Probability and Statistics

Probability and Statistics

Edited by
Nash Stokes

Larsen & Keller
www.larsen-keller.com

Probability and Statistics
Edited by Nash Stokes
ISBN: 978-1-63549-233-0 (Hardback)

© 2017 Larsen & Keller

▤ Larsen & Keller

Published by Larsen and Keller Education,
5 Penn Plaza,
19th Floor,
New York, NY 10001, USA

Cataloging-in-Publication Data

Probability and statistics / edited by Nash Stokes .
    p. cm.
Includes bibliographical references and index.
ISBN 978-1-63549-233-0
1. Probabilities. 2. Statistics. 3. Mathematical statistics.
I. Stokes, Nash.
QA273 .P76 2017
519--dc23

The publisher's policy is to use permanent paper from mills that operate a sustainable forestry policy. Furthermore, the publisher ensures that the text paper and cover boards used have met acceptable environmental accreditation standards.

Printed and bound in the United States of America.

For more information regarding Larsen and Keller Education and its products, please visit the publisher's website www.larsen-keller.com

Table of Contents

Preface   **VII**

Chapter 1 **Understanding Statistics and Probability**   **1**
    a. Statistics   1
    b. Probability   17

Chapter 2 **Key Concepts of Probability**   **26**
    a. Random Variable   26
    b. Event (Probability Theory)   35
    c. Law of Total Probability   45
    d. Venn Diagram   47
    e. Mutual Exclusivity   55
    f. Probability Axioms   57

Chapter 3 **Theory of Probability Distributions**   **61**
    a. Probability Distribution   61
    b. Probability Theory   70
    c. Probability Mass Function   76
    d. Probability Density Function   78
    e. Cumulative Distribution Function   88
    f. Quantile Function   93
    g. Expected Value   97
    h. Variance   108

Chapter 4 **Conditional Probability: A Comprehensive Study**   **126**
    a. Conditional Probability   126
    b. Conditional Expectation   134
    c. Conditional Probability Distribution   141
    d. Regular Conditional Probability   144
    e. Disintegration Theorem   145
    f. Bayes' Theorem   147
    g. Rule of Succession   157
    h. Conditional Independence   164

Chapter 5 **Interpretation of Probability**   **169**
    a. Probability Interpretations   169
    b. Classical Definition of Probability   176
    c. Frequentist Probability   178
    d. Probabilistic Logic   182
    e. Propensity Probability   184
    f. Bayesian Probability   186

Chapter 6 **Stochastic Process: An Overview**   **191**
    a. Stochastic Process   191

b.	Wiener Process	196
c.	Ornstein–Uhlenbeck Process	206
d.	Random Walk	211
e.	Poisson Point Process	223

Chapter 7	**Statistical Models: An Integrated Study**	**248**
a.	Statistical Model	248
b.	Regression Analysis	251
c.	Bayesian Hierarchical Modeling	259
d.	Errors-in-Variables Models	263
e.	Generalized Linear Model	272
f.	Vector Generalized Linear Model	280

Chapter 8	**Mathematical Statistics**	**290**
a.	Descriptive Statistics	294
b.	Nonparametric Statistics	295
c.	Probability Distribution	299

Chapter 9	**Statistical Inference and Hypothesis Testing**	**308**
a.	Statistical Inference	308
b.	Bayesian Inference	314
c.	Asymptotic Theory	326
d.	Estimation Theory	329
e.	Statistical Hypothesis Testing	335

Chapter 10	**Evolution of Probability and Statistics**	**356**
a.	History of Statistics	356
b.	History of Probability	368

Permissions

Index

Preface

Statistics refers to the examination of data on the basis of its presentation, organization, collection, interpretation and analysis. It uses probability as a major factor in analysis of data. Probability explores the classification of data into absolutes. This book is compiled in such a manner, that it will provide in-depth knowledge about the theory and practice of statistics and probability. It unfolds the innovative aspects of the area, which will be crucial for the holistic understanding of the subject matter. It is meant for students who are looking for an elaborate reference text on probability and statistics. Those in search of information to further their knowledge will be greatly assisted by this textbook.

A detailed account of the significant topics covered in this book is provided below:

Chapter 1- The study of the collection, analysis and the organization of data is termed as statistics. The two main statistical methods used in data analysis are descriptive statistics and inferential statistics. This is an introductory chapter that will introduce briefly all the significant aspects of statistics and probability.

Chapter 2- The key concepts of probability that have been explained in the section are random variable, event in probability, law of total probability, Venn diagram, mutual exclusivity etc. Random variable is a variable whose probable value depends on a set of random events. An event in probability and statistics is the result of a trial to which a probability is assigned. The section strategically encompasses and incorporates the major components and key concepts of probability, providing a complete understanding on the topic.

Chapter 3- Probability distribution is the mathematical explanation of any event in terms of the probabilities of events. The theories of probability distribution that have been elucidated are probability mass function, probability density function, cumulative distribution function, quantile function and expected value. The topics discussed in the chapter are of great importance to broaden the existing knowledge on probability distributions.

Chapter 4- Measuring any event by the probability of another event occurring is known as conditional probability. It is the most fundamental theory of probability. Conditional probability is used in statistical inference as a revision of the probability of an event; the information on which this is based should be new. Conditional expectation, conditional probability distribution, regular conditional probability and Bayes theorem are some of the aspects of conditional probability explained in the following section.

Chapter 5- Probability as a word has numerous ways of being used; it has two broad categories that can be called physical probability and evidential probability. Frequentist probability, probabilistic logic, propensity probability, Bayesian probability etc. are the interpretations of probability that has been discussed within this chapter.

Chapter 6- Stochastic process is an important model in probability. It is used to describe a time sequence of any system and helps in representing the evolution of that system by a variable. This section is an overview of the subject matter incorporating all the major aspects of stochastic process.

Chapter 7- A statistical model is a mathematical model that helps in the demonstration of a set of assumptions. These assumptions are concerned with the generation of sample data from a larger population. The aspects elucidated in this section are of vital importance and provide a better understanding of statistical models.

Chapter 8- The application of mathematics to statistics is known as mathematical statistics. The techniques used in this are mathematical analysis, linear algebra, stochastic analysis and differential equations. This chapter helps the readers in developing an in-depth understanding of mathematical statistics.

Chapter 9- The technique of analyzing properties of an underlying distribution by analysis of data is termed as statistical inference. The topics elucidated in the section are Bayesian inference, asymptotic theory, estimation theory and statistical hypothesis theory. The chapter serves as a source to understand the major categories related to statistical inference and hypothesis testing.

Chapter 10- The meaning of statistics has involved in the past few decades. Initially the meaning of the term was limited to information about states. The meaning has since then also included all the collections of information of all types and analysis and interpretation of such data. The evolution of probability and statistics helps the reader in understanding the growth of the subject.

It gives me an immense pleasure to thank our entire team for their efforts. Finally in the end, I would like to thank my family and colleagues who have been a great source of inspiration and support.

Editor

Understanding Statistics and Probability

The study of the collection, analysis and the organization of data is termed as statistics. The two main statistical methods used in data analysis are descriptive statistics and inferential statistics. This is an introductory chapter that will introduce briefly all the significant aspects of statistics and probability.

Statistics

Statistics is the study of the collection, analysis, interpretation, presentation, and organization of data. In applying statistics to, e.g., a scientific, industrial, or social problem, it is conventional to begin with a statistical population or a statistical model process to be studied. Populations can be diverse topics such as "all people living in a country" or "every atom composing a crystal". Statistics deals with all aspects of data including the planning of data collection in terms of the design of surveys and experiments.

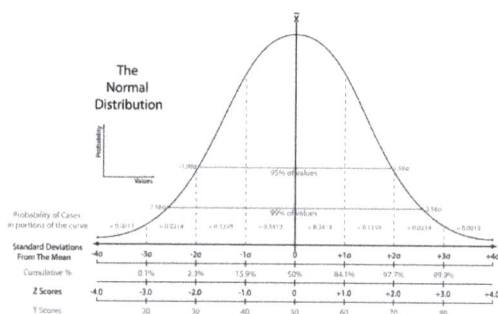

More probability density is found as one gets closer to the expected (mean) value in a normal distribution. Statistics used in standardized testing assessment are shown. The scales include standard deviations, cumulative percentages, percentile equivalents, Z-scores, T-scores, standard nines, and percentages in standard nines.

Some popular definitions are:

- Merriam-Webster dictionary defines statistics as "classified facts representing the conditions of a people in a state – especially the facts that can be stated in numbers or any other tabular or classified arrangement".

- Statistician Sir Arthur Lyon Bowley defines statistics as "Numerical statements of facts in any department of inquiry placed in relation to each other".

When census data cannot be collected, statisticians collect data by developing specific experiment designs and survey samples. Representative sampling assures that inferences and conclusions can safely extend from the sample to the population as a whole. An experimental study involves taking measurements of the system under study, manipulating the system, and then taking additional measurements using the same procedure to determine if the manipulation has modified the values of the measurements. In contrast, an observational study does not involve experimental manipulation.

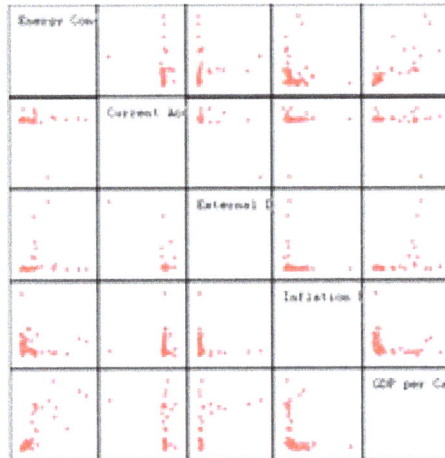

Scatter plots are used in descriptive statistics to show the observed relationships between different variables.

Two main statistical methodologies are used in data analysis: descriptive statistics, which summarizes data from a sample using indexes such as the mean or standard deviation, and inferential statistics, which draws conclusions from data that are subject to random variation (e.g., observational errors, sampling variation). Descriptive statistics are most often concerned with two sets of properties of a *distribution* (sample or population): *central tendency* (or *location*) seeks to characterize the distribution's central or typical value, while *dispersion* (or *variability*) characterizes the extent to which members of the distribution depart from its center and each other. Inferences on mathematical statistics are made under the framework of probability theory, which deals with the analysis of random phenomena.

A standard statistical procedure involves the test of the relationship between two statistical data sets, or a data set and a synthetic data drawn from idealized model. A hypothesis is proposed for the statistical relationship between the two data sets, and this is compared as an alternative to an idealized null hypothesis of no relationship between two data sets. Rejecting or disproving the null hypothesis is done using statistical tests that quantify the sense in which the null can be proven false, given the data that are used in the test. Working from a null hypothesis, two basic forms of error are recognized: Type I errors (null hypothesis is falsely rejected giving a "false positive") and Type II errors (null hypothesis fails to be rejected and an actual difference between populations is missed giving a "false negative"). Multiple problems have come to be associated with this framework: ranging from obtaining a sufficient sample size to specifying an adequate null hypothesis.

Measurement processes that generate statistical data are also subject to error. Many of these errors are classified as random (noise) or systematic (bias), but other types of errors (e.g., blunder, such as when an analyst reports incorrect units) can also be important. The presence of missing data and/or censoring may result in biased estimates and specific techniques have been developed to address these problems.

Statistics can be said to have begun in ancient civilization, going back at least to the 5th century BC, but it was not until the 18th century that it started to draw more heavily from calculus and probability theory. Statistics continues to be an area of active research, for example on the problem of how to analyze Big data.

Scope

Statistics is a mathematical body of science that pertains to the collection, analysis, interpretation or explanation, and presentation of data, or as a branch of mathematics. Some consider statistics to be a distinct mathematical science rather than a branch of mathematics. While many scientific investigations make use of data, statistics is concerned with the use of data in the context of uncertainty and decision making in the face of uncertainty.

Mathematical Statistics

Mathematical statistics is the application of mathematics to statistics, which was originally conceived as the science of the state — the collection and analysis of facts about a country: its economy, land, military, population, and so forth. Mathematical techniques used for this include mathematical analysis, linear algebra, stochastic analysis, differential equations, and measure-theoretic probability theory.

Overview

In applying statistics to a problem, it is common practice to start with a population or process to be studied. Populations can be diverse topics such as "all persons living in a country" or "every atom composing a crystal".

Ideally, statisticians compile data about the entire population (an operation called census). This may be organized by governmental statistical institutes. Descriptive statistics can be used to summarize the population data. Numerical descriptors include mean and standard deviation for continuous data types (like income), while frequency and percentage are more useful in terms of describing categorical data (like race).

When a census is not feasible, a chosen subset of the population called a sample is studied. Once a sample that is representative of the population is determined, data is collected for the sample members in an observational or experimental setting. Again, descriptive statistics can be used to summarize the sample data. However, the drawing of the sample has been subject to an element of randomness, hence the established numerical descriptors from the sample are also due to uncertainty. To still draw meaningful conclusions about the entire population, inferential statistics is needed. It uses patterns in the sample data to draw inferences about the population represented, accounting for randomness. These inferences may take the form of: answering yes/no questions about the data (hypothesis testing), estimating numerical characteristics of the data (estimation), describing associations within the data (correlation) and modeling relationships within the data (for example, using regression analysis). Inference can extend to forecasting, prediction and estimation of unobserved values either in or associated with the population being studied; it can include extrapolation and interpolation of time series or spatial data, and can also include data mining.

Data Collection

Sampling

When full census data cannot be collected, statisticians collect sample data by developing specific experiment designs and survey samples. Statistics itself also provides tools for prediction and forecasting the use of data through statistical models. To use a sample as a guide to an entire population, it is important that it truly represents the overall population. Representative sampling assures that inferences and conclusions can safely extend from the sample to the population as a whole. A major problem lies in determining the extent that the sample chosen is actually representative. Statistics offers methods to estimate and correct for any bias within the sample and data collection procedures. There are also methods of experimental design for experiments that can lessen these issues at the outset of a study, strengthening its capability to discern truths about the population.

Sampling theory is part of the mathematical discipline of probability theory. Probability is used in mathematical statistics to study the sampling distributions of sample statistics and, more generally, the properties of statistical procedures. The use of any statistical method is valid when the system or population under consideration satisfies the assumptions of the method. The difference in point of view between classic probability theory and sampling theory is, roughly, that probability theory starts from the given parameters of a total population to deduce probabilities that pertain to samples. Statistical inference, however, moves in the opposite direction—inductively inferring from samples to the parameters of a larger or total population.

Experimental and Observational Studies

A common goal for a statistical research project is to investigate causality, and in particular to draw a conclusion on the effect of changes in the values of predictors or independent variables on dependent variables. There are two major types of causal statistical studies: experimental studies and observational studies. In both types of studies, the effect of differences of an independent variable (or variables) on the behavior of the dependent variable are observed. The difference between the two types lies in how the study is actually conducted. Each can be very effective. An experimental study involves taking measurements of the system under study, manipulating the system, and then taking additional measurements using the same procedure to determine if the manipulation has modified the values of the measurements. In contrast, an observational study does not involve experimental manipulation. Instead, data are gathered and correlations between predictors and response are investigated. While the tools of data analysis work best on data from randomized studies, they are also applied to other kinds of data – like natural experiments and observational studies – for which a statistician would use a modified, more structured estimation method (e.g., Difference in differences estimation and instrumental variables, among many others) that produce consistent estimators.

Experiments

The basic steps of a statistical experiment are:

1. Planning the research, including finding the number of replicates of the study, using the following information: preliminary estimates regarding the size of treatment effects, alter-

native hypotheses, and the estimated experimental variability. Consideration of the selection of experimental subjects and the ethics of research is necessary. Statisticians recommend that experiments compare (at least) one new treatment with a standard treatment or control, to allow an unbiased estimate of the difference in treatment effects.

2. Design of experiments, using blocking to reduce the influence of confounding variables, and randomized assignment of treatments to subjects to allow unbiased estimates of treatment effects and experimental error. At this stage, the experimenters and statisticians write the *experimental protocol* that will guide the performance of the experiment and which specifies the *primary analysis* of the experimental data.

3. Performing the experiment following the experimental protocol and analyzing the data following the experimental protocol.

4. Further examining the data set in secondary analyses, to suggest new hypotheses for future study.

5. Documenting and presenting the results of the study.

Experiments on human behavior have special concerns. The famous Hawthorne study examined changes to the working environment at the Hawthorne plant of the Western Electric Company. The researchers were interested in determining whether increased illumination would increase the productivity of the assembly line workers. The researchers first measured the productivity in the plant, then modified the illumination in an area of the plant and checked if the changes in illumination affected productivity. It turned out that productivity indeed improved (under the experimental conditions). However, the study is heavily criticized today for errors in experimental procedures, specifically for the lack of a control group and blindness. The Hawthorne effect refers to finding that an outcome (in this case, worker productivity) changed due to observation itself. Those in the Hawthorne study became more productive not because the lighting was changed but because they were being observed.

Observational Study

An example of an observational study is one that explores the association between smoking and lung cancer. This type of study typically uses a survey to collect observations about the area of interest and then performs statistical analysis. In this case, the researchers would collect observations of both smokers and non-smokers, perhaps through a case-control study, and then look for the number of cases of lung cancer in each group.

Types of Data

Various attempts have been made to produce a taxonomy of levels of measurement. The psychophysicist Stanley Smith Stevens defined nominal, ordinal, interval, and ratio scales. Nominal measurements do not have meaningful rank order among values, and permit any one-to-one transformation. Ordinal measurements have imprecise differences between consecutive values, but have a meaningful order to those values, and permit any order-preserving transformation. Interval measurements have meaningful distances between measurements defined, but the zero value is arbitrary (as in the case with longitude and temperature measurements in Celsius or

Fahrenheit), and permit any linear transformation. Ratio measurements have both a meaningful zero value and the distances between different measurements defined, and permit any rescaling transformation.

Because variables conforming only to nominal or ordinal measurements cannot be reasonably measured numerically, sometimes they are grouped together as categorical variables, whereas ratio and interval measurements are grouped together as quantitative variables, which can be either discrete or continuous, due to their numerical nature. Such distinctions can often be loosely correlated with data type in computer science, in that dichotomous categorical variables may be represented with the Boolean data type, polytomous categorical variables with arbitrarily assigned integers in the integral data type, and continuous variables with the real data type involving floating point computation. But the mapping of computer science data types to statistical data types depends on which categorization of the latter is being implemented.

Other categorizations have been proposed. For example, Mosteller and Tukey (1977) distinguished grades, ranks, counted fractions, counts, amounts, and balances. Nelder (1990) described continuous counts, continuous ratios, count ratios, and categorical modes of data.

The issue of whether or not it is appropriate to apply different kinds of statistical methods to data obtained from different kinds of measurement procedures is complicated by issues concerning the transformation of variables and the precise interpretation of research questions. "The relationship between the data and what they describe merely reflects the fact that certain kinds of statistical statements may have truth values which are not invariant under some transformations. Whether or not a transformation is sensible to contemplate depends on the question one is trying to answer" (Hand, 2004).

Terminology and Theory of Inferential Statistics

Statistics, Estimators and Pivotal Quantities

Consider independent identically distributed (IID) random variables with a given probability distribution: standard statistical inference and estimation theory defines a random sample as the random vector given by the column vector of these IID variables. The population being examined is described by a probability distribution that may have unknown parameters.

A statistic is a random variable that is a function of the random sample, but *not a function of unknown parameters*. The probability distribution of the statistic, though, may have unknown parameters.

Consider now a function of the unknown parameter: an estimator is a statistic used to estimate such function. Commonly used estimators include sample mean, unbiased sample variance and sample covariance.

A random variable that is a function of the random sample and of the unknown parameter, but whose probability distribution *does not depend on the unknown parameter* is called a pivotal quantity or pivot. Widely used pivots include the z-score, the chi square statistic and Student's t-value.

Between two estimators of a given parameter, the one with lower mean squared error is said to be more efficient. Furthermore, an estimator is said to be unbiased if its expected value is equal to the true value of the unknown parameter being estimated, and asymptotically unbiased if its expected value converges at the limit to the true value of such parameter.

Other desirable properties for estimators include: UMVUE estimators that have the lowest variance for all possible values of the parameter to be estimated (this is usually an easier property to verify than efficiency) and consistent estimators which converges in probability to the true value of such parameter.

This still leaves the question of how to obtain estimators in a given situation and carry the computation, several methods have been proposed: the method of moments, the maximum likelihood method, the least squares method and the more recent method of estimating equations.

Null Hypothesis and Alternative Hypothesis

Interpretation of statistical information can often involve the development of a null hypothesis which is usually (but not necessarily) that no relationship exists among variables or that no change occurred over time.

The best illustration for a novice is the predicament encountered by a criminal trial. The null hypothesis, H_0, asserts that the defendant is innocent, whereas the alternative hypothesis, H_1, asserts that the defendant is guilty. The indictment comes because of suspicion of the guilt. The H_0 (status quo) stands in opposition to H_1 and is maintained unless H_1 is supported by evidence "beyond a reasonable doubt". However, "failure to reject H_0" in this case does not imply innocence, but merely that the evidence was insufficient to convict. So the jury does not necessarily *accept* H_0 but *fails to reject* H_0. While one can not "prove" a null hypothesis, one can test how close it is to being true with a power test, which tests for type II errors.

What statisticians call an alternative hypothesis is simply an hypothesis that contradicts the null hypothesis.

Error

Working from a null hypothesis, two basic forms of error are recognized:

- Type I errors where the null hypothesis is falsely rejected giving a "false positive".

- Type II errors where the null hypothesis fails to be rejected and an actual difference between populations is missed giving a "false negative".

Standard deviation refers to the extent to which individual observations in a sample differ from a central value, such as the sample or population mean, while Standard error refers to an estimate of difference between sample mean and population mean.

A statistical error is the amount by which an observation differs from its expected value, a residual is the amount an observation differs from the value the estimator of the expected value assumes on a given sample (also called prediction).

Mean squared error is used for obtaining efficient estimators, a widely used class of estimators. Root mean square error is simply the square root of mean squared error.

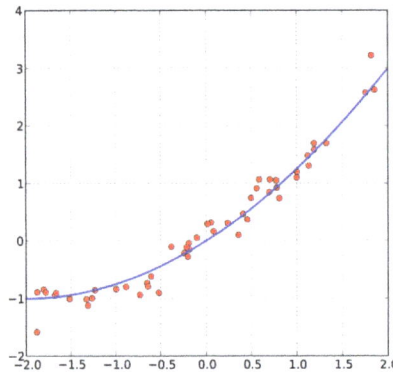

A least squares fit: in red the points to be fitted, in blue the fitted line.

Many statistical methods seek to minimize the residual sum of squares, and these are called "methods of least squares" in contrast to Least absolute deviations. The latter gives equal weight to small and big errors, while the former gives more weight to large errors. Residual sum of squares is also differentiable, which provides a handy property for doing regression. Least squares applied to linear regression is called ordinary least squares method and least squares applied to nonlinear regression is called non-linear least squares. Also in a linear regression model the non deterministic part of the model is called error term, disturbance or more simply noise. Both linear regression and non-linear regression are addressed in polynomial least squares, which also describes the variance in a prediction of the dependent variable (y axis) as a function of the independent variable (x axis) and the deviations (errors, noise, disturbances) from the estimated (fitted) curve.

Measurement processes that generate statistical data are also subject to error. Many of these errors are classified as random (noise) or systematic (bias), but other types of errors (e.g., blunder, such as when an analyst reports incorrect units) can also be important. The presence of missing data and/or censoring may result in biased estimates and specific techniques have been developed to address these problems.

Interval Estimation

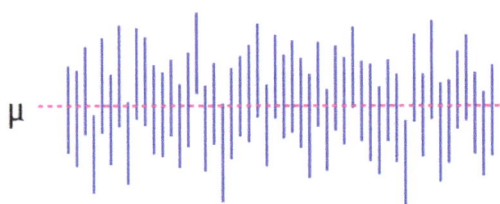

Confidence intervals: the red line is true value for the mean in this example, the blue lines are random confidence intervals for 100 realizations.

Most studies only sample part of a population, so results don't fully represent the whole population. Any estimates obtained from the sample only approximate the population value. Confidence intervals allow statisticians to express how closely the sample estimate matches the true value in the whole population. Often they are expressed as 95% confidence intervals. Formally, a 95%

confidence interval for a value is a range where, if the sampling and analysis were repeated under the same conditions (yielding a different dataset), the interval would include the true (population) value in 95% of all possible cases. This does *not* imply that the probability that the true value is in the confidence interval is 95%. From the frequentist perspective, such a claim does not even make sense, as the true value is not a random variable. Either the true value is or is not within the given interval. However, it is true that, before any data are sampled and given a plan for how to construct the confidence interval, the probability is 95% that the yet-to-be-calculated interval will cover the true value: at this point, the limits of the interval are yet-to-be-observed random variables. One approach that does yield an interval that can be interpreted as having a given probability of containing the true value is to use a credible interval from Bayesian statistics: this approach depends on a different way of interpreting what is meant by "probability", that is as a Bayesian probability.

In principle confidence intervals can be symmetrical or asymmetrical. An interval can be asymmetrical because it works as lower or upper bound for a parameter (left-sided interval or right sided interval), but it can also be asymmetrical because the two sided interval is built violating symmetry around the estimate. Sometimes the bounds for a confidence interval are reached asymptotically and these are used to approximate the true bounds.

Significance

Statistics rarely give a simple Yes/No type answer to the question under analysis. Interpretation often comes down to the level of statistical significance applied to the numbers and often refers to the probability of a value accurately rejecting the null hypothesis (sometimes referred to as the p-value).

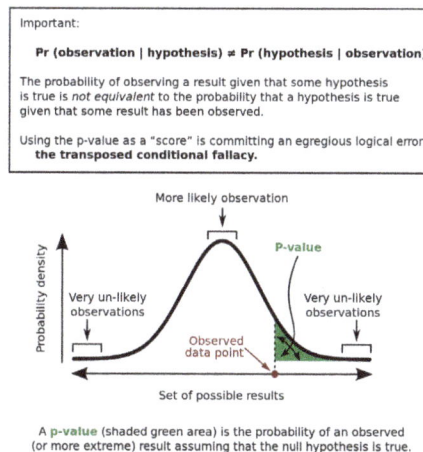

In this graph the black line is probability distribution for the test statistic, the critical region is the set of values to the right of the observed data point (observed value of the test statistic) and the p-value is represented by the green area.

The standard approach is to test a null hypothesis against an alternative hypothesis. A critical region is the set of values of the estimator that leads to refuting the null hypothesis. The probability of type I error is therefore the probability that the estimator belongs to the critical region given that null hypothesis is true (statistical significance) and the probability of type II error is the probability that the estimator doesn't belong to the critical region given that the alternative hypothesis is true. The statistical power of a test is the probability that it correctly rejects the null hypothesis when the null hypothesis is false.

Referring to statistical significance does not necessarily mean that the overall result is significant in real world terms. For example, in a large study of a drug it may be shown that the drug has a statistically significant but very small beneficial effect, such that the drug is unlikely to help the patient noticeably.

While in principle the acceptable level of statistical significance may be subject to debate, the p-value is the smallest significance level that allows the test to reject the null hypothesis. This is logically equivalent to saying that the p-value is the probability, assuming the null hypothesis is true, of observing a result at least as extreme as the test statistic. Therefore, the smaller the p-value, the lower the probability of committing type I error.

Some problems are usually associated with this framework:

- A difference that is highly statistically significant can still be of no practical significance, but it is possible to properly formulate tests to account for this. One response involves going beyond reporting only the significance level to include the p-value when reporting whether a hypothesis is rejected or accepted. The p-value, however, does not indicate the size or importance of the observed effect and can also seem to exaggerate the importance of minor differences in large studies. A better and increasingly common approach is to report confidence intervals. Although these are produced from the same calculations as those of hypothesis tests or p-values, they describe both the size of the effect and the uncertainty surrounding it.

- Fallacy of the transposed conditional, aka prosecutor's fallacy: criticisms arise because the hypothesis testing approach forces one hypothesis (the null hypothesis) to be favored, since what is being evaluated is probability of the observed result given the null hypothesis and not probability of the null hypothesis given the observed result. An alternative to this approach is offered by Bayesian inference, although it requires establishing a prior probability.

- Rejecting the null hypothesis does not automatically prove the alternative hypothesis.

- As everything in inferential statistics it relies on sample size, and therefore under fat tails p-values may be seriously mis-computed.

Examples

Some well-known statistical tests and procedures are:

- Analysis of variance (ANOVA)
- Chi-squared test
- Correlation
- Factor analysis
- Mann–Whitney U
- Mean square weighted deviation (MSWD)
- Pearson product-moment correlation coefficient

- Regression analysis

- Spearman's rank correlation coefficient

- Student's t-test

- Time series analysis

- Conjoint Analysis

Misuse

Misuse of statistics can produce subtle, but serious errors in description and interpretation—subtle in the sense that even experienced professionals make such errors, and serious in the sense that they can lead to devastating decision errors. For instance, social policy, medical practice, and the reliability of structures like bridges all rely on the proper use of statistics.

Even when statistical techniques are correctly applied, the results can be difficult to interpret for those lacking expertise. The statistical significance of a trend in the data—which measures the extent to which a trend could be caused by random variation in the sample—may or may not agree with an intuitive sense of its significance. The set of basic statistical skills (and skepticism) that people need to deal with information in their everyday lives properly is referred to as statistical literacy.

There is a general perception that statistical knowledge is all-too-frequently intentionally misused by finding ways to interpret only the data that are favorable to the presenter. A mistrust and misunderstanding of statistics is associated with the quotation, "There are three kinds of lies: lies, damned lies, and statistics". Misuse of statistics can be both inadvertent and intentional, and the book *How to Lie with Statistics* outlines a range of considerations. In an attempt to shed light on the use and misuse of statistics, reviews of statistical techniques used in particular fields are conducted (e.g. Warne, Lazo, Ramos, and Ritter (2012)).

Ways to avoid misuse of statistics include using proper diagrams and avoiding bias. Misuse can occur when conclusions are overgeneralized and claimed to be representative of more than they really are, often by either deliberately or unconsciously overlooking sampling bias. Bar graphs are arguably the easiest diagrams to use and understand, and they can be made either by hand or with simple computer programs. Unfortunately, most people do not look for bias or errors, so they are not noticed. Thus, people may often believe that something is true even if it is not well represented. To make data gathered from statistics believable and accurate, the sample taken must be representative of the whole. According to Huff, "The dependability of a sample can be destroyed by [bias]... allow yourself some degree of skepticism."

To assist in the understanding of statistics Huff proposed a series of questions to be asked in each case:

- Who says so? (Does he/she have an axe to grind?)

- How does he/she know? (Does he/she have the resources to know the facts?)

- What's missing? (Does he/she give us a complete picture?)

- Did someone change the subject? (Does he/she offer us the right answer to the wrong problem?)

- Does it make sense? (Is his/her conclusion logical and consistent with what we already know?)

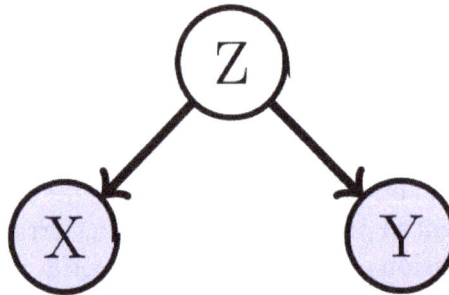

The confounding variable problem: X and Y may be correlated, not because there is causal relationship between them, but because both depend on a third variable Z. Z is called a confounding factor.

Misinterpretation: Correlation

The concept of correlation is particularly noteworthy for the potential confusion it can cause. Statistical analysis of a data set often reveals that two variables (properties) of the population under consideration tend to vary together, as if they were connected. For example, a study of annual income that also looks at age of death might find that poor people tend to have shorter lives than affluent people. The two variables are said to be correlated; however, they may or may not be the cause of one another. The correlation phenomena could be caused by a third, previously unconsidered phenomenon, called a lurking variable or confounding variable. For this reason, there is no way to immediately infer the existence of a causal relationship between the two variables.

History of Statistical Science

Gerolamo Cardano, the earliest pioneer on the mathematics of probability.

Statistical methods date back at least to the 5th century BC.

Some scholars pinpoint the origin of statistics to 1663, with the publication of *Natural and Political Observations upon the Bills of Mortality* by John Graunt. Early applications of statistical thinking revolved around the needs of states to base policy on demographic and economic data, hence its *stat-* etymology. The scope of the discipline of statistics broadened in the early 19th century to include the collection and analysis of data in general. Today, statistics is widely employed in government, business, and natural and social sciences.

Its mathematical foundations were laid in the 17th century with the development of the probability theory by Gerolamo Cardano, Blaise Pascal and Pierre de Fermat. Mathematical probability theory arose from the study of games of chance, although the concept of probability was already examined in medieval law and by philosophers such as Juan Caramuel. The method of least squares was first described by Adrien-Marie Legendre in 1805.

Karl Pearson, a founder of mathematical statistics.

The modern field of statistics emerged in the late 19th and early 20th century in three stages. The first wave, at the turn of the century, was led by the work of Francis Galton and Karl Pearson, who transformed statistics into a rigorous mathematical discipline used for analysis, not just in science, but in industry and politics as well. Galton's contributions included introducing the concepts of standard deviation, correlation, regression analysis and the application of these methods to the study of the variety of human characteristics – height, weight, eyelash length among others. Pearson developed the Pearson product-moment correlation coefficient, defined as a product-moment, the method of moments for the fitting of distributions to samples and the Pearson distribution, among many other things. Galton and Pearson founded *Biometrika* as the first journal of mathematical statistics and biostatistics (then called biometry), and the latter founded the world's first university statistics department at University College London.

Ronald Fisher coined the term null hypothesis during the Lady tasting tea experiment, which "is never proved or established, but is possibly disproved, in the course of experimentation".

The second wave of the 1910s and 20s was initiated by William Gosset, and reached its culmination in the insights of Ronald Fisher, who wrote the textbooks that were to define the academic discipline in universities around the world. Fisher's most important publications were his 1918 seminal paper *The Correlation between Relatives on the Supposition of Mendelian Inheritance*, which was the first to use the statistical term, variance, his classic 1925 work *Statistical Methods for Research Workers* and his 1935 *The Design of Experiments*, where he developed rigorous design of experiments models. He originated the concepts of sufficiency, ancillary statistics, Fisher's linear discriminator and Fisher information. In his 1930 book *The Genetical Theory of Natural Selection* he applied statistics to various biological concepts such as Fisher's principle). Nevertheless, A. W. F. Edwards has remarked that it is "probably the most celebrated argument in evolutionary biology". (about the sex ratio), the Fisherian runaway, a concept in sexual selection about a positive feedback runaway affect found in evolution.

The final wave, which mainly saw the refinement and expansion of earlier developments, emerged from the collaborative work between Egon Pearson and Jerzy Neyman in the 1930s. They introduced the concepts of "Type II" error, power of a test and confidence intervals. Jerzy Neyman in 1934 showed that stratified random sampling was in general a better method of estimation than purposive (quota) sampling.

Today, statistical methods are applied in all fields that involve decision making, for making accurate inferences from a collated body of data and for making decisions in the face of uncertainty based on statistical methodology. The use of modern computers has expedited large-scale statistical computations, and has also made possible new methods that are impractical to perform manually. Statistics continues to be an area of active research, for example on the problem of how to analyze Big data.

Applications

Applied Statistics, Theoretical Statistics and Mathematical Statistics

"Applied statistics" comprises descriptive statistics and the application of inferential statistics. *Theoretical statistics* concerns both the logical arguments underlying justification of approaches to statistical inference, as well encompassing *mathematical statistics*. Mathematical statistics includes not only the manipulation of probability distributions necessary for deriving results related to methods of estimation and inference, but also various aspects of computational statistics and the design of experiments.

Machine Learning and Data Mining

There are two applications for machine learning and data mining: data management and data analysis. Statistics tools are necessary for the data analysis.

Statistics in Society

Statistics is applicable to a wide variety of academic disciplines, including natural and social sciences, government, and business. Statistical consultants can help organizations and companies that don't have in-house expertise relevant to their particular questions.

Statistical Computing

The rapid and sustained increases in computing power starting from the second half of the 20th century have had a substantial impact on the practice of statistical science. Early statistical models were almost always from the class of linear models, but powerful computers, coupled with suitable numerical algorithms, caused an increased interest in nonlinear models (such as neural networks) as well as the creation of new types, such as generalized linear models and multilevel models.

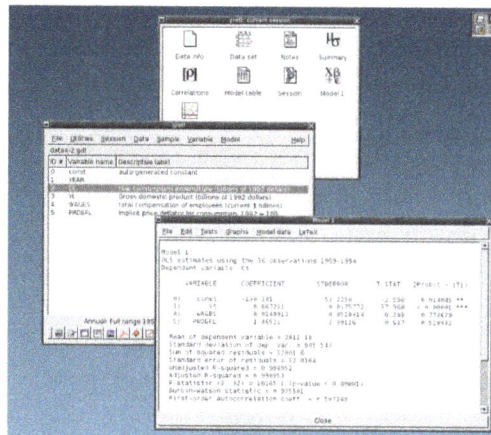

gretl, an example of an open source statistical package

Increased computing power has also led to the growing popularity of computationally intensive methods based on resampling, such as permutation tests and the bootstrap, while techniques such as Gibbs sampling have made use of Bayesian models more feasible. The computer revolution has implications for the future of statistics with new emphasis on "experimental" and "empirical" statistics. A large number of both general and special purpose statistical software are now available.

Statistics Applied to Mathematics or the Arts

Traditionally, statistics was concerned with drawing inferences using a semi-standardized methodology that was "required learning" in most sciences. This has changed with use of statistics in non-inferential contexts. What was once considered a dry subject, taken in many fields as a degree-requirement, is now viewed enthusiastically.Initially derided by some mathematical purists, it is now considered essential methodology in certain areas.

- In number theory, scatter plots of data generated by a distribution function may be transformed with familiar tools used in statistics to reveal underlying patterns, which may then lead to hypotheses.

- Methods of statistics including predictive methods in forecasting are combined with chaos theory and fractal geometry to create video works that are considered to have great beauty.

- The process art of Jackson Pollock relied on artistic experiments whereby underlying distributions in nature were artistically revealed.With the advent of computers, statistical methods were applied to formalize such distribution-driven natural processes to make and analyze moving video art.

- Methods of statistics may be used predicatively in performance art, as in a card trick based on a Markov process that only works some of the time, the occasion of which can be predicted using statistical methodology.

- Statistics can be used to predicatively create art, as in the statistical or stochastic music invented by Iannis Xenakis, where the music is performance-specific. Though this type of artistry does not always come out as expected, it does behave in ways that are predictable and tunable using statistics.

Specialized Disciplines

Statistical techniques are used in a wide range of types of scientific and social research, including: biostatistics, computational biology, computational sociology, network biology, social science, sociology and social research. Some fields of inquiry use applied statistics so extensively that they have specialized terminology. These disciplines include:

- Actuarial science (assesses risk in the insurance and finance industries)
- Applied information economics
- Astrostatistics (statistical evaluation of astronomical data)
- Biostatistics
- Business statistics
- Chemometrics (for analysis of data from chemistry)
- Data mining (applying statistics and pattern recognition to discover knowledge from data)
- Data science
- Demography
- Econometrics (statistical analysis of economic data)
- Energy statistics
- Engineering statistics
- Epidemiology (statistical analysis of disease)
- Geography and Geographic Information Systems, specifically in Spatial analysis
- Image processing
- Medical Statistics
- Political Science
- Psychological statistics
- Reliability engineering
- Social statistics
- Statistical Mechanics

In addition, there are particular types of statistical analysis that have also developed their own specialised terminology and methodology:

- Bootstrap / Jackknife resampling

- Multivariate statistics

- Statistical classification

- Structured data analysis (statistics)

- Structural equation modelling

- Survey methodology

- Survival analysis

- Statistics in various sports, particularly baseball - known as Sabermetrics - and cricket

Statistics form a key basis tool in business and manufacturing as well. It is used to understand measurement systems variability, control processes (as in statistical process control or SPC), for summarizing data, and to make data-driven decisions. In these roles, it is a key tool, and perhaps the only reliable tool.

Probability

Probability is the measure of the likelihood that an event will occur. Probability is quantified as a number between 0 and 1 (where 0 indicates impossibility and 1 indicates certainty). The higher the probability of an event, the more certain that the event will occur. A simple example is the tossing of a fair (unbiased) coin. Since the coin is unbiased, the two outcomes ("head" and "tail") are both equally probable; the probability of "head" equals the probability of "tail." Since no other outcomes are possible, the probability is 1/2 (or 50%), of either "head" or "tail". In other words, the probability of "head" is 1 out of 2 outcomes and the probability of "tail" is also 1 out of 2 outcomes, expressed as 0.5 when converted to decimal, with the above-mentioned quantification system. This type of probability is also called a priori probability.

These concepts have been given an axiomatic mathematical formalization in probability theory, which is used widely in such areas of study as mathematics, statistics, finance, gambling, science (in particular physics), artificial intelligence/machine learning, computer science, game theory, and philosophy to, for example, draw inferences about the expected frequency of events. Probability theory is also used to describe the underlying mechanics and regularities of complex systems.

Interpretations

When dealing with experiments that are random and well-defined in a purely theoretical setting (like tossing a fair coin), probabilities can be numerically described by the number of desired outcomes divided by the total number of all outcomes. For example, tossing a fair coin twice will yield "head-head", "head-tail", "tail-head", and "tail-tail" outcomes. The probability of getting an outcome of "head-head" is 1 out of 4 outcomes or 1/4 or 0.25 (or 25%). When it comes to practical application however, there are two major competing categories of probability interpretations, whose adherents possess different views about the fundamental nature of probability:

1. Objectivists assign numbers to describe some objective or physical state of affairs. The most popular version of objective probability is frequentist probability, which claims that the probability of a random event denotes the *relative frequency of occurrence* of an experiment's outcome, when repeating the experiment. This interpretation considers probability to be the relative frequency "in the long run" of outcomes. A modification of this is propensity probability, which interprets probability as the tendency of some experiment to yield a certain outcome, even if it is performed only once.

2. Subjectivists assign numbers per subjective probability, i.e., as a degree of belief. The degree of belief has been interpreted as, "the price at which you would buy or sell a bet that pays 1 unit of utility if E, 0 if not E." The most popular version of subjective probability is Bayesian probability, which includes expert knowledge as well as experimental data to produce probabilities. The expert knowledge is represented by some (subjective) prior probability distribution. These data are incorporated in a likelihood function. The product of the prior and the likelihood, normalized, results in a posterior probability distribution that incorporates all the information known to date. Starting from arbitrary, subjective probabilities for a group of agents, some Bayesians claim that all agents will eventually have sufficiently similar assessments of probabilities, given enough evidence.

Etymology

The word *probability* derives from the Latin *probabilitas*, which can also mean "probity", a measure of the authority of a witness in a legal case in Europe, and often correlated with the witness's nobility. In a sense, this differs much from the modern meaning of *probability*, which, in contrast, is a measure of the weight of empirical evidence, and is arrived at from inductive reasoning and statistical inference.

History

The scientific study of probability is a modern development of mathematics. Gambling shows that there has been an interest in quantifying the ideas of probability for millennia, but exact mathematical descriptions arose much later. There are reasons of course, for the slow development of the mathematics of probability. Whereas games of chance provided the impetus for the mathematical study of probability, fundamental issues are still obscured by the superstitions of gamblers.

Christiaan Huygens probably published the first book on probability

According to Richard Jeffrey, "Before the middle of the seventeenth century, the term 'probable' (Latin *probabilis*) meant *approvable*, and was applied in that sense, univocally, to opinion and to action. A probable action or opinion was one such as sensible people would undertake or hold, in the circumstances." However, in legal contexts especially, 'probable' could also apply to propositions for which there was good evidence.

Gerolamo Cardano

The sixteenth century Italian polymath Gerolamo Cardano demonstrated the efficacy of defining odds as the ratio of favourable to unfavourable outcomes (which implies that the probability of an event is given by the ratio of favourable outcomes to the total number of possible outcomes). Aside from the elementary work by Cardano, the doctrine of probabilities dates to the correspondence of Pierre de Fermat and Blaise Pascal (1654). Christiaan Huygens (1657) gave the earliest known scientific treatment of the subject. Jakob Bernoulli's *Ars Conjectandi* (posthumous, 1713) and Abraham de Moivre's *Doctrine of Chances* (1718) treated the subject as a branch of mathematics. *The Emergence of Probability* and James Franklin's *The Science of Conjecture* for histories of the early development of the very concept of mathematical probability.

The theory of errors may be traced back to Roger Cotes's *Opera Miscellanea* (posthumous, 1722), but a memoir prepared by Thomas Simpson in 1755 (printed 1756) first applied the theory to the discussion of errors of observation. The reprint (1757) of this memoir lays down the axioms that positive and negative errors are equally probable, and that certain assignable limits define the range of all errors. Simpson also discusses continuous errors and describes a probability curve.

The first two laws of error that were proposed both originated with Pierre-Simon Laplace. The first law was published in 1774 and stated that the frequency of an error could be expressed as an exponential function of the numerical magnitude of the error, disregarding sign. The second law of error was proposed in 1778 by Laplace and stated that the frequency of the error is an exponential function of the square of the error. The second law of error is called the normal distribution or the Gauss law. "It is difficult historically to attribute that law to Gauss, who in spite of his well-known precocity had probably not made this discovery before he was two years old."

Daniel Bernoulli (1778) introduced the principle of the maximum product of the probabilities of a system of concurrent errors.

Carl Friedrich Gauss

Adrien-Marie Legendre (1805) developed the method of least squares, and introduced it in his *Nouvelles méthodes pour la détermination des orbites des comètes* (*New Methods for Determining the Orbits of Comets*). In ignorance of Legendre's contribution, an Irish-American writer, Robert Adrain, editor of "The Analyst" (1808), first deduced the law of facility of error,

$$\phi(x) = ce^{-h^2 x^2},$$

where h is a constant depending on precision of observation, and c is a scale factor ensuring that the area under the curve equals 1. He gave two proofs, the second being essentially the same as John Herschel's (1850).Gauss gave the first proof that seems to have been known in Europe (the third after Adrain's) in 1809. Further proofs were given by Laplace (1810, 1812), Gauss (1823), James Ivory (1825, 1826), Hagen (1837), Friedrich Bessel (1838), W. F. Donkin (1844, 1856), and Morgan Crofton (1870). Other contributors were Ellis (1844), De Morgan (1864), Glaisher (1872), and Giovanni Schiaparelli (1875). Peters's (1856) formula for r, the probable error of a single observation, is well known.

In the nineteenth century authors on the general theory included Laplace, Sylvestre Lacroix (1816), Littrow (1833), Adolphe Quetelet (1853), Richard Dedekind (1860), Helmert (1872), Hermann Laurent (1873), Liagre, Didion, and Karl Pearson. Augustus De Morgan and George Boole improved the exposition of the theory.

Andrey Markov introduced the notion of Markov chains (1906), which played an important role in stochastic processes theory and its applications. The modern theory of probability based on the measure theory was developed by Andrey Kolmogorov (1931).

On the geometric side contributors to *The Educational Times* were influential (Miller, Crofton, McColl, Wolstenholme, Watson, and Artemas Martin).

Theory

Like other theories, the theory of probability is a representation of probabilistic concepts in formal terms—that is, in terms that can be considered separately from their meaning. These formal terms are manipulated by the rules of mathematics and logic, and any results are interpreted or translated back into the problem domain.

There have been at least two successful attempts to formalize probability, namely the Kolmogorov formulation and the Cox formulation. In Kolmogorov's formulation, sets are interpreted as events and probability itself as a measure on a class of sets. In Cox's theorem, probability is taken as a primitive (that is, not further analyzed) and the emphasis is on constructing a consistent assignment of probability values to propositions. In both cases, the laws of probability are the same, except for technical details.

There are other methods for quantifying uncertainty, such as the Dempster–Shafer theory or possibility theory, but those are essentially different and not compatible with the laws of probability as usually understood.

Applications

Probability theory is applied in everyday life in risk assessment and modeling. The insurance industry and markets use actuarial science to determine pricing and make trading decisions. Governments apply probabilistic methods in environmental regulation, entitlement analysis (Reliability theory of aging and longevity), and financial regulation.

A good example of the use of probibility theory in equity trading is the effect of the perceived probability of any widespread Middle East conflict on oil prices, which have ripple effects in the economy as a whole. An assessment by a commodity trader that a war is more likely can send that commodity's prices up or down, and signals other traders of that opinion. Accordingly, the probabilities are neither assessed independently nor necessarily very rationally. The theory of behavioral finance emerged to describe the effect of such groupthink on pricing, on policy, and on peace and conflict.

In addition to financial assessment, probability can be used to analyze trends in biology (e.g. disease spread) as well as ecology (e.g. biological Punnett squares). As with finance, risk assessment can be used as a statistical tool to calculate the likelihood of undesirable events occurring and can assist with implementing protocols to avoid encountering such circumstances. Probability is used to design games of chance so that casinos can make a guaranteed profit, yet provide payouts to players that are frequent enough to encourage continued play.

The discovery of rigorous methods to assess and combine probability assessments has changed society.It is important for most citizens to understand how probability assessments are made, and how they contribute to decisions.

Another significant application of probability theory in everyday life is reliability. Many consumer products, such as automobiles and consumer electronics, use reliability theory in product design to reduce the probability of failure. Failure probability may influence a manufacturer's decisions on a product's warranty.

The cache language model and other statistical language models that are used in natural language processing are also examples of applications of probability theory.

Mathematical Treatment

Consider an experiment that can produce a number of results. The collection of all possible results is called the sample space of the experiment. The power set of the sample space is formed by considering all different collections of possible results. For example, rolling a dice can produce six possible results. One collection of possible results gives an odd number on the dice. Thus, the subset {1,3,5} is an element of the power set of the sample space of dice rolls. These collections are called "events." In this case, {1,3,5} is the event that the dice falls on some odd number. If the results that actually occur fall in a given event, the event is said to have occurred.

A probability is a way of assigning every event a value between zero and one, with the requirement that the event made up of all possible results (in our example, the event {1,2,3,4,5,6}) is assigned a value of one. To qualify as a probability, the assignment of values must satisfy the requirement that if you look at a collection of mutually exclusive events (events with no common results, e.g., the events {1,6}, {3}, and {2,4} are all mutually exclusive), the probability that at least one of the events will occur is given by the sum of the probabilities of all the individual events.

The probability of an event A is written as $P(A)$, $p(A)$, or $\Pr(A)$. This mathematical definition of probability can extend to infinite sample spaces, and even uncountable sample spaces, using the concept of a measure.

The *opposite* or *complement* of an event A is the event [not A] (that is, the event of A not occurring), often denoted as $\bar{A}, A^C, \neg A$, or $\sim A$; its probability is given by $P(\text{not } A) = 1 - P(A)$. As an example, the chance of not rolling a six on a six-sided die is $1 - (\text{chance of rolling a six}) = 1 - \frac{1}{6} = \frac{5}{6}$.

If two events A and B occur on a single performance of an experiment, this is called the intersection or joint probability of A and B, denoted as $P(A \cap B)$..

Independent Events

If two events, A and B are independent then the joint probability is

$$P(A \text{ and } B) = P(A \cap B) = P(A)P(B),$$

for example, if two coins are flipped the chance of both being heads is $\frac{1}{2} \times \frac{1}{2} = \frac{1}{4}$.

Mutually Exclusive Events

If either event A or event B occurs on a single performance of an experiment this is called the union of the events A and B denoted as $P(A \cup B)$. If two events are mutually exclusive then the probability of either occurring is

$$P(A \text{ or } B) = P(A \cup B) = P(A) + P(B).$$

For example, the chance of rolling a 1 or 2 on a six-sided die is $P(1 \text{ or } 2) = P(1) + P(2) = \frac{1}{6} + \frac{1}{6} = \frac{1}{3}$.

Not Mutually Exclusive Events

If the events are not mutually exclusive then

$$P(A \text{ or } B) = P(A) + P(B) - P(A \text{ and } B).$$

For example, when drawing a single card at random from a regular deck of cards, the chance of getting a heart or a face card (J,Q,K) (or one that is both) is $\frac{13}{52} + \frac{12}{52} - \frac{3}{52} = \frac{11}{26}$, because of the 52 cards of a deck 13 are hearts, 12 are face cards, and 3 are both: here the possibilities included in the "3 that are both" are included in each of the "13 hearts" and the "12 face cards" but should only be counted once.

Conditional Probability

Conditional probability is the probability of some event A, given the occurrence of some other event B. Conditional probability is written $P(A|B)$ and is read "the probability of A, given B". It is defined by

$$P(A|B) = \frac{P(A \cap B)}{P(B)}.$$

If $P(B) = 0$ then $P(A|B)$ is formally undefined by this expression. However, it is possible to define a conditional probability for some zero-probability events using a σ-algebra of such events (such as those arising from a continuous random variable).

For example, in a bag of 2 red balls and 2 blue balls (4 balls in total), the probability of taking a red ball is $1/2$; however, when taking a second ball, the probability of it being either a red ball or a blue ball depends on the ball previously taken, such as, if a red ball was taken, the probability of picking a red ball again would be $1/3$ since only 1 red and 2 blue balls would have been remaining.

Inverse Probability

In probability theory and applications, Bayes' rule relates the odds of event A_1 to event A_2, before (prior to) and after (posterior to) conditioning on another event B. The odds on A_1 to event A_2 is simply the ratio of the probabilities of the two events. When arbitrarily many events A are of interest, not just two, the rule can be rephrased as posterior is proportional to prior times likelihood, $P(A|B) \propto P(A)P(B|A)$ where the proportionality symbol means that the left hand side is proportional to (i.e., equals a constant times) the right hand side as A varies, for fixed or given B (Lee, 2012; Bertsch McGrayne, 2012). In this form it goes back to Laplace (1774) and to Cournot (1843).

Summary of Probabilities

Summary of probabilities	
Event	**Probability**
A	$P(A) \in [0,1]$
not A	$P(A^c) = 1 - P(A)$
A or B	$P(A \cup B) = P(A) + P(B) - P(A \cap B)$ $P(A \cup B) = P(A) + P(B)$ if A and B are mutually exclusive

A and B	$P(A \cap B) = P(A\,\vert\,B)P(B) = P(B\,\vert\,A)P(A)$ $P(A \cap B) = P(A)P(B)$ if A and B are independent
A given B	$P(A\,\vert\,B) = \dfrac{P(A \cap B)}{P(B)} = \dfrac{P(B\,\vert\,A)P(A)}{P(B)}$

Relation to Randomness

In a deterministic universe, based on Newtonian concepts, there would be no probability if all conditions were known (Laplace's demon), (but there are situations in which sensitivity to initial conditions exceeds our ability to measure them, i.e. know them). In the case of a roulette wheel, if the force of the hand and the period of that force are known, the number on which the ball will stop would be a certainty (though as a practical matter, this would likely be true only of a roulette wheel that had not been exactly levelled — as Thomas A. Bass' Newtonian Casino revealed). Of course, this also assumes knowledge of inertia and friction of the wheel, weight, smoothness and roundness of the ball, variations in hand speed during the turning and so forth. A probabilistic description can thus be more useful than Newtonian mechanics for analyzing the pattern of outcomes of repeated rolls of a roulette wheel. Physicists face the same situation in kinetic theory of gases, where the system, while deterministic *in principle*, is so complex (with the number of molecules typically the order of magnitude of Avogadro constant $6.02 \cdot 10^{23}$) that only a statistical description of its properties is feasible.

Probability theory is required to describe quantum phenomena. A revolutionary discovery of early 20th century physics was the random character of all physical processes that occur at sub-atomic scales and are governed by the laws of quantum mechanics. The objective wave function evolves deterministically but, according to the Copenhagen interpretation, it deals with probabilities of observing, the outcome being explained by a wave function collapse when an observation is made. However, the loss of determinism for the sake of instrumentalism did not meet with universal approval. Albert Einstein famously remarked in a letter to Max Born: "I am convinced that God does not play dice". Like Einstein, Erwin Schrödinger, who discovered the wave function, believed quantum mechanics is a statistical approximation of an underlying deterministic reality. In some modern interpretations of the statistical mechanics of measurement, quantum decoherence is invoked to account for the appearance of subjectively probabilistic experimental outcomes.

References

- Chance, Beth L.; Rossman, Allan J. (2005). "Preface". Investigating Statistical Concepts, Applications, and Methods (PDF). Duxbury Press. ISBN 978-0-495-05064-3.

- Lakshmikantham,, ed. by D. Kannan,... V. (2002). Handbook of stochastic analysis and applications. New York: M. Dekker. ISBN 0824706609.

- Everitt, Brian (1998). The Cambridge Dictionary of Statistics. Cambridge, UK New York: Cambridge University Press. ISBN 0521593468.

- Drennan, Robert D. (2008). "Statistics in archaeology". In Pearsall, Deborah M. Encyclopedia of Archaeology. Elsevier Inc. pp. 2093–2100. ISBN 978-0-12-373962-9.

- Anderson, D.R.; Sweeney, D.J.; Williams, T.A. (1994) Introduction to Statistics: Concepts and Applications, pp. 5–9. West Group. ISBN 978-0-314-03309-3

- "Kendall's Advanced Theory of Statistics, Volume 1: Distribution Theory", Alan Stuart and Keith Ord, 6th Ed, (2009), ISBN 9780534243128

- Hogg, Robert V.; Craig, Allen; McKean, Joseph W. (2004). Introduction to Mathematical Statistics (6th ed.). Upper Saddle River: Pearson. ISBN 0-13-008507-3.

- Hacking, I. (2006) The Emergence of Probability: A Philosophical Study of Early Ideas about Probability, Induction and Statistical Inference, Cambridge University Press, ISBN 978-0-521-68557-3

- Ivancevic, Vladimir G.; Ivancevic, Tijana T. (2008). Quantum leap : from Dirac and Feynman, across the universe, to human body and mind. Singapore ; Hackensack, NJ: World Scientific. p. 16. ISBN 978-981-281-927-7.

- Franklin, James (2001). The Science of Conjecture: Evidence and Probability Before Pascal. Johns Hopkins University Press. ISBN 0801865697.

- Seneta, Eugene William. ""Adrien-Marie Legendre" (version 9)". StatProb: The Encyclopedia Sponsored by Statistics and Probability Societies. Retrieved 27 January 2016.

- Hájek, Alan. "Interpretations of Probability". The Stanford Encyclopedia of Philosophy (Winter 2012 Edition), Edward N. Zalta (ed.). Retrieved 22 April 2013.

- Ioannidis, J. P. A. (2005). "Why Most Published Research Findings Are False". PLoS Medicine. 2 (8): e124. doi:10.1371/journal.pmed.0020124. PMC 1182327. PMID 16060722.

- Warne, R. Lazo; Ramos, T.; Ritter, N. (2012). "Statistical Methods Used in Gifted Education Journals, 2006–2010". Gifted Child Quarterly. 56 (3): 134–149. doi:10.1177/0016986212444122.

Key Concepts of Probability

The key concepts of probability that have been explained in the section are random variable, event in probability, law of total probability, Venn diagram, mutual exclusivity etc. Random variable is a variable whose probable value depends on a set of random events. An event in probability and statistics is the result of a trial to which a probability is assigned. The section strategically encompasses and incorporates the major components and key concepts of probability, providing a complete understanding on the topic.

Random Variable

In probability and statistics, a random variable, random quantity, aleatory variable, or stochastic variable is a variable (i.e., not necessarily fixed) quantity whose possible values depend, in some clearly-defined way, on a set of random events. When the random variable is discrete, it can take on a value from a discrete set of possible different values, each with an associated probability. Like a traditional mathematical variable, its value is thus unknown a priori (before the outcome of the events is known).

In a more abstract sense, a random variable is defined as a function that maps outcomes (that is, points in a probability space) to mathematically convenient outcome labels, usually real numbers. In this sense, it is a procedure for assigning a number to an outcome, and, contrary to its name, this procedure itself is neither random nor variable. The function which characterizes a random variable must also be measurable, which rules out certain pathological cases such as those in which the random variable's quantity is infinitely sensitive to any small change in the outcome.

A random variable's possible values might represent the possible outcomes of a yet-to-be-performed experiment, or the possible outcomes of a past experiment whose already-existing value is uncertain (for example, due to imprecise measurements or quantum uncertainty). They may also conceptually represent either the results of an "objectively" random process (such as rolling a die) or the "subjective" randomness that results from incomplete knowledge of a quantity. The meaning of the probabilities assigned to the potential values of a random variable is not part of probability theory itself but is instead related to philosophical arguments over the interpretation of probability. The mathematics works the same regardless of the particular interpretation in use.

A random variable has a probability distribution, which specifies the probability that its value falls in any given interval. Random variables can be discrete, that is, taking any of a specified finite or countable list of values, endowed with a probability mass function characteristic of the random variable's probability distribution; or continuous, taking any numerical value in an interval or collection of intervals, via a probability density function that is characteristic of the random variable's probability distribution; or a mixture of both types. Two random variables with the same probability distribution can still differ in terms of their associations with, or independence from, other random variables. The realizations of a random variable, that is, the results of randomly choosing values according to the variable's probability distribution function, are called random variates.

The formal mathematical treatment of random variables is a topic in probability theory. In that context, a random variable is understood as a function defined on a sample space whose outputs are numerical values.

Definition

A *random variable* $X : \Omega \to E$ is a measurable function from the set of possible outcomes Ω to some set E. The technical axiomatic definition requires $\grave{U}$ to be a probability space and E to be a measurable space.

Note that although X is usually a real-valued function ($E = \mathbb{R}$), it does *not* return a probability. The probabilities of different outcomes or sets of outcomes (events) are already given by the probability measure P with which Ω is equipped. Rather, X describes some numerical property that outcomes in Ω may have — e.g., the number of heads in a random collection of coin flips, or the height of a random person. The probability that X takes value ≤ 3 is the probability of the set of outcomes $\{\omega \in \Omega : X(\omega) \leq 3\}$, denoted $P(X \leq 3)$.

Standard Case

In many cases, $E = \mathbb{R}$. In some contexts, the term *random element* is used to denote a random variable not of this form.

When the image (or range) of X is finite or countably infinite, the random variable is called a discrete random variable and its distribution can be described by a probability mass function which assigns a probability to each value in the image of X. If the image is uncountably infinite then X is called a continuous random variable. In the special case that it is absolutely continuous, its distribution can be described by a probability density function, which assigns probabilities to intervals; in particular, each individual point must necessarily have probability zero for an absolutely continuous random variable. Not all continuous random variables are absolutely continuous, for example a mixture distribution. Such random variables cannot be described by a probability density or a probability mass function.

Any random variable can be described by its cumulative distribution function, which describes the probability that the random variable will be less than or equal to a certain value.

Extensions

The term "random variable" in statistics is traditionally limited to the real-valued case ($E = \mathbb{R}$). This ensures that it is possible to define quantities such as the expected value and variance of a random variable, its cumulative distribution function, and the moments of its distribution.

However, the definition above is valid for any measurable space E of values. Thus one can consider random elements of other sets E, such as random boolean values, categorical values, complex numbers, vectors, matrices, sequences, trees, sets, shapes, manifolds, and functions. One may then specifically refer to a *random variable of type E*, or an *E − valued random variable*.

This more general concept of a random element is particularly useful in disciplines such as graph theory, machine learning, natural language processing, and other fields in discrete mathematics and computer science, where one is often interested in modeling the random variation of non-numerical data structures. In some cases, it is nonetheless convenient to represent each element of E using one or more real numbers. In this case, a random element may optionally be represented as a vector of real-valued random variables (all defined on the same underlying probability space Ω, which allows the different random variables to covary). For example:

- A random word may be represented as a random integer that serves as an index into the vocabulary of possible words. Alternatively, it can be represented as a random indicator vector whose length equals the size of the vocabulary, where the only values of positive probability are $(1\,0\,0\,0\cdots)$, $(0\,1\,0\,0\cdots)$, $(0\,0\,1\,0\cdots)$ and the position of the 1 indicates the word.

- A random sentence of given length N may be represented as a vector of N random words.

- A random graph on N given vertices may be represented as a $N \times N$ matrix of random variables, whose values specify the adjacency matrix of the random graph.

- A random function F may be represented as a collection of random variables $F(x)$, giving the function's values at the various points x in the function's domain. The $F(x)$ are ordinary real-valued random variables provided that the function is real-valued. For example, a stochastic process is a random function of time, a random vector is a random function of some index set such as $1, 2, \ldots$, and random field is a random function on any set (typically time, space, or a discrete set).

Examples

Discrete Random Variable

In an experiment a person may be chosen at random, and one random variable may be the person's height. Mathematically, the random variable is interpreted as a function which maps the person to the person's height. Associated with the random variable is a probability distribution that allows the computation of the probability that the height is in any subset of possible values, such as the probability that the height is between 180 and 190 cm, or the probability that the height is either less than 150 or more than 200 cm.

Another random variable may be the person's number of children; this is a discrete random variable with non-negative integer values. It allows the computation of probabilities for individual integer values – the probability mass function (PMF) – or for sets of values, including infinite sets. For example, the event of interest may be "an even number of children". For both finite and infinite event sets, their probabilities can be found by adding up the PMFs of the elements; that is, the probability of an even number of children is the infinite sum $\mathrm{PMF}(0) + \mathrm{PMF}(2) + \mathrm{PMF}(4) + \cdots$

In examples such as these, the sample space (the set of all possible persons) is often suppressed, since it is mathematically hard to describe, and the possible values of the random variables are then treated as a sample space. But when two random variables are measured on the same sample space of outcomes, such as the height and number of children being computed on the same random persons, it is easier to track their relationship if it is acknowledged that both height and number of children come from the same random person, for example so that questions of whether such random variables are correlated or not can be posed.

Coin Toss

The possible outcomes for one coin toss can be described by the sample space $\Omega = \{\text{heads}, \text{tails}\}$. We can introduce a real-valued random variable Y that models a \$1 payoff for a successful bet on heads as follows:

$$Y(\omega) = \begin{cases} 1, & \text{if } \omega = \text{heads}, \\ 0, & \text{if } \omega = \text{tails}. \end{cases}$$

If the coin is a fair coin, Y has a probability mass function f_Y given by:

$$f_Y(y) = \begin{cases} \frac{1}{2}, & \text{if } y = 1, \\ \frac{1}{2}, & \text{if } y = 0, \end{cases}$$

Dice Roll

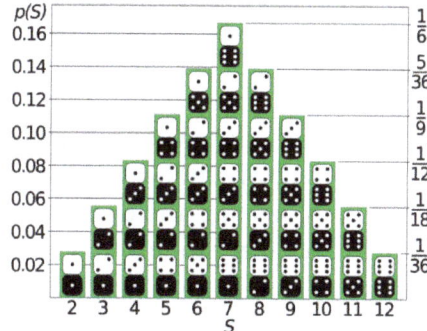

If the sample space is the set of possible numbers rolled on two dice, and the random variable of interest is the sum S of the numbers on the two dice, then S is a discrete random variable whose distribution is described by the probability mass function plotted as the height of picture columns here.

A random variable can also be used to describe the process of rolling dice and the possible outcomes. The most obvious representation for the two-dice case is to take the set of pairs of numbers n_1 and n_2 from $\{1, 2, 3, 4, 5, 6\}$ (representing the numbers on the two dice) as the sample space. The total number rolled (the sum of the numbers in each pair) is then a random variable X given by the function that maps the pair to the sum:

$$X((n_1, n_2)) = n_1 + n_2$$

and (if the dice are fair) has a probability mass function f_X given by:

$$f_X(S) = \frac{\min(S-1, 13-S)}{36}, \text{for } S \in \{2,3,4,5,6,7,8,9,10,11,12\}$$

Continuous Random Variable

An example of a continuous random variable would be one based on a spinner that can choose a horizontal direction. Then the values taken by the random variable are directions. We could represent these directions by North, West, East, South, Southeast, etc. However, it is commonly more convenient to map the sample space to a random variable which takes values which are real numbers. This can be done, for example, by mapping a direction to a bearing in degrees clockwise from North. The random variable then takes values which are real numbers from the interval [0, 360), with all parts of the range being "equally likely". In this case, X = the angle spun. Any real number has probability zero of being selected, but a positive probability can be assigned to any *range* of values. For example, the probability of choosing a number in [0, 180] is $\frac{1}{2}$. Instead of speaking of a probability mass function, we say that the probability *density* of X is 1/360. The probability of a subset of [0, 360) can be calculated by multiplying the measure of the set by 1/360. In general, the probability of a set for a given continuous random variable can be calculated by integrating the density over the given set.

Mixed Type

An example of a random variable of mixed type would be based on an experiment where a coin is flipped and the spinner is spun only if the result of the coin toss is heads. If the result is tails, $X = -1$; otherwise X = the value of the spinner as in the preceding example. There is a probability of $\frac{1}{2}$ that this random variable will have the value -1. Other ranges of values would have half the probabilities of the last example.

Measure-theoretic Definition

The most formal, axiomatic definition of a random variable involves measure theory. Continuous random variables are defined in terms of sets of numbers, along with functions that map such sets to probabilities. Because of various difficulties (e.g. the Banach–Tarski paradox) that arise if such sets are insufficiently constrained, it is necessary to introduce what is termed a sigma-algebra to constrain the possible sets over which probabilities can be defined. Normally, a particular such sigma-algebra is used, the Borel σ-algebra, which allows for probabilities to be defined over any sets that can be derived either directly from continuous intervals of numbers or by a finite or countably infinite number of unions and/or intersections of such intervals.

The measure-theoretic definition is as follows.

Let $(\Omega, \mathcal{F}, P)$ be a probability space and $(E, \mathcal{E})$ a measurable space. Then an $(E, \mathcal{E})$-valued random variable is a function $X : \Omega \to E$ which is $(\mathcal{F}, \mathcal{E})$-measurable. The latter means that, for every subset $B \in \mathcal{E}$, its preimage $X^{-1}(B) \in \mathcal{F}$ where $X^{-1}(B) = \{\omega : X(\omega) \in B\}$.. This definition enables us to measure any subset $B \in \mathcal{E}$ in the target space by looking at its preimage, which by assumption is measurable.

In more intuitive terms, Ω represents the "outcome", $\mathcal{F}$ represents the measurable subsets of possible outcomes, P represents the function that gives the probability of any such subset, E represents the kind of quantity the random value should take (such as real numbers), and $\mathcal{E}$ represents all the "well-behaved" (measurable) subsets of E (those for which you might want to find the probability). The random variable is then a function from any outcome to a quantity, such that the outcomes leading to any useful subset of quantities for the random variable have a well-defined probability.

When E is a topological space, then the most common choice for the σ-algebra $\mathcal{E}$ is the Borel σ-algebra $\mathcal{B}(E)$, which is the σ-algebra generated by the collection of all open sets in E. In such case the $(E, \mathcal{E})$-valued random variable is called the E-valued random variable. Moreover, when space E is the real line $\mathbb{R}$, then such a real-valued random variable is called simply the random variable.

Real-valued Random Variables

In this case the observation space is the set of real numbers. Recall, $(\Omega, \mathcal{F}, P)$ is the probability space. For real observation space, the function $X : \Omega \to \mathbb{R}$ is a real-valued random variable if

$$\{\omega : X(\omega) \le r\} \in \mathcal{F} \qquad \forall r \in \mathbb{R}.$$

This definition is a special case of the above because the set $\{(-\infty, r] : r \in \mathbb{R}\}$ generates the Borel σ-algebra on the set of real numbers, and it suffices to check measurability on any generating set. Here we can prove measurability on this generating set by using the fact that $\{\omega : X(\omega) \le r\} = X^{-1}((-\infty, r])$.

Distribution Functions of Random Variables

If a random variable $X : \Omega \to \mathbb{R}$ defined on the probability space $(\Omega, \mathcal{F}, P)$ is given, we can ask questions like "How likely is it that the value of X is equal to 2?". This is the same as the probability of the event $\{\omega : X(\omega) = 2\}$ which is often written as $P(X = 2)$ or $p_X(2)$ for short.

Recording all these probabilities of output ranges of a real-valued random variable X yields the probability distribution of X. The probability distribution "forgets" about the particular probability space used to define X and only records the probabilities of various values of X. Such a probability distribution can always be captured by its cumulative distribution function

$$F_X(x) = P(X \le x)$$

and sometimes also using a probability density function, p_X. In measure-theoretic terms, we use the random variable X to "push-forward" the measure P on Ω to a measure p_X on $\mathbb{R}$. The underlying probability space Ω is a technical device used to guarantee the existence of random variables, sometimes to construct them, and to define notions such as correlation and dependence or independence based on a joint distribution of two or more random variables on the same probability space. In practice, one often disposes of the space $\grave{U}$ altogether and just puts a measure on $\mathbb{R}$ that assigns measure 1 to the whole real line, i.e., one works with probability distributions instead of random variables.

Moments

The probability distribution of a random variable is often characterised by a small number of parameters, which also have a practical interpretation. For example, it is often enough to know what its "average value" is. This is captured by the mathematical concept of expected value of a random variable, denoted $E[X]$, and also called the first moment. In general, $E[f(X)]$ is not equal to

$f(\mathrm{E}[X])$. Once the "average value" is known, one could then ask how far from this average value the values of X typically are, a question that is answered by the variance and standard deviation of a random variable. $\mathrm{E}[X]$ can be viewed intuitively as an average obtained from an infinite population, the members of which are particular evaluations of X.

Mathematically, this is known as the (generalised) problem of moments: for a given class of random variables X, find a collection $\{f_i\}$ of functions such that the expectation values $\mathrm{E}[f_i(X)]$ fully characterise the distribution of the random variable X.

Moments can only be defined for real-valued functions of random variables (or complex-valued, etc.). If the random variable is itself real-valued, then moments of the variable itself can be taken, which are equivalent to moments of the identity function $f(X)=X$ of the random variable. However, even for non-real-valued random variables, moments can be taken of real-valued functions of those variables. For example, for a categorical random variable X that can take on the nominal values "red", "blue" or "green", the real-valued function $[X=\mathrm{green}]$ can be constructed; this uses the Iverson bracket, and has the value 1 if X has the value "green", 0 otherwise. Then, the expected value and other moments of this function can be determined.

Functions of Random Variables

A new random variable Y can be defined by applying a real Borel measurable function $g:\mathbb{R}\to\mathbb{R}$ to the outcomes of a real-valued random variable X. The cumulative distribution function of Y is

$$F_Y(y)=\mathrm{P}(g(X)\le y).$$

If function g is invertible, i.e., g^{-1} exists, and is either increasing or decreasing, then the previous relation can be extended to obtain

$$F_Y(y)=\mathrm{P}(g(X)\le y)=\begin{cases}\mathrm{P}(X\le g^{-1}(y))=F_X(g^{-1}(y)), & \text{if } g^{-1} \text{ increasing,}\\[2mm]\mathrm{P}(X\ge g^{-1}(y))=1-F_X(g^{-1}(y)), & \text{if } g^{-1} \text{ decreasing.}\end{cases}$$

and, again with the same hypotheses of invertibility of g, assuming also differentiability, we can find the relation between the probability density functions by differentiating both sides with respect to y, in order to obtain

$$f_Y(y)=f_X(g^{-1}(y))\left|\frac{dg^{-1}(y)}{dy}\right|.$$

If there is no invertibility of g but each y admits at most a countable number of roots (i.e., a finite, or countably infinite, number of x_i such that $y=g(x_i)$) then the previous relation between the probability density functions can be generalized with

$$f_Y(y)=\sum_i f_X(g_i^{-1}(y))\left|\frac{dg_i^{-1}(y)}{dy}\right|$$

where $x_i = g_i^{-1}(y)$. The formulas for densities do not demand g to be increasing.

In the measure-theoretic, axiomatic approach to probability, if we have a random variable X on Ω and a Borel measurable function $g : \mathbb{R} \to \mathbb{R}$, then $Y = g(X)$ will also be a random variable on Ω, since the composition of measurable functions is also measurable. (However, this is not true if g is Lebesgue measurable.) The same procedure that allowed one to go from a probability space (Ω, P) to $(\mathbb{R}, dF_X)$ can be used to obtain the distribution of Y .

Example 1

Let X be a real-valued, continuous random variable and let $Y = X^2$.

$$F_Y(y) = P(X^2 \le y).$$

If $y < 0$, then $P(X^2 \le y) = 0$, so

$$F_Y(y) = 0 \quad \text{if} \quad y < 0.$$

If $y \ge 0$, then

$$P(X^2 \le y) = P(|X| \le \sqrt{y}) = P(-\sqrt{y} \le X \le \sqrt{y}),$$

so

$$F_Y(y) = F_X(\sqrt{y}) - F_X(-\sqrt{y}) \quad \text{if} \quad y \ge 0.$$

Example 2

Suppose X is a random variable with a cumulative distribution

$$F_X(x) = P(X \le x) = \frac{1}{(1+e^{-x})^\theta}$$

where $\theta > 0$ is a fixed parameter. Consider the random variable $Y = \log(1 + e^{-X})$. Then,

$$F_Y(y) = P(Y \le y) = P(\log(1+e^{-X}) \le y) = P(X > -\log(e^y - 1)).$$

The last expression can be calculated in terms of the cumulative distribution of X, so

$$F_Y(y) = 1 - F_X(-\log(e^y - 1))$$

$$= 1 - \frac{1}{(1 + e^{\log(e^y-1)})^\theta}$$

$$= 1 - \frac{1}{(1 + e^y - 1)^\theta}$$

$$= 1 - e^{-y\theta}.$$

which is the cdf of an exponential distribution.

Example 3

Suppose X is a random variable with a standard normal distribution, whose density is

$$f_X(x) = \frac{1}{\sqrt{2\pi}} e^{-x^2/2}.$$

Consider the random variable $Y = X^2$. We can find the density using the above formula for a change of variables:

$$f_Y(y) = \sum_i f_X(g_i^{-1}(y)) \left| \frac{dg_i^{-1}(y)}{dy} \right|.$$

In this case the change is not monotonic, because every value of Y has two corresponding values of X (one positive and negative). However, because of symmetry, both halves will transform identically, i.e.,

$$f_Y(y) = 2 f_X(g^{-1}(y)) \left| \frac{dg^{-1}(y)}{dy} \right|.$$

The inverse transformation is

$$x = g^{-1}(y) = \sqrt{y}$$

and its derivative is

$$\frac{dg^{-1}(y)}{dy} = \frac{1}{2\sqrt{y}}.$$

Then,

$$f_Y(y) = 2 \frac{1}{\sqrt{2\pi}} e^{-y/2} \frac{1}{2\sqrt{y}} = \frac{1}{\sqrt{2\pi y}} e^{-y/2}.$$

This is a chi-squared distribution with one degree of freedom.

Equivalence of Random Variables

There are several different senses in which random variables can be considered to be equivalent. Two random variables can be equal, equal almost surely, or equal in distribution.

In increasing order of strength, the precise definition of these notions of equivalence is given below.

Equality in Distribution

If the sample space is a subset of the real line, random variables X and Y are *equal in distribution* (denoted $X \stackrel{d}{=} Y$) if they have the same distribution functions:

$$P(X \leq x) = P(Y \leq x) \quad \text{for all} \quad x.$$

Two random variables having equal moment generating functions have the same distribution. This provides, for example, a useful method of checking equality of certain functions of i.i.d. random variables. However, the moment generating function exists only for distributions that have a defined Laplace transform.

Almost Sure Equality

Two random variables X and Y are *equal almost surely* if, and only if, the probability that they are different is zero:

$$P(X \neq Y) = 0.$$

For all practical purposes in probability theory, this notion of equivalence is as strong as actual equality. It is associated to the following distance:

$$d_\infty(X,Y) = \operatorname*{ess\,sup}_\omega | X(\omega) - Y(\omega) |,$$

where "ess sup" represents the essential supremum in the sense of measure theory.

Equality

Finally, the two random variables X and Y are *equal* if they are equal as functions on their measurable space:

$$X(\omega) = Y(\omega) \qquad \textit{for all } \omega.$$

Convergence

A significant theme in mathematical statistics consists of obtaining convergence results for certain sequences of random variables; for instance the law of large numbers and the central limit theorem.

There are various senses in which a sequence (X_n) of random variables can converge to a random variable X. These are explained in the article on convergence of random variables.

Event (Probability Theory)

In probability theory, an event is a set of outcomes of an experiment (a subset of the sample space) to which a probability is assigned. A single outcome may be an element of many different events, and different events in an experiment are usually not equally likely, since they may include very different groups of outcomes. An event defines a complementary event, namely the complementary set (the event *not* occurring), and together these define a Bernoulli trial: did the event occur or not?

Typically, when the sample space is finite, any subset of the sample space is an event (*i.e.* all elements of the power set of the sample space are defined as events). However, this approach does not work well in cases where the sample space is uncountably infinite, most notably when the outcome is a real number. So, when defining a probability space it is possible, and often necessary, to exclude certain subsets of the sample space from being events.

A Simple Example

If we assemble a deck of 52 playing cards with no jokers, and draw a single card from the deck, then the sample space is a 52-element set, as each card is a possible outcome. An event, however, is any subset of the sample space, including any singleton set (an elementary event), the empty set (an impossible event, with probability zero) and the sample space itself (a certain event, with probability one). Other events are proper subsets of the sample space that contain multiple elements. So, for example, potential events include:

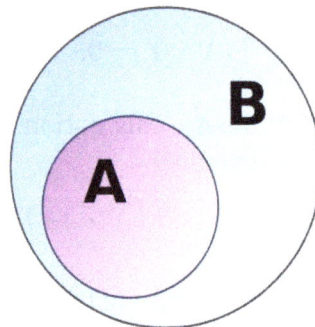

A Venn diagram of an event. B is the sample space and A is an event. By the ratio of their areas, the probability of A is approximately 0.4.

- "Red and black at the same time without being a joker" (0 elements),
- "The 5 of Hearts" (1 element),
- "A King" (4 elements),
- "A Face card" (12 elements),
- "A Spade" (13 elements),
- "A Face card or a red suit" (32 elements),
- "A card" (52 elements).

Since all events are sets, they are usually written as sets (e.g. {1, 2, 3}), and represented graphically using Venn diagrams. In the situation where each outcome in the sample space Ω is equally likely, the probability P of an event A is the following formula:

$$P(A) = \frac{|A|}{|\Omega|} \left(\text{alternatively: } \Pr(A) = \frac{|A|}{|\Omega|} \right)$$

This rule can readily be applied to each of the example events above.

Events in Probability Spaces

Defining all subsets of the sample space as events works well when there are only finitely many outcomes, but gives rise to problems when the sample space is infinite. For many standard probability distributions, such as the normal distribution, the sample space is the set of real numbers or some subset of the real numbers. Attempts to define probabilities for all subsets of the real numbers run into difficulties when one considers 'badly behaved' sets, such as those that are nonmeasurable. Hence, it is necessary to restrict attention to a more limited family of subsets. For the standard tools of probability theory, such as joint and conditional probabilities, to work, it is necessary to use a σ-algebra, that is, a family closed under complementation and countable unions of its members. The most natural choice is the Borel measurable set derived from unions and intersections of intervals. However, the larger class of Lebesgue measurable sets proves more useful in practice.

In the general measure-theoretic description of probability spaces, an event may be defined as an element of a selected σ-algebra of subsets of the sample space. Under this definition, any subset of the sample space that is not an element of the σ-algebra is not an event, and does not have a probability. With a reasonable specification of the probability space, however, all *events of interest* are elements of the σ-algebra.

A Note on Notation

Even though events are subsets of some sample space Ω, they are often written as propositional formulas involving random variables. For example, if X is a real-valued random variable defined on the sample space Ω, the event

$$\{\omega \in \Omega \mid u < X(\omega) \le v\}$$

can be written more conveniently as, simply,

$$u < X \le v.$$

This is especially common in formulas for a probability, such as

$$\Pr(u < X \le v) = F(v) - F(u).$$

The set $u < X \le v$ is an example of an inverse image under the mapping X because $\omega \in X^{-1}((u,v])$ if and only if $u < X(\omega) \le v$.

Types of Event (Probability Theory)

Elementary Event

In probability theory, an elementary event (also called an atomic event or simple event) is an event which contains only a single outcome in the sample space. Using set theory terminology, an elementary event is a singleton. Elementary events and their corresponding outcomes are often written interchangeably for simplicity, as such an event corresponds to precisely one outcome.

The following are examples of elementary events:

- All sets $\{k\}$, where $k \in \mathbb{N}$ if objects are being counted and the sample space is $S = \{0, 1, 2, 3, ...\}$ (the natural numbers).

- $\{HH\}$, $\{HT\}$, $\{TH\}$ and $\{TT\}$ if a coin is tossed twice. $S = \{HH, HT, TH, TT\}$. H stands for heads and T for tails.

- All sets $\{x\}$, where x is a real number. Here X is a random variable with a normal distribution and $S = (-\infty, +\infty)$. This example shows that, because the probability of each elementary event is zero, the probabilities assigned to elementary events do not determine a continuous probability distribution.

Probability of an Elementary Event

Elementary events may occur with probabilities that are between zero and one (inclusively). In a discrete probability distribution whose sample space is finite, each elementary event is assigned a particular probability. In contrast, in a continuous distribution, individual elementary events must all have a probability of zero because there are infinitely many of them— then non-zero probabilities can only be assigned to non-elementary events.

Some "mixed" distributions contain both stretches of continuous elementary events and some discrete elementary events; the discrete elementary events in such distributions can be called atoms or atomic events and can have non-zero probabilities.

Under the measure-theoretic definition of a probability space, the probability of an elementary event need not even be defined. In particular, the set of events on which probability is defined may be some σ-algebra on S and not necessarily the full power set.

Complementary Event

In probability theory, the complement of any event A is the event [not A], i.e. the event that A does not occur. The event A and its complement [not A] are mutually exclusive and exhaustive. Generally, there is only one event B such that A and B are both mutually exclusive and exhaustive; that event is the complement of A. The complement of an event A is usually denoted as A', A^c or $\overline{A}$. Given an event, the event and its complementary event define a Bernoulli trial: did the event occur or not?

For example, if a typical coin is tossed and one assumes that it cannot land on its edge, then it can either land showing "heads" or "tails." Because these two outcomes are mutually exclusive (i.e. the coin cannot simultaneously show both heads and tails) and collectively exhaustive (i.e. there are no other possible outcomes not represented between these two), they are therefore each other's complements. This means that [heads] is logically equivalent to [not tails], and [tails] is equivalent to [not heads].

Complement Rule

In a random experiment, the probabilities of all possible events (the sample space) must total to 1— that is, some outcome must occur on every trial. For two events to be complements, they must be collectively exhaustive, together filling the entire sample space. Therefore, the probability of an event's complement must be unity minus the probability of the event. That is, for an event A,

$$P(A^c) = 1 - P(A).$$

Equivalently, the probabilities of an event and its complement must always total to 1. This does not, however, mean that *any* two events whose probabilities total to 1 are each other's complements; complementary events must also fulfill the condition of mutual exclusivity.

Example of the Utility of this Concept

Suppose one throws an ordinary six-sided die eight times. What is the probability that one sees a "1" at least once?

It may be tempting to say that

Pr(["1" on 1st trial] or ["1" on second trial] or … or ["1" on 8th trial])

= Pr("1" on 1st trial) + Pr("1" on second trial) + … + P("1" on 8th trial)

= 1/6 + 1/6 + … + 1/6.

= 8/6 = 1.3333… (…and this is clearly wrong.)

That cannot be right because a probability cannot be more than 1. The technique is wrong because the eight events whose probabilities got added are not mutually exclusive.

One may resolve this overlap by the principle of inclusion-exclusion, or in this case one may instead more simply find the probability of the complementary event and subtract it from 1, thus:

Pr(at least one "1") = 1 − Pr(no "1"s)

= 1 − Pr([no "1" on 1st trial] and [no "1" on 2nd trial] and … and [no "1" on 8th trial])

= 1 − Pr(no "1" on 1st trial) × Pr(no "1" on 2nd trial) × … × Pr(no "1" on 8th trial)

= 1 −(5/6) × (5/6) × … × (5/6)

= $1 - (5/6)^8$

= 0.7674…

Independence (Probability Theory)

In probability theory, two events are independent, statistically independent, or stochastically independent if the occurrence of one does not affect the probability of the other. Similarly, two random variables are independent if the realization of one does not affect the probability distribution of the other.

The concept of independence extends to dealing with collections of more than two events or random variables, in which case the events are pairwise independent if each pair are independent of each other, and the events are mutually independent if each event is independent of each other combination of events.

Definition

Two Events

Two events A and B are independent (often written as $A \perp B$ or $A \perp\!\!\!\perp B$) if their joint probability equals the product of their probabilities:

$$P(A \cap B) = P(A)P(B).$$

Why this defines independence is made clear by rewriting with conditional probabilities:

$$P(A \cap B) = P(A)P(B) \Leftrightarrow P(A) = \frac{P(A)P(B)}{P(B)} = \frac{P(A \cap B)}{P(B)} = P(A \mid B).$$

and similarly

$$P(A \cap B) = P(A)P(B) \Leftrightarrow P(B) = P(B \mid A).$$

Thus, the occurrence of B does not affect the probability of A, and vice versa. Although the derived expressions may seem more intuitive, they are not the preferred definition, as the conditional probabilities may be undefined if $P(A)$ or $P(B)$ are o. Furthermore, the preferred definition makes clear by symmetry that when A is independent of B, B is also independent of A.

More than Two Events

A finite set of events $\{A_i\}$ is pairwise independent if and only if every pair of events is independent—that is, if and only if for all distinct pairs of indices m, k,

$$P(A_m \cap A_k) = P(A_m)P(A_k).$$

A finite set of events is mutually independent if and only if every event is independent of any intersection of the other events—that is, if and only if for every n-element subset $\{A_i\}$,

$$P\left(\bigcap_{i=1}^{n} A_i\right) = \prod_{i=1}^{n} P(A_i).$$

This is called the *multiplication rule* for independent events. Note that it is not a single condition involving only the product of all the probabilities of all single events it must hold true for all subset of events.

For more than two events, a mutually independent set of events is (by definition) pairwise independent; but the converse is not necessarily true.

For Random Variables

Two Random Variables

Two random variables X and Y are independent if and only if (iff) the elements of the π-system generated by them are independent; that is to say, for every a and b, the events $\{X \le a\}$ and $\{Y \le b\}$ are independent events (as defined above). That is, X and Y with cumulative distribution functions $F_X(x)$ and $F_Y(y)$, and probability densities $f_X(x)$ and $f_Y(y)$, are independent iff the combined random variable (X, Y) has a joint cumulative distribution function

$$F_{X,Y}(x, y) = F_X(x)F_Y(y),$$

or equivalently, if the joint density exists,

$$f_{X,Y}(x, y) = f_X(x)f_Y(y).$$

More than Two Random Variables

A set of random variables is pairwise independent if and only if every pair of random variables is independent.

A set of random variables is mutually independent if and only if for any finite subset $X_1, \dots, X_n$ and any finite sequence of numbers $a_1, \dots, a_n$, the events $\{X_1 \le a_1\}, \dots, \{X_n \le a_n\}$ are mutually independent events (as defined above).

The measure-theoretically inclined may prefer to substitute events $\{X \in A\}$ for events $\{X \le a\}$ in the above definition, where A is any Borel set. That definition is exactly equivalent to the one above when the values of the random variables are real numbers. It has the advantage of working also for complex-valued random variables or for random variables taking values in any measurable space (which includes topological spaces endowed by appropriate σ-algebras).

Conditional Independence

Intuitively, two random variables X and Y are conditionally independent given Z if, once Z is known, the value of Y does not add any additional information about X. For instance, two measurements X and Y of the same underlying quantity Z are not independent, but they are conditionally independent given Z (unless the errors in the two measurements are somehow connected).

The formal definition of conditional independence is based on the idea of conditional distributions. If X, Y, and Z are discrete random variables, then we define X and Y to be *conditionally independent given Z* if

$$P(X \le x, Y \le y \mid Z = z) = P(X \le x \mid Z = z) \cdot P(Y \le y \mid Z = z)$$

for all x, y and z such that $P(Z = z) > 0$. On the other hand, if the random variables are continuous and have a joint probability density function p, then X and Y are conditionally independent given Z if

$$p_{XY|Z}(x, y \mid z) = p_{X|Z}(x \mid z) \cdot p_{Y|Z}(y \mid z)$$

for all real numbers x, y and z such that $p_Z(z) > 0$.

If X and Y are conditionally independent given Z, then

$$P(X = x \mid Y = y, Z = z) = P(X = x \mid Z = z)$$

for any x, y and z with $P(Z = z) > 0$. That is, the conditional distribution for X given Y and Z is the same as that given Z alone. A similar equation holds for the conditional probability density functions in the continuous case.

Independence can be seen as a special kind of conditional independence, since probability can be seen as a kind of conditional probability given no events.

Independent σ-Algebras

The definitions above are both generalized by the following definition of independence for σ-algebras. Let $(\Omega, \Sigma, \Pr)$ be a probability space and let A and B be two sub-σ-algebras of Σ. A and B are said to be independent if, whenever $A \in A$ and $B \in B$,

$$P(A \cap B) = P(A)P(B).$$

Likewise, a finite family of σ-algebras $(\tau_i)_{i \in I}$ is said to be independent if and only if

$$\forall (A_i)_{i \in I} \in \prod_{i \in I} \tau_i : P\left(\bigcap_{i \in I} A_i\right) = \prod_{i \in I} P(A_i)$$

and an infinite family of σ-algebras is said to be independent if all its finite subfamilies are independent.

The new definition relates to the previous ones very directly:

- Two events are independent (in the old sense) if and only if the σ-algebras that they generate are independent (in the new sense). The σ-algebra generated by an event $E \in \Sigma$ is, by definition,

$$\sigma(\{E\}) = \{\varnothing, E, \Omega \setminus E, \Omega\}.$$

- Two random variables X and Y defined over Ω are independent (in the old sense) if and only if the σ-algebras that they generate are independent (in the new sense). The σ-algebra generated by a random variable X taking values in some measurable space S consists, by definition, of all subsets of Ω of the form $X^{-1}(U)$, where U is any measurable subset of S.

Using this definition, it is easy to show that if X and Y are random variables and Y is constant, then X and Y are independent, since the σ-algebra generated by a constant random variable is the trivial σ-algebra $\{\varnothing, \Omega\}$. Probability zero events cannot affect independence so independence also holds if Y is only Pr-almost surely constant.

Properties

Self-independence

An event is independent of itself if and only if

$$P(A) = P(A \cap A) = P(A) \cdot P(A) \Leftrightarrow P(A) = 0 \text{ or} 1.$$

Thus an event is independent of itself if and only if it almost surely occurs or its complement almost surely occurs.

Expectation and Covariance

If X and Y are independent, then the expectation operator E has the property

$$E[XY] = E[X]E[Y],$$

and the covariance cov(X, Y) is zero, since we have

$$\text{cov}[X, Y] = E[XY] - E[X]E[Y].$$

(The converse of these, i.e. the proposition that if two random variables have a covariance of 0 they must be independent, is not true.)

Characteristic Function

Two random variables X and Y are independent if and only if the characteristic function of the random vector (X, Y) satisfies

$$\varphi_{(X,Y)}(t,s) = \varphi_X(t) \cdot \varphi_Y(s).$$

In particular the characteristic function of their sum is the product of their marginal characteristic functions:

$$\varphi_{X+Y}(t) = \varphi_X(t) \cdot \varphi_Y(t),$$

though the reverse implication is not true. Random variables that satisfy the latter condition are called subindependent.

Examples

Rolling Dice

The event of getting a 6 the first time a die is rolled and the event of getting a 6 the second time are *independent*. By contrast, the event of getting a 6 the first time a die is rolled and the event that the sum of the numbers seen on the first and second trials is 8 are *not* independent.

Drawing Cards

If two cards are drawn *with* replacement from a deck of cards, the event of drawing a red card on the first trial and that of drawing a red card on the second trial are *independent*. By contrast, if two cards are drawn *without* replacement from a deck of cards, the event of drawing a red card on the first trial and that of drawing a red card on the second trial are again *not* independent.

Pairwise and Mutual Independence

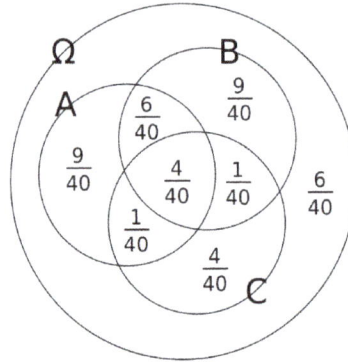

Pairwise independent, but not mutually independent, events.

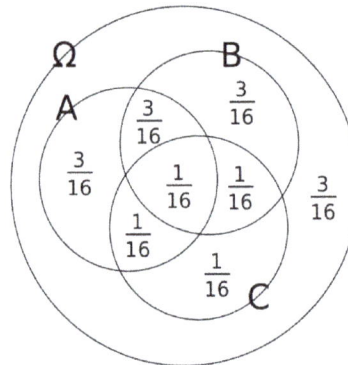

Mutually independent events.

Consider the two probability spaces shown. In both cases, $P(A) = P(B) = 1/2$ and $P(C) = 1/4$ The first space is pairwise independent but not mutually independent. The second space is both pairwise independent and mutually independent. To illustrate the difference, consider conditioning on two events. In the pairwise independent case, although any one event is independent of each of the other two individually, it is not independent of the intersection of the other two:

$$P(A \mid BC) = \frac{\dfrac{4}{40}}{\dfrac{4}{40} + \dfrac{1}{40}} = \tfrac{4}{5} \neq P(A)$$

$$P(B \mid AC) = \frac{\dfrac{4}{40}}{\dfrac{4}{40} + \dfrac{1}{40}} = \tfrac{4}{5} \neq P(B)$$

$$P(C \mid AB) = \frac{\dfrac{4}{40}}{\dfrac{4}{40} + \dfrac{6}{40}} = \tfrac{2}{5} \neq P(C)$$

In the mutually independent case however:

$$P(A \mid BC) = \frac{\frac{1}{16}}{\frac{1}{16} + \frac{1}{16}} = \tfrac{1}{2} = P(A)$$

$$P(B \mid AC) = \frac{\frac{1}{16}}{\frac{1}{16} + \frac{1}{16}} = \tfrac{1}{2} = P(B)$$

$$P(C \mid AB) = \frac{\frac{1}{16}}{\frac{1}{16} + \frac{3}{16}} = \tfrac{1}{4} = P(C)$$

Mutual Independence

It is possible to create a three-event example in which

$$P(A \cap B \cap C) = P(A)P(B)P(C),$$

and yet no two of the three events are pairwise independent (and hence the set of events are not mutually independent). This example shows that mutual independence involves requirements on the products of probabilities of all combinations of events, not just the single events as in this example.

Law of Total Probability

In probability theory, the law (or formula) of total probability is a fundamental rule relating marginal probabilities to conditional probabilities. It expresses the total probability of an outcome which can be realized via several distinct events - hence the name.

Statement

The law of total probability is the proposition that if $\{B_n : n = 1, 2, 3, \ldots\}$ is a finite or countably infinite partition of a sample space (in other words, a set of pairwise disjoint events whose union is the entire sample space) and each event B_n is measurable, then for any event A of the same probability space:

$$\Pr(A) = \sum_n \Pr(A \cap B_n)$$

or, alternatively,

$$\Pr(A) = \sum_n \Pr(A \mid B_n) \Pr(B_n),$$

where, for any n for which $\Pr(B_n) = 0$ these terms are simply omitted from the summation, because $\Pr(A \mid B_n)$ is finite.

The summation can be interpreted as a weighted average, and consequently the marginal probability, $\Pr(A)$, is sometimes called "average probability"; "overall probability" is sometimes used in less formal writings.

The law of total probability can also be stated for conditional probabilities. Taking the B_n as above, and assuming C is an event independent with any of the B_n:

$$\Pr(A\,|\,C) = \sum_n \Pr(A\,|\,C \cap B_n)\Pr(B_n\,|\,C) = \sum_n \Pr(A\,|\,C \cap B_n)\Pr(B_n)$$

Informal Formulation

The above mathematical statement might be interpreted as follows: *given an outcome A, with known conditional probabilities given any of the B_n events, each with a known probability itself, what is the total probability that A will happen?* The answer to this question is given by $\Pr(A)$.

Example

Suppose that two factories supply light bulbs to the market. Factory X's bulbs work for over 5000 hours in 99% of cases, whereas factory Y's bulbs work for over 5000 hours in 95% of cases. It is known that factory X supplies 60% of the total bulbs available. What is the chance that a purchased bulb will work for longer than 5000 hours?

Applying the law of total probability, we have:

$$\Pr(A) = \Pr(A\,|\,B_X)\cdot\Pr(B_X) + \Pr(A\,|\,B_Y)\cdot\Pr(B_Y) = \frac{99}{100}\cdot\frac{6}{10} + \frac{95}{100}\cdot\frac{4}{10} = \frac{594+380}{1000} = \frac{974}{1000}$$

where

- $\Pr(B_X) = \frac{6}{10}$ is the probability that the purchased bulb was manufactured by factory X;
- $\Pr(B_Y) = \frac{4}{10}$ is the probability that the purchased bulb was manufactured by factory Y;
- $\Pr(A\,|\,B_X) = \frac{99}{100}$ is the probability that a bulb manufactured by X will work for over 5000 hours;
- $\Pr(A\,|\,B_Y) = \frac{95}{100}$ is the probability that a bulb manufactured by Y will work for over 5000 hours.

Thus each purchased light bulb has a 97.4% chance to work for more than 5000 hours.

Applications

One common application of the law is where the events coincide with a discrete random variable X taking each value in its range, i.e. B_n is the event $X = x_n$. It follows that the probability of an event A is equal to the expected value of the conditional probabilities of A given $X = x_n$. That is,

$$\Pr(A) = \sum_n \Pr(A\,|\,X = x_n)\Pr(X = x_n) = E[\Pr(A\,|\,X)],$$

where $\Pr(A \mid X)$ is the conditional probability of A given the value of the random variable X. This conditional probability is a random variable in its own right, whose value depends on that of X. The conditional probability $\Pr(A \mid X = x)$ is simply a conditional probability given an event, $[X = x]$. It is a function of x, say $g(x) = \Pr(A \mid X = x)$. Then the conditional probability $\Pr(A \mid X)$ is $g(X)$, hence itself a random variable. This version of the law of total probability says that the expected value of this random variable is the same as $\Pr(A)$.

This result can be generalized to continuous random variables (via continuous conditional density), and the expression becomes

$$\Pr(A) = E[\Pr(A \mid \mathcal{F}_X)],$$

where $\mathcal{F}_X$ denotes the sigma-algebra generated by the random variable X.

Other Names

The term *law of total probability* is sometimes taken to mean the law of alternatives, which is a special case of the law of total probability applying to discrete random variables.One author even uses the terminology "continuous law of alternatives" in the continuous case. This result is given by Grimmett and Welsh as the partition theorem, a name that they also give to the related law of total expectation.

Venn Diagram

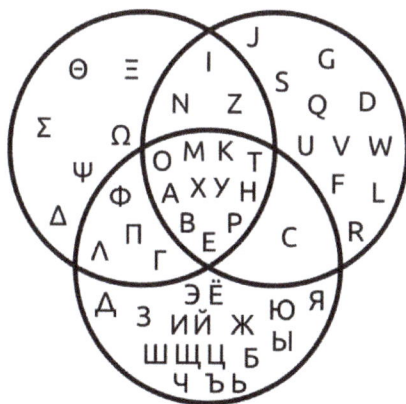

Venn diagram showing which uppercase letter glyphs are shared by the Greek, Latin and Russian alphabets

A Venn diagram (also called a set diagram or logic diagram) is a diagram that shows all possible logical relations between a finite collection of different sets. Typically overlapping shapes, usually circles, are used, and an area-proportional or scaled Venn diagram is one in which the area of the shape is proportional to the number of elements it contains. These diagrams represent elements as points in the plane, and sets as regions inside curves. An element is in a set S just in case the corresponding point is in the region for S. They are thus a special case of Euler diagrams, which do not necessarily show all relations. Venn diagrams were conceived around 1880 by John Venn. They are used to teach elementary set theory, as well as illustrate simple set relationships in probability, logic, statistics, linguistics and computer science.

Example

Sets A (creatures with two legs) and B (creatures that can fly)

This example involves two sets, A and B, represented here as coloured circles. The orange circle, set A, represents all living creatures that are two-legged. The blue circle, set B, represents the living creatures that can fly. Each separate type of creature can be imagined as a point somewhere in the diagram. Living creatures that both can fly *and* have two legs—for example, parrots—are then in both sets, so they correspond to points in the region where the blue and orange circles overlap. That region contains all such and only such living creatures.

Humans and penguins are bipedal, and so are then in the orange circle, but since they cannot fly they appear in the left part of the orange circle, where it does not overlap with the blue circle. Mosquitoes have six legs, and fly, so the point for mosquitoes is in the part of the blue circle that does not overlap with the orange one. Creatures that are not two-legged and cannot fly (for example, whales and spiders) would all be represented by points outside both circles.

The combined region of sets A and B is called the *union* of A and B, denoted by A ∪ B. The union in this case contains all living creatures that are either two-legged or that can fly (or both).

The region in both A and B, where the two sets overlap, is called the *intersection* of A and B, denoted by A ∩ B. For example, the intersection of the two sets is not empty, because there *are* points that represent creatures that are in *both* the orange and blue circles.

History

Venn diagrams were introduced in 1880 by John Venn in a paper entitled *On the Diagrammatic and Mechanical Representation of Propositions and Reasonings* in the "Philosophical Magazine and Journal of Science", about the different ways to represent propositions by diagrams. The use of these types of diagrams in formal logic, according to Ruskey and M. Weston, is "not an easy history to trace, but it is certain that the diagrams that are popularly associated with Venn, in fact, originated much earlier. They are rightly associated with Venn, however, because he comprehensively surveyed and formalized their usage, and was the first to generalize them".

Venn himself did not use the term "Venn diagram" and referred to his invention as "Eulerian Circles". For example, in the opening sentence of his 1880 article Venn writes, "Schemes of diagrammatic representation have been so familiarly introduced into logical treatises during the last century or so, that many readers, even those who have made no professional study of logic, may be supposed to be acquainted with the general nature and object of such devices. Of these schemes one only, viz. that commonly called 'Eulerian circles,' has met with any general acceptance..." The first to use the term "Venn diagram" was Clarence Irving Lewis in 1918, in his book "A Survey of Symbolic Logic".

Venn diagrams are very similar to Euler diagrams, which were invented by Leonhard Euler in the 18th century. M. E. Baron has noted that Leibniz (1646–1716) in the 17th century produced similar diagrams before Euler, but much of it was unpublished. She also observes even earlier Euler-like diagrams by Ramon Lull in the 13th Century.

In the 20th century, Venn diagrams were further developed. D.W. Henderson showed in 1963 that the existence of an n-Venn diagram with n-fold rotational symmetry implied that n was a prime number. He also showed that such symmetric Venn diagrams exist when n is 5 or 7. In 2002 Peter Hamburger found symmetric Venn diagrams for $n = 11$ and in 2003, Griggs, Killian, and Savage showed that symmetric Venn diagrams exist for all other primes. Thus rotationally symmetric Venn diagrams exist if and only if n is a prime number.

Venn diagrams and Euler diagrams were incorporated as part of instruction in set theory as part of the new math movement in the 1960s. Since then, they have also been adopted in the curriculum of other fields such as reading.

Overview

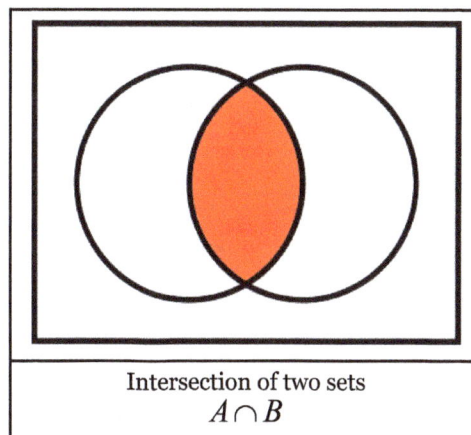

Intersection of two sets
$A \cap B$

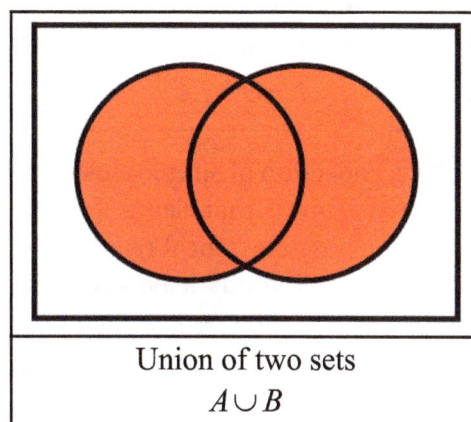

Union of two sets
$A \cup B$

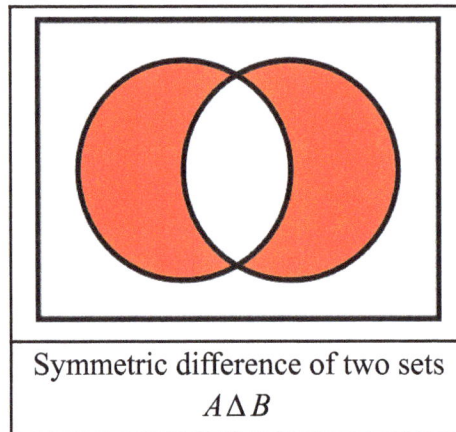

Symmetric difference of two sets
$A \triangle B$

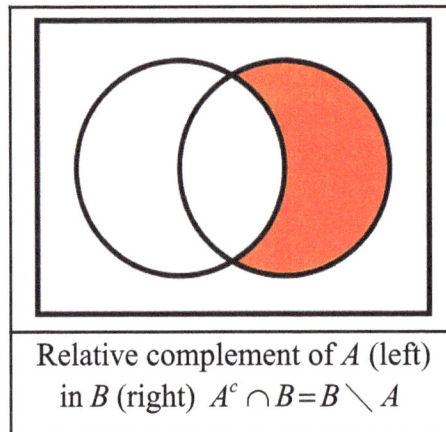

Relative complement of A (left)
in B (right) $A^c \cap B = B \setminus A$

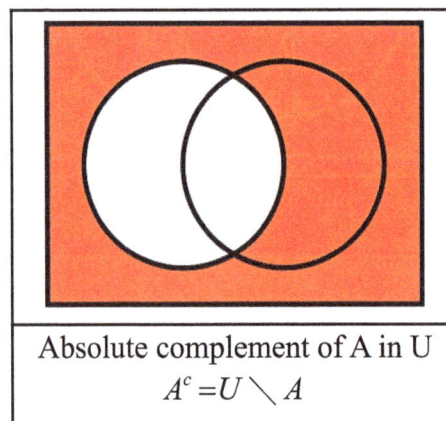

Absolute complement of A in U
$A^c = U \setminus A$

A Venn diagram is constructed with a collection of simple closed curves drawn in a plane. According to Lewis, the "principle of these diagrams is that classes [or *sets*] be represented by regions in such relation to one another that all the possible logical relations of these classes can be indicated in the same diagram. That is, the diagram initially leaves room for any possible relation of the classes, and the actual or given relation, can then be specified by indicating that some particular region is null or is not-null".

Venn diagrams normally comprise overlapping circles. The interior of the circle symbolically represents the elements of the set, while the exterior represents elements that are not members of the set. For instance, in a two-set Venn diagram, one circle may represent the group of all wooden objects, while another circle may represent the set of all tables. The overlapping region or *intersection* would then represent the set of all wooden tables. Shapes other than circles can be employed as shown below by Venn's own higher set diagrams. Venn diagrams do not generally contain information on the relative or absolute sizes (cardinality) of sets; i.e. they are schematic diagrams.

Venn diagrams are similar to Euler diagrams. However, a Venn diagram for n component sets must contain all 2^n hypothetically possible zones that correspond to some combination of inclusion or exclusion in each of the component sets. Euler diagrams contain only the actually possible zones in a given context. In Venn diagrams, a shaded zone may represent an empty zone, whereas in an Euler diagram the corresponding zone is missing from the diagram. For example, if one set represents *dairy products* and another *cheeses*, the Venn diagram contains a zone for cheeses that are not dairy products. Assuming that in the context *cheese* means some type of dairy product, the Euler diagram has the cheese zone entirely contained within the dairy-product zone—there is no zone for (non-existent) non-dairy cheese. This means that as the number of contours increases, Euler diagrams are typically less visually complex than the equivalent Venn diagram, particularly if the number of non-empty intersections is small.

The difference between Euler and Venn diagrams can be seen in the following example. Take the three sets:

- $A = \{1,2,5\}$

- $B = \{1,6\}$

- $C = \{4,7\}$

The Venn and the Euler diagram of those sets are:

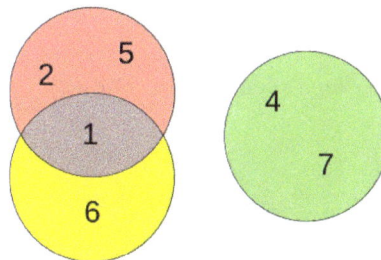

Euler diagram

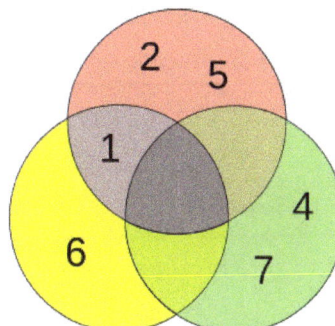

Venn diagram

Extensions to Higher Numbers of Sets

Venn diagrams typically represent two or three sets, but there are forms that allow for higher numbers. Shown below, four intersecting spheres form the highest order Venn diagram that has the symmetry of a simplex and can be visually represented. The 16 intersections correspond to the vertices of a tesseract (or the cells of a 16-cell respectively).

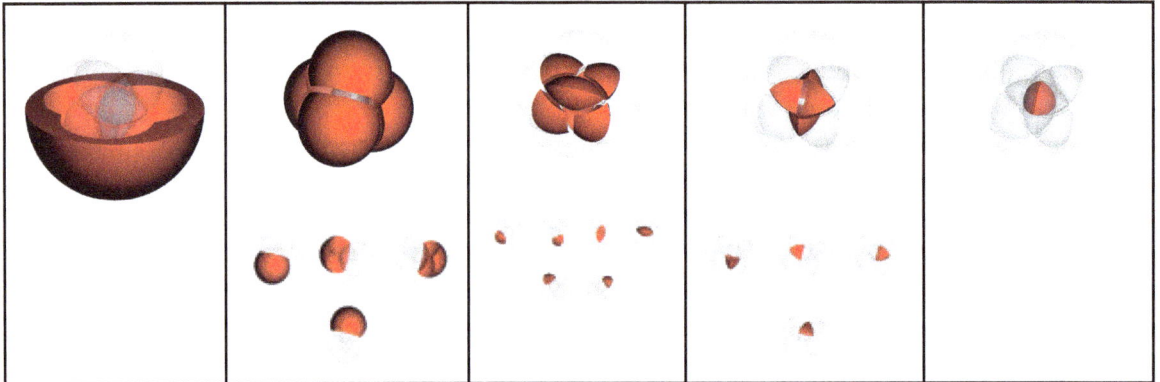

For higher numbers of sets, some loss of symmetry in the diagrams is unavoidable. Venn was keen to find "symmetrical figures...elegant in themselves," that represented higher numbers of sets, and he devised a four-set diagram using ellipses. He also gave a construction for Venn diagrams for *any* number of sets, where each successive curve that delimits a set interleaves with previous curves, starting with the three-circle diagram.

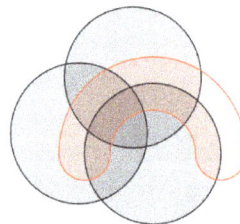

Venn's construction for 4 sets

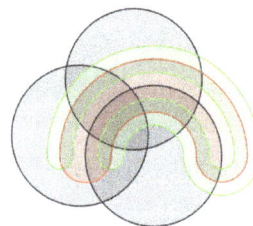

Venn's construction for 5 sets

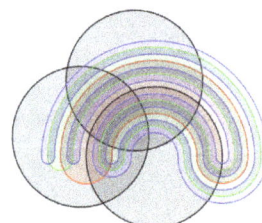

Venn's construction for 6 sets

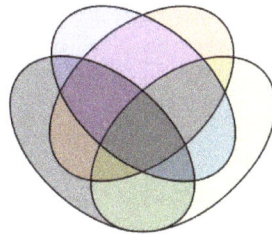

Venn's four-set diagram using ellipses

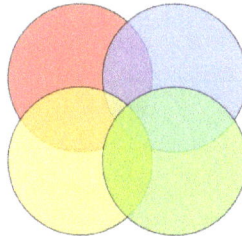

Non-example: This Euler diagram is not a Venn diagram for four sets as it has only 13 regions (excluding the outside); there is no region where only the yellow and blue, or only the red and green circles meet.

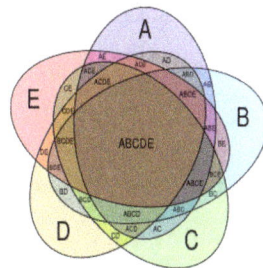

Five-set Venn diagram using congruent ellipses in a 5-fold rotationally symmetrical arrangement devised by Branko Grünbaum. Labels have been simplified for greater readability; for example, A denotes A ∩ B^c ∩ C^c ∩ D^c ∩ E^c, while BCE denotes A^c ∩ B ∩ C ∩ D^c ∩ E.

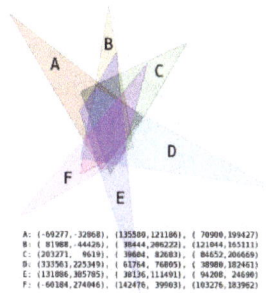

Six-set Venn diagram made of only triangles (interactive version)

Edwards' Venn Diagrams

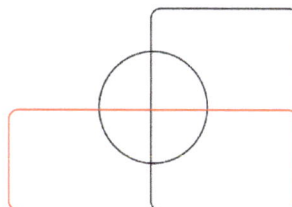

Three sets

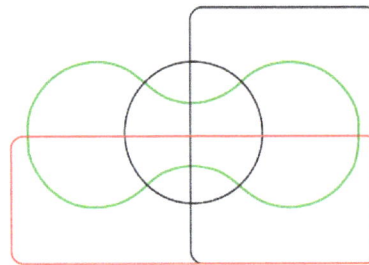

Four sets

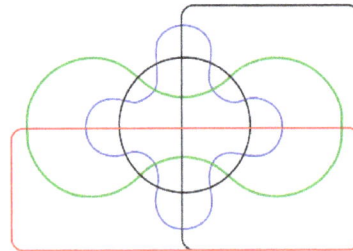

Five sets

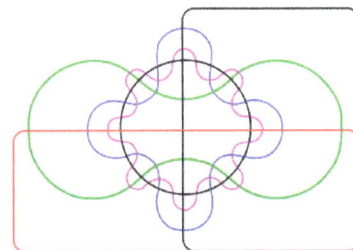

Six sets

A. W. F. Edwards constructed a series of Venn diagrams for higher numbers of sets by segmenting the surface of a sphere. For example, three sets can be easily represented by taking three hemispheres of the sphere at right angles ($x = 0$, $y = 0$ and $z = 0$). A fourth set can be added to the representation by taking a curve similar to the seam on a tennis ball, which winds up and down around the equator, and so on. The resulting sets can then be projected back to a plane to give *cogwheel* diagrams with increasing numbers of teeth. These diagrams were devised while designing a stained-glass window in memory of Venn.

Other Diagrams

Edwards' Venn diagrams are topologically equivalent to diagrams devised by Branko Grünbaum, which were based around intersecting polygons with increasing numbers of sides. They are also 2-dimensional representations of hypercubes.

Henry John Stephen Smith devised similar n-set diagrams using sine curves with the series of equations

$$y_i = \frac{\sin(2^i x)}{2i} \text{ where } 0 \le i \le n-2 \text{ and } i \in \mathbb{N}.$$

Charles Lutwidge Dodgson devised a five-set diagram.

Related Concepts

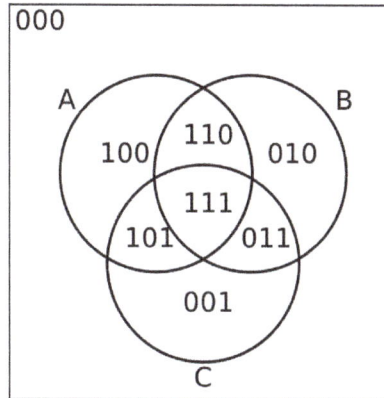

Venn diagram as a truth table

Venn diagrams correspond to truth tables for the propositions $x \in A$, $x \in B$, etc., in the sense that each region of Venn diagram corresponds to one row of the truth table. Another way of representing sets is with R-Diagrams.

Mutual Exclusivity

In logic and probability theory, two propositions (or events) are mutually exclusive or disjoint if they cannot both be true (occur). A clear example is the set of outcomes of a single coin toss, which can result in either heads or tails, but not both.

In the coin-tossing example, both outcomes are, in theory, jointly exhaustive, which means that at least one of the outcomes must happen, so these two possibilities together exhaust all the possibilities. However, not all mutually exclusive events are collectively exhaustive. For example, the outcomes 1 and 4 of a single roll of a six-sided die are mutually exclusive (both cannot happen at the same time) but not collectively exhaustive (there are other possible outcomes; 2,3,5,6).

Logic

In logic, two mutually exclusive propositions are propositions that logically cannot be true in the same sense at the same time. To say that more than two propositions are mutually exclusive, depending on context, means that one cannot be true if the other one is true, or at least one of them cannot be true. The term *pairwise mutually exclusive* always means that two of them cannot be true simultaneously.

Probability

In probability theory, events E_1, E_2, ..., E_n are said to be mutually exclusive if the occurrence of any one of them implies the non-occurrence of the remaining $n - 1$ events. Therefore, two mutually exclusive events cannot both occur. Formally said, the intersection of each two of them is empty (the null event): $A \cap B = \emptyset$. In consequence, mutually exclusive events have the property: $P(A \cap B) = 0$.

For example, it is impossible to draw a card that is both red and a club because clubs are always black. If just one card is drawn from the deck, either a red card (heart or diamond) or a black card (club or spade) will be drawn. When A and B are mutually exclusive, $P(A \cup B) = P(A) + P(B)$. To find the probability of drawing a red card or a club, for example, add together the probability of drawing a red card and the probability of drawing a club. In a standard 52-card deck, there are twenty-six red cards and thirteen clubs: $26/52 + 13/52 = 39/52$ or $3/4$.

One would have to draw at least two cards in order to draw both a red card and a club. The probability of doing so in two draws depends on whether the first card drawn were replaced before the second drawing, since without replacement there is one fewer card after the first card was drawn. The probabilities of the individual events (red, and club) are multiplied rather than added. The probability of drawing a red and a club in two drawings without replacement is then $26/52 \times 13/51 \times 2 = 676/2652$, or $13/51$. With replacement, the probability would be $26/52 \times 13/52 \times 2 = 676/2704$, or $13/52$.

In probability theory, the word *or* allows for the possibility of both events happening. The probability of one or both events occurring is denoted $P(A \cup B)$ and in general it equals $P(A) + P(B) - P(A \cap B)$. Therefore, in the case of drawing a red card or a king, drawing any of a red king, a red non-king, or a black king is considered a success. In a standard 52-card deck, there are twenty-six red cards and four kings, two of which are red, so the probability of drawing a red or a king is $26/52 + 4/52 - 2/52 = 28/52$.

Events are collectively exhaustive if all the possibilities for outcomes are exhausted by those possible events, so at least one of those outcomes must occur. The probability that at least one of the events will occur is equal to one. For example, there are theoretically only two possibilities for flipping a coin. Flipping a head and flipping a tail are collectively exhaustive events, and there is a probability of one of flipping either a head or a tail. Events can be both mutually exclusive and collectively exhaustive. In the case of flipping a coin, flipping a head and flipping a tail are also mutually exclusive events. Both outcomes cannot occur for a single trial (i.e., when a coin is flipped only once). The probability of flipping a head and the probability of flipping a tail can be added to yield a probability of 1: $1/2 + 1/2 = 1$.

Statistics

In statistics and regression analysis, an independent variable that can take on only two possible values is called a dummy variable. For example, it may take on the value 0 if an observation is of a male subject or 1 if the observation is of a female subject. The two possible categories associated with the two possible values are mutually exclusive, so that no observation falls into more than one category, and the categories are exhaustive, so that every observation falls into some category. Sometimes there are three or more possible categories, which are pairwise mutually exclusive and are collectively exhaustive — for example, under 18 years of age, 18 to 64 years of age, and age 65 or above. In this case a set of dummy variables is constructed, each dummy variable having two mutually exclusive and jointly exhaustive categories — in this example, one dummy variable (called D_1) would equal 1 if age is less than 18, and would equal 0 *otherwise*; a second dummy variable (called D_2) would equal 1 if age is in the range 18-64, and 0 otherwise. In this set-up, the dummy variable pairs (D_1, D_2) can have the values (1,0) (under 18), (0,1) (between 18 and 64), or (0,0) (65 or older) (but not (1,1), which would nonsensically imply that an observed subject is both under 18 and between 18 and 64). Then the dummy variables can be included as independent (explanatory) variables in a regression. Note that the number of dummy variables is always one less than the number of categories: with the two categories male and female there is a single dummy variable to distinguish them, while with the three age categories two dummy variables are needed to distinguish them.

Such qualitative data can also be used for dependent variables. For example, a researcher might want to predict whether someone goes to college or not, using family income, a gender dummy variable, and so forth as explanatory variables. Here the variable to be explained is a dummy variable that equals 0 if the observed subject does not go to college and equals 1 if the subject does go to college. In such a situation, ordinary least squares (the basic regression technique) is widely seen as inadequate; instead probit regression or logistic regression is used. Further, sometimes there are three or more categories for the dependent variable — for example, no college, community college, and four-year college. In this case, the multinomial probit or multinomial logit technique is used.

Probability Axioms

In Kolmogorov's probability theory, the probability P of some event E, denoted $P(E)$, is usually defined such that P satisfies the Kolmogorov axioms, named after the Russian mathematician Andrey Kolmogorov.

These assumptions can be summarised as follows: Let (Ω, F, P) be a measure space with $P(\Omega) = 1$. Then (Ω, F, P) is a probability space, with sample space Ω, event space F and probability measure P.

An alternative approach to formalising probability, favoured by some Bayesians, is given by Cox's theorem.

Axioms

First Axiom

The probability of an event is a non-negative real number:

$$P(E) \in \mathbb{R}, P(E) \geq 0 \qquad \forall E \in F$$

where F is the event space. In particular, $P(E)$ is always finite, in contrast with more general measure theory. Theories which assign negative probability relax the first axiom.

Second Axiom

This is the assumption of unit measure: that the probability that at least one of the elementary events in the entire sample space will occur is 1.

$$P(\Omega) = 1.$$

Third Axiom

This is the assumption of σ-additivity:

> Any countable sequence of disjoint sets (synonymous with *mutually exclusive* events) $E_1, E_2, \ldots$ satisfies

$$P\left(\bigcup_{i=1}^{\infty} E_i\right) = \sum_{i=1}^{\infty} P(E_i).$$

Some authors consider merely finitely additive probability spaces, in which case one just needs an algebra of sets, rather than a σ-algebra. Quasiprobability distributions in general relax the third axiom.

Consequences

From the Kolmogorov axioms, one can deduce other useful rules for calculating probabilities.

The Probability of the Empty Set

$$P(\varnothing) = 0.$$

In some cases, $\varnothing$ is not the only event with probability 0.

Monotonicity

$$\text{if} \quad A \subseteq B \quad \text{then} \quad P(A) \leq P(B).$$

If A is a subset of, or equal to B, then the probability of A is less than, or equal to the probability of B.

The Numeric Bound

It immediately follows from the monotonicity property that

$$0 \leq P(E) \leq 1 \qquad \forall E \in F.$$

Proofs

The proofs of these properties are both interesting and insightful. They illustrate the power of the third axiom, and its interaction with the remaining two axioms. When studying axiomatic probability theory, many deep consequences follow from merely these three axioms. In order to verify the monotonicity property, we set $E_1 = A$ and $E_2 = B \setminus A$, where $A \subseteq B$ and $E_i = \varnothing$ for $i \geq 3$. It is easy to see that the sets E_i are pairwise disjoint and $E_1 \cup E_2 \cup \cdots = B$. Hence, we obtain from the third axiom that

$$P(A) + P(B \setminus A) + \sum_{i=3}^{\infty} P(\varnothing) = P(B).$$

Since the left-hand side of this equation is a series of non-negative numbers, and that it converges to $P(B)$ which is finite, we obtain both $P(A) \leq P(B)$ and $P(\emptyset) = 0$. The second part of the statement is seen by contradiction: if $P(\emptyset) = a$ then the left hand side is not less than infinity

$$\sum_{i=3}^{\infty} P(E_i) = \sum_{i=3}^{\infty} P(\emptyset) = \sum_{i=3}^{\infty} a = \begin{cases} 0 & \text{if } a = 0, \\ \infty & \text{if } a > 0. \end{cases}$$

If $a > 0$ then we obtain a contradiction, because the sum does not exceed $P(B)$ which is finite. Thus, $a = 0$. We have shown as a byproduct of the proof of monotonicity that $P(\emptyset) = 0$.

Further Consequences

Another important property is:

$$P(A \cup B) = P(A) + P(B) - P(A \cap B).$$

This is called the addition law of probability, or the sum rule. That is, the probability that A or B will happen is the sum of the probabilities that A will happen and that B will happen, minus the probability that both A and B will happen. The proof of this is as follows:

$$P(A \cup B) = P(A) + P(B \setminus (A \cap B)) \text{ (by Axiom 3)}$$

now, $P(B) = P(B \setminus (A \cap B)) + P(A \cap B)$. Eliminating $P(B \setminus (A \cap B))$ from both equations gives us the desired result.

An extension of the addition law to any number of sets is the principle of inclusion-exclusion.

Setting B to the complement A^c of A in the addition law gives

$$P\left(A^c\right) = P(\Omega \setminus A) = 1 - P(A)$$

That is, the probability that any event will *not* happen (or the event's complement) is 1 minus the probability that it will.

Simple Example: Coin Toss

Consider a single coin-toss, and assume that the coin will either land heads (H) or tails (T) (but not both). No assumption is made as to whether the coin is fair.

We may define:

$$\Omega = \{H, T\}$$

$$F = \{\emptyset, \{H\}, \{T\}, \{H, T\}\}$$

Kolmogorov's axioms imply that:

$$P(\emptyset) = 0$$

The probability of *neither* heads *nor* tails, is 0.

$$P(\{H,T\}) = 1$$

The probability of *either* heads *or* tails, is 1.

$$P(\{H\}) + P(\{T\}) = 1$$

The sum of the probability of heads and the probability of tails, is 1.

References

- Yates, Daniel S.; Moore, David S; Starnes, Daren S. (2003). The Practice of Statistics (2nd ed.). New York: Freeman. ISBN 978-0-7167-4773-4.

- Foerster, Paul A. (2006). Algebra and trigonometry: Functions and applications, Teacher's edition (Classics ed.). Upper Saddle River, NJ: Prentice Hall. p. 634. ISBN 0-13-165711-9.

- Wackerly, Denniss; William Mendenhall; Richard Scheaffer. Mathematical Statistics with Applications. Duxbury. ISBN 0-534-37741-6.

- Pfeiffer, Paul E. (1978) Concepts of probability theory. Dover Publications. ISBN 978-0-486-63677-1 (online copy, p. 18, at Google Books)

- Robert R. Johnson, Patricia J. Kuby: Elementary Statistics. Cengage Learning 2007, ISBN 978-0-495-38386-4, p. 229 (restricted online copy, p. 229, at Google Books)

- Yates, Daniel S.; Moore, David S; Starnes, Daren S. (2003). The Practice of Statistics (2nd ed.). New York: Freeman. ISBN 978-0-7167-4773-4.

- Russell, Stuart; Norvig, Peter (2002). Artificial Intelligence: A Modern Approach. Prentice Hall. p. 478. ISBN 0-13-790395-2.

- Zwillinger, D., Kokoska, S. (2000) CRC Standard Probability and Statistics Tables and Formulae, CRC Press. ISBN 1-58488-059-7 page 31.

- Johnson, D. L. (2001). "3.3 Laws". Elements of logic via numbers and sets. Springer Undergraduate Mathematics Series. Berlin: Springer-Verlag. p. 62. ISBN 3-540-76123-3.

- Steigerwald, Douglas G. "Economics 245A – Introduction to Measure Theory" (PDF). University of California, Santa Barbara. Retrieved April 26, 2013.

- L. Castañeda; V. Arunachalam & S. Dharmaraja (2012). Introduction to Probability and Stochastic Processes with Applications. Wiley. p. 67.

Theory of Probability Distributions

Probability distribution is the mathematical explanation of any event in terms of the probabilities of events. The theories of probability distribution that have been elucidated are probability mass function, probability density function, cumulative distribution function, quantile function and expected value. The topics discussed in the chapter are of great importance to broaden the existing knowledge on probability distributions.

Probability Distribution

In probability and statistics, a probability distribution is a mathematical description of a random phenomenon in terms of the probabilities of events. Examples of random phenomena include the results of an experiment or survey. A probability distribution is defined in terms of an underlying sample space, which is the set of all possible outcomes of the random phenomenon being observed. The sample space may be the set of real numbers or a higher-dimensional vector space, or it may be a list of non-numerical values; for example, the sample space of a coin flip would be . Probability distributions are generally divided into two classes. A discrete probability distribution can be encoded by a list of the probabilities of the outcomes, known as a probability mass function. On the other hand, in a continuous probability distribution, the probability of any individual outcome is 0. Continuous probability distributions can often be described by probability density functions; however, more complex experiments, such as those involving stochastic processes defined in continuous time, may demand the use of more general probability measures.

In applied probability, a probability distribution can be specified in a number of different ways, often chosen for mathematical convenience:

- by supplying a valid probability mass function or probability density function

- by supplying a valid cumulative distribution function or survival function

- by supplying a valid hazard function

- by supplying a valid characteristic function

- by supplying a rule for constructing a new random variable from other random variables whose joint probability distribution is known.

A probability distribution whose sample space is the set of real numbers is called univariate, while a distribution whose sample space is a vector space is called multivariate. A univariate distribution gives the probabilities of a single random variable taking on various alternative values; a multivariate distribution (a joint probability distribution) gives the probabilities of a random vector—a list of two or more random variables—taking on various combinations of values. Important and commonly encountered univariate probability distributions include the binomial distribution, the hypergeometric distribution, and the normal distribution. The multivariate normal distribution is a commonly encountered multivariate distribution.

Introduction

To define probability distributions for the simplest cases, one needs to distinguish between discrete and continuous random variables. In the discrete case, it is sufficient to specify a probability mass function assigning a probability to each possible outcome: for example, when throwing a fair dice, each of the six values *1* to *6* has the probability 1/6. The probability of an event is then defined to be the sum of the probabilities of the outcomes that satisfy the event; for example, the probability of the event "the die rolls an even value" is $\text{Prob}(2) + \text{Prob}(4) + \text{Prob}(6) = 1/6 + 1/6 + 1/6 = 1/2$.

In contrast, when a random variable takes values from a continuum then typically, any individual outcome has probability zero and only events that include infinitely many outcomes, such as intervals, can have positive probability. For example, the probability that a given object weighs *exactly* 500 g is zero, because the probability of measuring exactly 500 g tends to zero as the accuracy of our measuring instruments increases. Nevertheless, in quality control one might demand that the probability of a "500 g" package containing between 490 g and 510 g should be no less than 98%, and this demand is less sensitive to the accuracy of our instruments.

Continuous probability distributions can be described in several ways. The probability density function describes the infinitesimal probability of any given value, and the probability that the outcome lies in a given interval can be computed by integrating the probability density function over that interval. On the other hand, the cumulative distribution function describes the probability that the random variable is no larger than a given value; the probability that the outcome lies in a given interval can be computed by taking the difference between the values of the cumulative distribution function at the endpoints of the interval. The cumulative distribution function is the antiderivative of the probability density function provided that the latter function exists.

Terminology

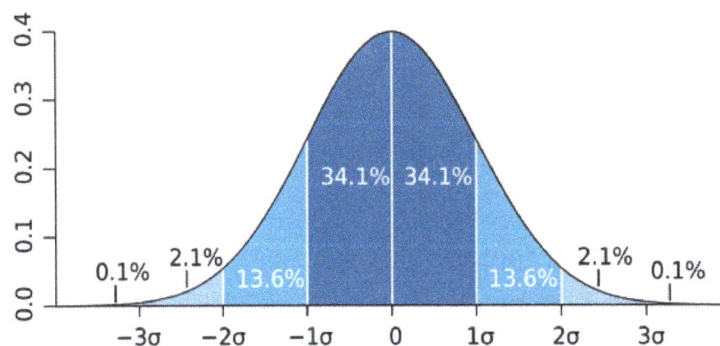

The probability density function (pdf) of the normal distribution, also called Gaussian or "bell curve", the most important continuous random distribution. As notated on the figure, the probabilities of intervals of values correspond to the area under the curve.

As probability theory is used in quite diverse applications, terminology is not uniform and sometimes confusing. The following terms are used for non-cumulative probability distribution functions:

- Probability mass, Probability mass function, p.m.f.: for discrete random variables.

- Categorical distribution: for discrete random variables with a finite set of values.

- Probability density, Probability density function, p.d.f.: most often reserved for continuous random variables.

The following terms are somewhat ambiguous as they can refer to non-cumulative or cumulative distributions, depending on authors' preferences:

- Probability distribution function: continuous or discrete, non-cumulative or cumulative.

- Probability function: even more ambiguous, can mean any of the above or other things.

Finally,

- Probability distribution: sometimes the same as *probability distribution function*, but usually refers to the more complete assignment of probabilities to all measurable subsets of outcomes (i.e. the corresponding probability measure), not just to specific outcomes or ranges of outcomes.

Basic Terms

- Mode: for a discrete random variable, the value with highest probability (the location at which the probability mass function has its peak); for a continuous random variable, the location at which the probability density function has its peak.

- Support: the smallest closed set whose complement has probability zero.

- Head: the range of values where the pmf or pdf is relatively high.

- Tail: the complement of the head within the support; the large set of values where the pmf or pdf is relatively low.

- Expected value or mean: the weighted average of the possible values, using their probabilities as their weights; or the continuous analog thereof.

- Median: the value such that the set of values less than the median has a probability of one-half.

- Variance: the second moment of the pmf or pdf about the mean; an important measure of the dispersion of the distribution.

- Standard deviation: the square root of the variance, and hence another measure of dispersion.

- Symmetry: a property of some distributions in which the portion of the distribution to the left of a specific value is a mirror image of the portion to its right.

- Skewness: a measure of the extent to which a pmf or pdf "leans" to one side of its mean. The third standardized moment of the distribution.

- Kurtosis: a measure of the "fatness" of the tails of a pmf or pdf. The fourth standardized moment of the distribution.

Cumulative Distribution Function

Because a probability distribution Pr on the real line is determined by the probability of a scalar random variable X being in a half-open interval $(-\infty, x]$, the probability distribution is completely characterized by its cumulative distribution function:

$$F(x) = \Pr[X \le x] \qquad \text{for all } x \in \mathbb{R}.$$

Discrete Probability Distribution

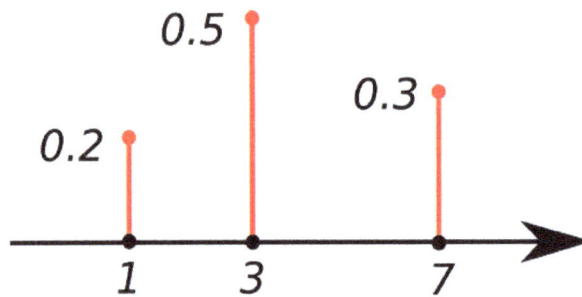

The probability mass function of a discrete probability distribution. The probabilities of the singletons {1}, {3}, and {7} are respectively 0.2, 0.5, 0.3. A set not containing any of these points has probability zero.

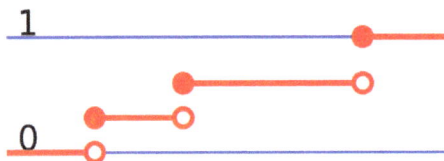

The cdf of a discrete probability distribution, ...

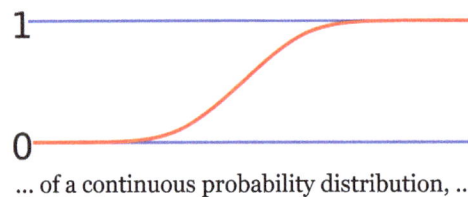

... of a continuous probability distribution, ...

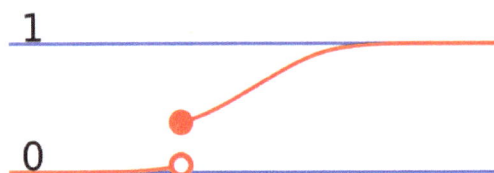

... of a distribution which has both a continuous part and a discrete part.

A discrete probability distribution is a probability distribution characterized by a probability mass function. Thus, the distribution of a random variable X is discrete, and X is called a discrete random variable, if

$$\sum_u \Pr(X = u) = 1$$

as u runs through the set of all possible values of X. A discrete random variable can assume only a finite or countably infinite number of values. For the number of potential values to be countably infinite, even though their probabilities sum to 1, the probabilities have to decline to zero fast enough. For example, if $\Pr(X = n) = \frac{1}{2^n}$ for n = 1, 2, ..., we have the sum of probabilities $1/2 + 1/4 + 1/8 + ... = 1$.

Well-known discrete probability distributions used in statistical modeling include the Poisson distribution, the Bernoulli distribution, the binomial distribution, the geometric distribution, and the negative binomial distribution. Additionally, the discrete uniform distribution is commonly used in computer programs that make equal-probability random selections between a number of choices.

Measure Theoretic Formulation

A measurable function $X : A \rightarrow B$ between a probability space $(A, \mathcal{A}, P)$ and a measurable space $(B, \mathcal{B})$ is called a discrete random variable provided its image is a countable set and the pre-image of singleton sets are measurable, i.e., $X^{-1}(b) \in \mathcal{A}$ for all $b \in B$.. The latter requirement induces a probability mass function $f_X : X(A) \rightarrow \mathbb{R}$ via $f_X(b) := P(X^{-1}(b))$. Since the pre-images of disjoint sets are disjoint

$$\sum_{b \in X(A)} f_X(b) = \sum_{b \in X(A)} P(X^{-1}(b)) = P\left(\bigcup_{b \in X(A)} X^{-1}(b) \right) = P(A) = 1.$$

This recovers the definition given above.

Cumulative Density

Equivalently to the above, a discrete random variable can be defined as a random variable whose cumulative distribution function (cdf) increases only by jump discontinuities—that is, its cdf increases only where it "jumps" to a higher value, and is constant between those jumps. The points where jumps occur are precisely the values which the random variable may take.

Delta-function Representation

Consequently, a discrete probability distribution is often represented as a generalized probability density function involving Dirac delta functions, which substantially unifies the treatment of continuous and discrete distributions. This is especially useful when dealing with probability distributions involving both a continuous and a discrete part.

Indicator-function Representation

For a discrete random variable X, let u_0, u_1, ... be the values it can take with non-zero probability. Denote

$$\Omega_i = X^{-1}(u_i) = \{\omega : X(\omega) = u_i\}, i = 0, 1, 2, \ldots$$

These are disjoint sets, and by formula (1)

$$\Pr\left(\bigcup_i \Omega_i\right) = \sum_i \Pr(\Omega_i) = \sum_i \Pr(X = u_i) = 1.$$

It follows that the probability that X takes any value except for u_0, u_1, ... is zero, and thus one can write X as

$$X(\omega) = \sum_i u_i 1_{\Omega_i}(\omega)$$

except on a set of probability zero, where 1_A is the indicator function of A. This may serve as an alternative definition of discrete random variables.

Continuous Probability Distribution

A continuous probability distribution is a *probability distribution* that has a cumulative distribution function that is continuous. Most often they are generated by having a probability density function. Mathematicians call distributions with probability density functions absolutely continuous, since their cumulative distribution function is absolutely continuous with respect to the Lebesgue measure λ. If the distribution of X is continuous, then X is called a continuous random variable. There are many examples of continuous probability distributions: normal, uniform, chi-squared, and others.

Intuitively, a continuous random variable is the one which can take a continuous range of values—as opposed to a discrete distribution, where the set of possible values for the random variable is at most countable. While for a discrete distribution an event with probability zero is impossible (e.g., rolling 31/2 on a standard dice is impossible, and has probability zero), this is not so in the case of a continuous random variable. For example, if one measures the width of an oak leaf, the result of 3½ cm is possible; however, it has probability zero because uncountably many other potential values exist even between 3 cm and 4 cm. Each of these individual outcomes has probability zero, yet the probability that the outcome will fall into the interval (3 cm, 4 cm) is nonzero. This apparent paradox is resolved by the fact that the probability that X attains some value within an infinite set, such as an interval, cannot be found by naively adding the probabilities for individual values. Formally, each value has an infinitesimally small probability, which statistically is equivalent to zero.

Formally, if X is a continuous random variable, then it has a probability density function $f(x)$, and therefore its probability of falling into a given interval, say $[a, b]$ is given by the integral

$$\Pr[a \leq X \leq b] = \int_a^b f(x)dx$$

In particular, the probability for X to take any single value a (that is $a \leq X \leq a$) is zero, because an integral with coinciding upper and lower limits is always equal to zero.

The definition states that a continuous probability distribution must possess a density, or equivalently, its cumulative distribution function be absolutely continuous. This requirement is stronger than simple continuity of the cumulative distribution function, and there is a special class of distributions, singular distributions, which are neither continuous nor discrete nor a mixture of those. An example is given by the Cantor distribution. Such singular distributions however are never encountered in practice.

Note on terminology: some authors use the term "continuous distribution" to denote the distribution with continuous cumulative distribution function. Thus, their definition includes both the (absolutely) continuous and singular distributions.

By one convention, a probability distribution μ is called *continuous* if its cumulative distribution function $F(x) = \mu(-\infty, x]$ is continuous and, therefore, the probability measure of singletons $\mu\{x\}=0$ for all x.

Another convention reserves the term *continuous probability distribution* for absolutely continuous distributions. These distributions can be characterized by a probability density function: a non-negative Lebesgue integrable function f defined on the real numbers such that

$$F(x) = \mu(-\infty, x] = \int_{-\infty}^{x} f(t)dt.$$

Discrete distributions and some continuous distributions (like the Cantor distribution) do not admit such a density.

Some Properties

- The probability distribution of the sum of two independent random variables is the convolution of each of their distributions.

- Probability distributions are not a vector space—they are not closed under linear combinations, as these do not preserve non-negativity or total integral 1—but they are closed under convex combination, thus forming a convex subset of the space of functions (or measures).

Kolmogorov Definition

In the measure-theoretic formalization of probability theory, a random variable is defined as a measurable function X from a probability space $(\Omega, \mathcal{F}, P)$ to measurable space $(\mathcal{X}, \mathcal{A})$. A probability distribution of X is the pushforward measure X_*P of X, which is a probability measure on $(\mathcal{X}, \mathcal{A})$ satisfying $X_*P = PX^{-1}$.

Random Number Generation

A frequent problem in statistical simulations (the Monte Carlo method) is the generation of pseudo-random numbers that are distributed in a given way. Most algorithms are based on a pseudo-random number generator that produces numbers X that are uniformly distributed in the interval [0,1). These random variates X are then transformed via some algorithm to create a new random variate having the required probability distribution.

Applications

The concept of the probability distribution and the random variables which they describe underlies the mathematical discipline of probability theory, and the science of statistics. There is spread or variability in almost any value that can be measured in a population (e.g. height of people, durability of a metal, sales growth, traffic flow, etc.); almost all measurements are made with some intrinsic error; in physics many processes are described probabilistically,from the kinetic properties of gases to the quantum mechanical description of fundamental particles. For these and many other reasons, simple numbers are often inadequate for describing a quantity, while probability distributions are often more appropriate.

As a more specific example of an application, the cache language models and other statistical language models used in natural language processing to assign probabilities to the occurrence of particular words and word sequences do so by means of probability distributions.

Common Probability Distributions

The following is a list of some of the most common probability distributions, grouped by the type of process that they are related to. For a more complete list, list of probability distributions, which groups by the nature of the outcome being considered (discrete, continuous, multivariate, etc.)

Note also that all of the univariate distributions below are singly peaked; that is, it is assumed that the values cluster around a single point. In practice, actually observed quantities may cluster around multiple values. Such quantities can be modeled using a mixture distribution.

Related to Real-valued Quantities that Grow Linearly (e.g. Errors, Offsets)

- Normal distribution (Gaussian distribution), for a single such quantity; the most common continuous distribution

Related to Positive real-valued Quantities that Grow Exponentially (e.g. Prices, incomes, Populations)

- Log-normal distribution, for a single such quantity whose log is normally distributed

- Pareto distribution, for a single such quantity whose log is exponentially distributed; the prototypical power law distribution

Related to Real-valued Quantities that Are Assumed to Be Uniformly Distributed Over a (Possibly Unknown) Region

- Discrete uniform distribution, for a finite set of values (e.g. the outcome of a fair die)

- Continuous uniform distribution, for continuously distributed values

Related to Bernoulli Trials (Yes/no Events, With a Given Probability)

- Basic distributions:

- o Bernoulli distribution, for the outcome of a single Bernoulli trial (e.g. success/failure, yes/no)

- o Binomial distribution, for the number of "positive occurrences" (e.g. successes, yes votes, etc.) given a fixed total number of independent occurrences

- o Negative binomial distribution, for binomial-type observations but where the quantity of interest is the number of failures before a given number of successes occurs

- o Geometric distribution, for binomial-type observations but where the quantity of interest is the number of failures before the first success; a special case of the negative binomial distribution

- Related to sampling schemes over a finite population:

- o Hypergeometric distribution, for the number of "positive occurrences" (e.g. successes, yes votes, etc.) given a fixed number of total occurrences, using sampling without replacement

- o Beta-binomial distribution, for the number of "positive occurrences" (e.g. successes, yes votes, etc.) given a fixed number of total occurrences, sampling using a Polya urn scheme (in some sense, the "opposite" of sampling without replacement)

Related to Categorical Outcomes (Events With K Possible Outcomes, With a Given Probability for Each Outcome)

- Categorical distribution, for a single categorical outcome (e.g. yes/no/maybe in a survey); a generalization of the Bernoulli distribution

- Multinomial distribution, for the number of each type of categorical outcome, given a fixed number of total outcomes; a generalization of the binomial distribution

- Multivariate hypergeometric distribution, similar to the multinomial distribution, but using sampling without replacement; a generalization of the hypergeometric distribution

Related to Events in a Poisson Process (Events that Occur Independently With a Given Rate)

- Poisson distribution, for the number of occurrences of a Poisson-type event in a given period of time

- Exponential distribution, for the time before the next Poisson-type event occurs

- Gamma distribution, for the time before the next k Poisson-type events occur

Related to the Absolute Values of Vectors With Normally Distributed Components

- Rayleigh distribution, for the distribution of vector magnitudes with Gaussian distributed orthogonal components. Rayleigh distributions are found in RF signals with Gaussian real and imaginary components.

- Rice distribution, a generalization of the Rayleigh distributions for where there is a stationary background signal component. Found in Rician fading of radio signals due to multipath propagation and in MR images with noise corruption on non-zero NMR signals.

Related to Normally Distributed Quantities Operated With Sum of Squares (for Hypothesis Testing)

- Chi-squared distribution, the distribution of a sum of squared standard normal variables; useful e.g. for inference regarding the sample variance of normally distributed samples

- Student's t distribution, the distribution of the ratio of a standard normal variable and the square root of a scaled chi squared variable; useful for inference regarding the mean of normally distributed samples with unknown variance

- F-distribution, the distribution of the ratio of two scaled chi squared variables; useful e.g. for inferences that involve comparing variances or involving R-squared (the squared correlation coefficient)

Useful as Conjugate Prior Distributions in Bayesian Inference

- Beta distribution, for a single probability (real number between 0 and 1); conjugate to the Bernoulli distribution and binomial distribution

- Gamma distribution, for a non-negative scaling parameter; conjugate to the rate parameter of a Poisson distribution or exponential distribution, the precision (inverse variance) of a normal distribution, etc.

- Dirichlet distribution, for a vector of probabilities that must sum to 1; conjugate to the categorical distribution and multinomial distribution; generalization of the beta distribution

- Wishart distribution, for a symmetric non-negative definite matrix; conjugate to the inverse of the covariance matrix of a multivariate normal distribution; generalization of the gamma distribution

Probability Theory

Probability theory is the branch of mathematics concerned with probability, the analysis of random phenomena. The central objects of probability theory are random variables, stochastic processes, and events: mathematical abstractions of non-deterministic events or measured quantities that may either be single occurrences or evolve over time in an apparently random fashion.

It is not possible to predict precisely results of random events. However, if a sequence of individual events, such as coin flipping or the roll of dice, is influenced by other factors, such as friction, it will exhibit certain patterns, which can be studied and predicted. Two representative mathematical results describing such patterns are the law of large numbers and the central limit theorem.

As a mathematical foundation for statistics, probability theory is essential to many human activities that involve quantitative analysis of large sets of data. Methods of probability theory also apply to descriptions of complex systems given only partial knowledge of their state, as in statistical mechanics. A great discovery of twentieth century physics was the probabilistic nature of physical phenomena at atomic scales, described in quantum mechanics.

History

The mathematical theory of probability has its roots in attempts to analyze games of chance by Gerolamo Cardano in the sixteenth century, and by Pierre de Fermat and Blaise Pascal in the seventeenth century (for example the "problem of points"). Christiaan Huygens published a book on the subject in 1657 and in the 19th century, Pierre Laplace completed what is today considered the classic interpretation.

Initially, probability theory mainly considered discrete events, and its methods were mainly combinatorial. Eventually, analytical considerations compelled the incorporation of continuous variables into the theory.

This culminated in modern probability theory, on foundations laid by Andrey Nikolaevich Kolmogorov. Kolmogorov combined the notion of sample space, introduced by Richard von Mises, and measure theory and presented his axiom system for probability theory in 1933. This became the mostly undisputed axiomatic basis for modern probability theory; but, alternatives exist, such as the adoption of finite rather than countable additivity by Bruno de Finetti.

Treatment

Most introductions to probability theory treat discrete probability distributions and continuous probability distributions separately. The more mathematically advanced measure theory-based treatment of probability covers the discrete, continuous, a mix of the two, and more.

Motivation

Consider an experiment that can produce a number of outcomes. The set of all outcomes is called the *sample space* of the experiment. The *power set* of the sample space (or equivalently, the event space) is formed by considering all different collections of possible results. For example, rolling an honest die produces one of six possible results. One collection of possible results corresponds to getting an odd number. Thus, the subset {1,3,5} is an element of the power set of the sample space of die rolls. These collections are called *events*. In this case, {1,3,5} is the event that the die falls on some odd number. If the results that actually occur fall in a given event, that event is said to have occurred.

Probability is a way of assigning every "event" a value between zero and one, with the requirement that the event made up of all possible results (in our example, the event {1,2,3,4,5,6}) be assigned a value of one. To qualify as a probability distribution, the assignment of values must satisfy the requirement that if you look at a collection of mutually exclusive events (events that contain no common results, e.g., the events {1,6}, {3}, and {2,4} are all mutually exclusive), the probability that one of the events will occur is given by the sum of the probabilities of the individual events.

The probability that any one of the events {1,6}, {3}, or {2,4} will occur is 5/6. This is the same as saying that the probability of event {1,2,3,4,6} is 5/6. This event encompasses the possibility of any number except five being rolled. The mutually exclusive event {5} has a probability of 1/6, and the event {1,2,3,4,5,6} has a probability of 1, that is, absolute certainty.

Discrete Probability Distributions

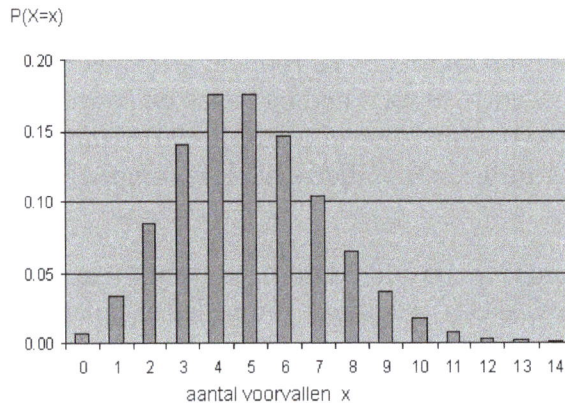

The Poisson distribution, a discrete probability distribution.

Discrete probability theory deals with events that occur in countable sample spaces.

Examples: Throwing dice, experiments with decks of cards, random walk, and tossing coins

Classical definition: Initially the probability of an event to occur was defined as number of cases favorable for the event, over the number of total outcomes possible in an equiprobable sample space.

For example, if the event is "occurrence of an even number when a die is rolled", the probability is given by $\frac{3}{6} = \frac{1}{2}$, since 3 faces out of the 6 have even numbers and each face has the same probability of appearing.

Modern definition: The modern definition starts with a finite or countable set called the sample space, which relates to the set of all *possible outcomes* in classical sense, denoted by . It is then assumed that for each element $\Omega.$, an intrinsic "probability" value $x \in \Omega,$ is attached, which satisfies the following properties:

1. $f(x) \in [0,1]$ *for all* $x \in \Omega$;

2. $\sum_{x \in U} f(x) = 1.$

That is, the probability function $f(x)$ lies between zero and one for every value of x in the sample space Ω, and the sum of $f(x)$ over all values x in the sample space Ω is equal to 1. An event is defined as any subset E of the sample space Ω. The probability of the event E is defined as

$$P(E) = \sum_{x \in E} f(x).$$

So, the probability of the entire sample space is 1, and the probability of the null event is 0.

The function $f(x)$ mapping a point in the sample space to the "probability" value is called a probability mass function abbreviated as pmf. The modern definition does not try to answer how probability mass functions are obtained; instead it builds a theory that assumes their existence.

Continuous Probability Distributions

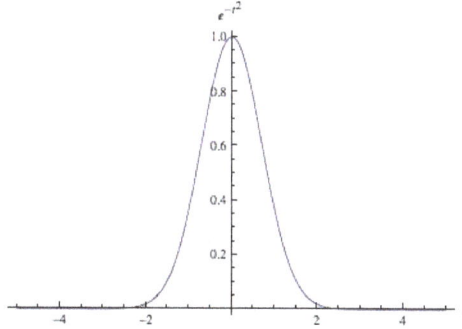

The normal distribution, a continuous probability distribution.

Continuous probability theory deals with events that occur in a continuous sample space.

Classical definition: The classical definition breaks down when confronted with the continuous case.

Modern definition: If the outcome space of a random variable X is the set of real numbers ($\mathbb{R}$) or a subset thereof, then a function called the cumulative distribution function (or cdf) F exists, defined by $F(x) = P(X \leq x)$. That is, $F(x)$ returns the probability that X will be less than or equal to x.

The cdf necessarily satisfies the following properties.

1. F is a monotonically non-decreasing, right-continuous function;

2. $\lim\limits_{x \to -\infty} F(x) = 0$;

3. $\lim\limits_{x \to \infty} F(x) = 1$.

If F is absolutely continuous, i.e., its derivative exists and integrating the derivative gives us the cdf back again, then the random variable X is said to have a probability density function or pdf or simply density $f(x) = \dfrac{dF(x)}{dx}$.

For a set $E \subseteq \mathbb{R}$, the probability of the random variable X being in E is

$$P(X \in E) = \int_{x \in E} dF(x).$$

In case the probability density function exists, this can be written as

$$P(X \in E) = \int_{x \in E} f(x)dx.$$

Whereas the *pdf* exists only for continuous random variables, the *cdf* exists for all random variables (including discrete random variables) that take values in $\mathbb{R}$.

These concepts can be generalized for multidimensional cases on $\mathbb{R}^n$ and other continuous sample spaces.

Measure-theoretic Probability Theory

The *raison d'être* of the measure-theoretic treatment of probability is that it unifies the discrete and the continuous cases, and makes the difference a question of which measure is used. Furthermore, it covers distributions that are neither discrete nor continuous nor mixtures of the two.

An example of such distributions could be a mix of discrete and continuous distributions—for example, a random variable that is 0 with probability 1/2, and takes a random value from a normal distribution with probability 1/2. It can still be studied to some extent by considering it to have a pdf of $(\delta[x] + \varphi(x))/2$, , where $\delta[x]$ is the Dirac delta function.

Other distributions may not even be a mix, for example, the Cantor distribution has no positive probability for any single point, neither does it have a density. The modern approach to probability theory solves these problems using measure theory to define the probability space:

Given any set Ω, (also called sample space) and a σ-algebra $\mathcal{F}$ on it, a measure P defined on $\mathcal{F}$ is called a probability measure if $P(\Omega) = 1$.

If $\mathcal{F}$ is the Borel σ-algebra on the set of real numbers, then there is a unique probability measure on $\mathcal{F}$ for any cdf, and vice versa. The measure corresponding to a cdf is said to be induced by the cdf. This measure coincides with the pmf for discrete variables and pdf for continuous variables, making the measure-theoretic approach free of fallacies.

The *probability* of a set E in the σ-algebra $\mathcal{F}$ is defined as

$$P(E) = \int_{\omega \in E} \mu_F(d\omega)$$

where the integration is with respect to the measure μ_F induced by F.

Along with providing better understanding and unification of discrete and continuous probabilities, measure-theoretic treatment also allows us to work on probabilities outside $\mathbb{R}^n$, as in the theory of stochastic processes. For example, to study Brownian motion, probability is defined on a space of functions.

When it's convenient to work with a dominating measure, the Radon-Nikodym theorem is used to define a density as the Radon-Nikodym derivative of the probability distribution of interest with respect to this dominating measure. Discrete densities are usually defined as this derivative with respect to a counting measure over the set of all possible outcomes. Densities for absolutely continuous distributions are usually defined as this derivative with respect to the Lebesgue measure. If a theorem can be proved in this general setting, it holds for both discrete and continuous distributions as well as others; separate proofs are not required for discrete and continuous distributions.

Classical Probability Distributions

Certain random variables occur very often in probability theory because they well describe many natural or physical processes. Their distributions therefore have gained *special importance* in probability theory. Some fundamental *discrete distributions* are the discrete uniform, Bernoulli, binomial, negative binomial, Poisson and geometric distributions. Important *continuous distributions* include the continuous uniform, normal, exponential, gamma and beta distributions.

Convergence of Random Variables

In probability theory, there are several notions of convergence for random variables. They are listed below in the order of strength, i.e., any subsequent notion of convergence in the list implies convergence according to all of the preceding notions.

Weak convergence

A sequence of random variables $X_1, X_2, \ldots$, converges weakly to the random variable X if their respective cumulative *distribution functions* $F_1, F_2, \ldots$ converge to the cumulative distribution function F of X, wherever F is continuous. Weak convergence is also called convergence in distribution.

Most common shorthand notation: $X_n \xrightarrow{D} X$

Convergence in probability

The sequence of random variables $X_1, X_2, \ldots$ is said to converge towards the random variable X in probability if $\lim_{n \to \infty} P(|X_n - X| \geq \varepsilon) = 0$ for every $\varepsilon > 0$.

Most common shorthand notation: $X_n \xrightarrow{P} X$

Strong convergence

The sequence of random variables $X_1, X_2, \ldots$ is said to converge towards the random variable X strongly if $P(\lim_{n \to \infty} X_n = X) = 1$. Strong convergence is also known as almost sure convergence.

Most common shorthand notation: $X_n \xrightarrow{a.s.} X$

As the names indicate, weak convergence is weaker than strong convergence. In fact, strong convergence implies convergence in probability, and convergence in probability implies weak convergence. The reverse statements are not always true.

Law of Large Numbers

Common intuition suggests that if a fair coin is tossed many times, then *roughly* half of the time it will turn up *heads*, and the other half it will turn up *tails*. Furthermore, the more often the coin is tossed, the more likely it should be that the ratio of the number of *heads* to the number of *tails* will approach unity. Modern probability theory provides a formal version of this intuitive idea, known as the law of large numbers. This law is remarkable because it is not assumed in the foundations of probability theory, but instead emerges from these foundations as a theorem. Since it links theoretically derived probabilities to their actual frequency of occurrence in the real world, the law of large numbers is considered as a pillar in the history of statistical theory and has had widespread influence.

The law of large numbers (LLN) states that the sample average

$$\overline{X}_n = \frac{1}{n}\sum_{k=1}^{n} X_k$$

of a sequence of independent and identically distributed random variables X_k converges towards their common expectation μ, provided that the expectation of $|X_k|$ is finite.

It is in the different forms of convergence of random variables that separates the *weak* and the *strong* law of large numbers

Weak law: $\overline{X}_n \xrightarrow{P} \mu$ for $n \to \infty$

Strong law: $\overline{X}_n \xrightarrow{a.s.} \mu$ for $n \to \infty$.

It follows from the LLN that if an event of probability p is observed repeatedly during independent experiments, the ratio of the observed frequency of that event to the total number of repetitions converges towards p.

For example, if $Y_1, Y_2, \dots$ are independent Bernoulli random variables taking values 1 with probability p and 0 with probability 1-p, then $E(Y_i) = p$ for all i, so that $\overline{Y}_n$ converges to p almost surely.

Central Limit Theorem

"The central limit theorem (CLT) is one of the great results of mathematics." It explains the ubiquitous occurrence of the normal distribution in nature.

The theorem states that the average of many independent and identically distributed random variables with finite variance tends towards a normal distribution *irrespective* of the distribution followed by the original random variables. Formally, let $X_1, X_2, \dots$ be independent random variables with mean μ and variance $\sigma^2 > 0$. Then the sequence of random variables

$$Z_n = \frac{\sum_{i=1}^{n}(X_i - \mu)}{\sigma\sqrt{n}}$$

converges in distribution to a standard normal random variable.

For some classes of random variables the classic central limit theorem works rather fast, for example the distributions with finite first, second and third moment from the exponential family; on the other hand, for some random variables of the heavy tail and fat tail variety, it works very slowly or may not work at all: in such cases one may use the Generalized Central Limit Theorem (GCLT).

Probability Mass Function

In probability theory and statistics, a probability mass function (pmf) is a function that gives the probability that a discrete random variable is exactly equal to some value. The probability mass function is often the primary means of defining a discrete probability distribution, and such functions exist for either scalar or multivariate random variables whose domain is discrete.

A probability mass function differs from a probability density function (pdf) in that the latter is associated with continuous rather than discrete random variables; the values of the latter are not probabilities as such: a pdf must be integrated over an interval to yield a probability.

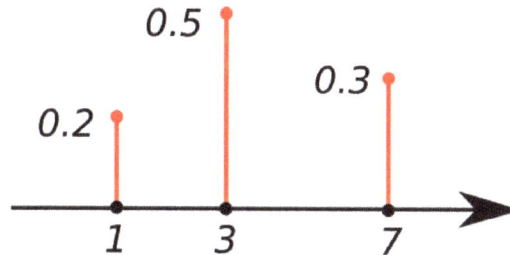

The graph of a probability mass function. All the values of this function must be non-negative and sum up to 1.

Formal Definition

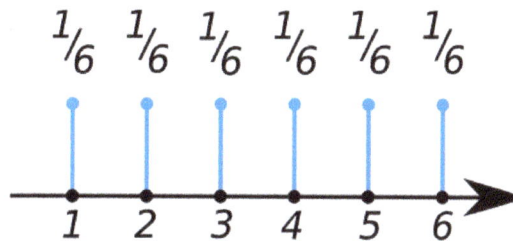

The probability mass function of a fair die. All the numbers on the die have an equal chance of appearing on top when the die stops rolling.

Suppose that $X: S \to A$ ($A \subseteq R$) is a discrete random variable defined on a sample space S. Then the probability mass function $f_X: A \to [0, 1]$ for X is defined as

$$f_X(x) = \Pr(X = x) = \Pr(\{s \in S : X(s) = x\}).$$

Thinking of probability as mass helps to avoid mistakes since the physical mass is conserved as is the total probability for all hypothetical outcomes x:

$$\sum_{x \in A} f_X(x) = 1$$

When there is a natural order among the hypotheses x, it may be convenient to assign numerical values to them (or n-tuples in case of a discrete multivariate random variable) and to consider also values not in the image of X. That is, f_X may be defined for all real numbers and $f_X(x) = 0$ for all $x \notin X(S)$.

Since the image of X is countable, the probability mass function $f_X(x)$ is zero for all but a countable number of values of x. The discontinuity of probability mass functions is related to the fact that the cumulative distribution function of a discrete random variable is also discontinuous. Where it is differentiable, the derivative is zero, just as the probability mass function is zero at all such points.

Measure Theoretic Formulation

A probability mass function of a discrete random variable X can be seen as a special case of two more general measure theoretic constructions: the distribution of X and the probability density function of X with respect to the counting measure. We make this more precise below.

Suppose that $(A, \mathcal{A}, P)$ is a probability space and that $(B, \mathcal{B})$ is a measurable space whose underlying σ-algebra is discrete, so in particular contains singleton sets of B. In this setting, a random variable $X : A \to B$ is discrete provided its image is countable. The pushforward measure $X_*(P)$ ---called a distribution of X in this context---is a probability measure on B whose restriction to singleton sets induces a probability mass function $f_X : B \to \mathbb{R}$ since $f_X(b) = P(X^{-1}(b)) = [X_*(P)](\{b\})$ for each b in B.

Now suppose that $(B, \mathcal{B}, \mu)$ is a measure space equipped with the counting measure μ. The probability density function f of X with respect to the counting measure, if it exists, is the Radon-Nikodym derivative of the pushforward measure of X (with respect to the counting measure), so $f = dX_*P / d\mu$ and f is a function from B to the non-negative reals. As a consequence, for any b in B we have

$$P(X = b) = P(X^{-1}(\{b\})) := \int_{X^{-1}(\{b\})} dP = \int_{\{b\}} f d\mu = f(b),$$

demonstrating that f is in fact a probability mass function.

Examples

Suppose that S is the sample space of all outcomes of a single toss of a fair coin, and X is the random variable defined on S assigning 0 to "tails" and 1 to "heads". Since the coin is fair, the probability mass function is

$$f_X(x) = \begin{cases} \dfrac{1}{2}, & x \in \{0,1\}, \\ 0, & x \notin \{0,1\}. \end{cases}$$

This is a special case of the binomial distribution, the Bernoulli distribution.

An example of a multivariate discrete distribution, and of its probability mass function, is provided by the multinomial distribution.

Probability Density Function

In probability theory, a probability density function (PDF), or density of a continuous random variable, is a function that describes the relative likelihood for this random variable to take on a given value. The probability of the random variable falling within a particular range of values is given by the integral of this variable's density over that range—that is, it is given by the area under the density function but above the horizontal axis and between the lowest and greatest values of the range. The probability density function is nonnegative everywhere, and its integral over the entire space is equal to one.

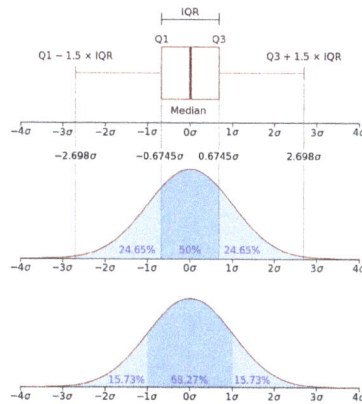

Boxplot and probability density function of a normal distribution $N(0, \sigma^2)$.

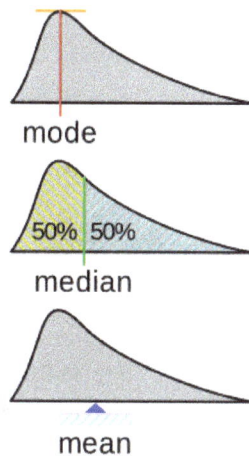

Geometric visualisation of the mode, median and mean of an arbitrary probability density function.

The terms *"probability distribution function"* and *"probability function"* have also sometimes been used to denote the probability density function. However, this use is not standard among probabilists and statisticians. In other sources, "probability distribution function" may be used when the probability distribution is defined as a function over general sets of values, or it may refer to the cumulative distribution function, or it may be a probability mass function (PMF) rather than the density. Further confusion of terminology exists because *density function* has also been used for what is here called the "probability mass function" (PMF). In general though, the PMF is used in the context of discrete random variables (random variables that take values on a discrete set), while PDF is used in the context of continuous random variables.

Example

Suppose a species of bacteria typically lives 4 to 6 hours. What is the probability that a bacterium lives *exactly* 5 hours? The answer is 0%. A lot of bacteria live for *approximately* 5 hours, but there is no chance that any given bacterium dies at *exactly* 5.0000000000... hours.

Instead we might ask: What is the probability that the bacterium dies between 5 hours and 5.01 hours? Let's say the answer is 0.02 (i.e., 2%). Next: What is the probability that the bacterium dies between 5 hours and 5.001 hours? The answer is probably around 0.002, since this is 1/10th of the previous interval. The probability that the bacterium dies between 5 hours and 5.0001 hours is probably about 0.0002, and so on.

In these three examples, the ratio (probability of dying during an interval) / (duration of the interval) is approximately constant, and equal to 2 per hour (or 2 hour^{-1}). For example, there is 0.02 probability of dying in the 0.01-hour interval between 5 and 5.01 hours, and (0.02 probability / 0.01 hours) = 2 hour^{-1}. This quantity 2 hour^{-1} is called the *probability density* for dying at around 5 hours.

Therefore, in response to the question "What is the probability that the bacterium dies at 5 hours?", a literally correct but unhelpful answer is "0", but a better answer can be written as (2 hour^{-1}) dt. This is the probability that the bacterium dies within a small (infinitesimal) window of time around 5 hours, where dt is the duration of this window.

For example, the probability that it lives longer than 5 hours, but shorter than (5 hours + 1 nanosecond), is (2 hour^{-1})×(1 nanosecond) $\simeq$ 6×10^{-13} (using the unit conversion 3.6×10^{12} nanoseconds = 1 hour).

There is a *probability density function* f with f(5 hours) = 2 hour^{-1}. The integral of f over any window of time (not only infinitesimal windows but also large windows) is the probability that the bacterium dies in that window.

Absolutely Continuous Univariate Distributions

A probability density function is most commonly associated with absolutely continuous univariate distributions. A random variable X has density f_X, where f_X is a non-negative Lebesgue-integrable function, if:

$$\Pr[a \leq X \leq b] = \int_a^b f_X(x)dx.$$

Hence, if F_X is the cumulative distribution function of X, then:

$$F_X(x) = \int_{-\infty}^x f_X(u)du,$$

and (if f_X is continuous at x)

$$f_X(x) = \frac{d}{dx}F_X(x).$$

Intuitively, one can think of $f_X(x)$ dx as being the probability of X falling within the infinitesimal interval [x, x + dx].

Formal Definition

A random variable X with values in a measurable space $(\mathcal{X}, \mathcal{A})$ (usually R^n with the Borel sets as measurable subsets) has as probability distribution the measure $X_* P$ on $(\mathcal{X}, \mathcal{A})$ the density of X with respect to a reference measure μ on $(\mathcal{X}, \mathcal{A})$ is the Radon–Nikodym derivative:

$$f = \frac{dX_*P}{d\mu}.$$

That is, f is any measurable function with the property that:

$$\Pr[X \in A] = \int_{X^{-1}A} dP = \int_A f \, d\mu$$

for any measurable set $A \in \mathcal{A}$.

Discussion

In the continuous univariate case above, the reference measure is the Lebesgue measure. The probability mass function of a discrete random variable is the density with respect to the counting measure over the sample space (usually the set of integers, or some subset thereof).

Note that it is not possible to define a density with reference to an arbitrary measure (e.g. one can't choose the counting measure as a reference for a continuous random variable). Furthermore, when it does exist, the density is almost everywhere unique.

Further Details

Unlike a probability, a probability density function can take on values greater than one; for example, the uniform distribution on the interval [0, ½] has probability density $f(x) = 2$ for $0 \le x \le ½$ and $f(x) = 0$ elsewhere.

The standard normal distribution has probability density

$$f(x) = \frac{1}{\sqrt{2\pi}} e^{-x^2/2}.$$

If a random variable X is given and its distribution admits a probability density function f, then the expected value of X (if the expected value exists) can be calculated as

$$E[X] = \int_{-\infty}^{\infty} x f(x) dx.$$

Not every probability distribution has a density function: the distributions of discrete random variables do not; nor does the Cantor distribution, even though it has no discrete component, i.e., does not assign positive probability to any individual point.

A distribution has a density function if and only if its cumulative distribution function $F(x)$ is absolutely continuous. In this case: F is almost everywhere differentiable, and its derivative can be used as probability density:

$$\frac{d}{dx} F(x) = f(x).$$

If a probability distribution admits a density, then the probability of every one-point set $\{a\}$ is zero; the same holds for finite and countable sets.

Two probability densities f and g represent the same probability distribution precisely if they differ only on a set of Lebesgue measure zero.

In the field of statistical physics, a non-formal reformulation of the relation above between the derivative of the cumulative distribution function and the probability density function is generally used as the definition of the probability density function. This alternate definition is the following:

If dt is an infinitely small number, the probability that X is included within the interval $(t, t + dt)$ is equal to $f(t)\, dt$, or:

$$\Pr(t < X < t + dt) = f(t)dt.$$

Link Between Discrete and Continuous Distributions

It is possible to represent certain discrete random variables as well as random variables involving both a continuous and a discrete part with a generalized probability density function, by using the Dirac delta function. For example, let us consider a binary discrete random variable having the Rademacher distribution—that is, taking −1 or 1 for values, with probability ½ each. The density of probability associated with this variable is:

$$f(t) = \frac{1}{2}(\delta(t+1) + \delta(t-1)).$$

More generally, if a discrete variable can take n different values among real numbers, then the associated probability density function is:

$$f(t) = \sum_{i=1}^{n} p_i \delta(t - x_i),$$

where $x_1, ..., x_n$ are the discrete values accessible to the variable and $p_1, ..., p_n$ are the probabilities associated with these values.

This substantially unifies the treatment of discrete and continuous probability distributions. For instance, the above expression allows for determining statistical characteristics of such a discrete variable (such as its mean, its variance and its kurtosis), starting from the formulas given for a continuous distribution of the probability.

Families of Densities

It is common for probability density functions (and probability mass functions) to be parametrized—that is, to be characterized by unspecified parameters. For example, the normal distribution is parametrized in terms of the mean and the variance, denoted by μ and σ^2 respectively, giving the family of densities

$$f(x; \mu, \sigma^2) = \frac{1}{\sigma\sqrt{2\pi}} e^{-\frac{1}{2}\left(\frac{x-\mu}{\sigma}\right)^2}.$$

It is important to keep in mind the difference between the domain of a family of densities and the parameters of the family. Different values of the parameters describe different distributions of different random variables on the same sample space (the same set of all possible values of the variable); this sample space is the domain of the family of random variables that this family of distributions describes. A given set of parameters describes a single distribution within the family sharing the functional form of the density. From the perspective of a given distribution, the parameters are constants, and terms in a density function that contain only parameters, but not variables, are part of the normalization factor of a distribution (the multiplicative factor that ensures that the area under the density—the probability of *something* in the domain occurring— equals 1). This normalization factor is outside the kernel of the distribution.

Since the parameters are constants, reparametrizing a density in terms of different parameters, to give a characterization of a different random variable in the family, means simply substituting the new parameter values into the formula in place of the old ones. Changing the domain of a probability density, however, is trickier and requires more work.

Densities Associated With Multiple Variables

For continuous random variables $X_1, ..., X_n$, it is also possible to define a probability density function associated to the set as a whole, often called joint probability density function. This density function is defined as a function of the n variables, such that, for any domain D in the n-dimensional space of the values of the variables $X_1, ..., X_n$, the probability that a realisation of the set variables falls inside the domain D is

$$\Pr\left(X_1, \cdots, X_n \in D\right) = \int_D f_{X_1, \cdots, X_n}(x_1, \cdots, x_n) dx_1 \cdots dx_n.$$

If $F(x_1, ..., x_n) = \Pr(X_1 \le x_1, ..., X_n \le x_n)$ is the cumulative distribution function of the vector $(X_1, ..., X_n)$, then the joint probability density function can be computed as a partial derivative

$$f(x) = \frac{\partial^n F}{\partial x_1 \cdots \partial x_n}\bigg|_x$$

Marginal Densities

For $i=1, 2, ...,n$, let $f_{Xi}(x_i)$ be the probability density function associated with variable X_i alone. This is called the "marginal" density function, and can be deduced from the probability density associated with the random variables $X_1, ..., X_n$ by integrating on all values of the $n - 1$ other variables:

$$f_{X_i}(x_i) = \int f(x_1, \cdots, x_n) dx_1 \cdots dx_{i-1} dx_{i+1} \cdots dx_n.$$

Independence

Continuous random variables $X_1, ..., X_n$ admitting a joint density are all independent from each other if and only if

$$f_{X_1, \cdots, X_n}(x_1, \cdots, x_n) = f_{X_1}(x_1) \cdots f_{X_n}(x_n).$$

Corollary

If the joint probability density function of a vector of n random variables can be factored into a product of n functions of one variable

$$f_{X_1,\cdots,X_n}(x_1,\cdots,x_n) = f_1(x_1)\cdots f_n(x_n),$$

(where each f_i is not necessarily a density) then the n variables in the set are all independent from each other, and the marginal probability density function of each of them is given by

$$f_{X_i}(x_i) = \frac{f_i(x_i)}{\int f_i(x)dx}.$$

Example

This elementary example illustrates the above definition of multidimensional probability density functions in the simple case of a function of a set of two variables. Let us call $\vec{R}$ a 2-dimensional random vector of coordinates (X, Y): the probability to obtain $\vec{R}$ in the quarter plane of positive x and y is

$$\Pr(X > 0, Y > 0) = \int_0^\infty \int_0^\infty f_{X,Y}(x, y)dxdy.$$

Dependent Variables and Change of Variables

If the probability density function of a random variable X is given as $f_X(x)$, it is possible to calculate the probability density function of some variable $Y = g(X)$. This is also called a "change of variable" and is in practice used to generate a random variable of arbitrary shape $f_{g(X)} = f_Y$ using a known (for instance uniform) random number generator.

If the function g is monotonic, then the resulting density function is

$$f_Y(y) = \left| \frac{d}{dy}(g^{-1}(y)) \right| \cdot f_X(g^{-1}(y)).$$

Here g^{-1} denotes the inverse function.

This follows from the fact that the probability contained in a differential area must be invariant under change of variables. That is,

$$\left| f_Y(y)dy \right| = \left| f_X(x)dx \right|,$$

or

$$f_Y(y) = \left| \frac{dx}{dy} \right| f_X(x) = \left| \frac{d}{dy}(x) \right| f_X(x) = \left| \frac{d}{dy}(g^{-1}(y)) \right| f_X(g^{-1}(y)) = \frac{f_X(g^{-1}(y))}{|g'(g^{-1}(y))|}.$$

For functions which are not monotonic the probability density function for y is

$$\sum_{k=1}^{n(y)} \left| \frac{d}{dy} g_k^{-1}(y) \right| \cdot f_X(g_k^{-1}(y))$$

where $n(y)$ is the number of solutions in x for the equation $g(x) = y$, and $g^{-1}{}_k(y)$ are these solutions.

It is tempting to think that in order to find the expected value $E(g(X))$ one must first find the probability density $f_{g(X)}$ of the new random variable $Y = g(X)$. However, rather than computing

$$E(g(X)) = \int_{-\infty}^{\infty} y f_{g(X)}(y) dy,$$

one may find instead

$$E(g(X)) = \int_{-\infty}^{\infty} g(x) f_X(x) dx.$$

The values of the two integrals are the same in all cases in which both X and $g(X)$ actually have probability density functions. It is not necessary that g be a one-to-one function. In some cases the latter integral is computed much more easily than the former.

Multiple Variables

The above formulas can be generalized to variables (which we will again call y) depending on more than one other variable. $f(x_1, ..., x_n)$ shall denote the probability density function of the variables that y depends on, and the dependence shall be $y = g(x_1, ..., x_n)$. Then, the resulting density function is

$$\int_{y=g(x_1,\cdots,x_n)} \frac{f(x_1,\cdots,x_n)}{\sqrt{\sum_{j=1}^{n} \frac{\partial g}{\partial x_j}(x_1,\cdots,x_n)^2}} \, dV$$

where the integral is over the entire $(n-1)$-dimensional solution of the subscripted equation and the symbolic dV must be replaced by a parametrization of this solution for a particular calculation; the variables $x_1, ..., x_n$ are then of course functions of this parametrization.

This derives from the following, perhaps more intuitive representation: Suppose x is an n-dimensional random variable with joint density f. If $y = H(x)$, where H is a bijective, differentiable function, then y has density g:

$$g(\mathbf{y}) = f(\mathbf{x}) \left| \det\left(\frac{d\mathbf{x}}{d\mathbf{y}} \right) \right|$$

with the differential regarded as the Jacobian of the inverse of H, evaluated at y.

Using the delta-function (and assuming independence) the same result is formulated as follows.

If the probability density function of independent random variables X_i, $i = 1, 2, ...n$ are given as $f_{Xi}(x_i)$, it is possible to calculate the probability density function of some variable $Y = G(X_1, X_2, ... X_n)$. The following formula establishes a connection between the probability density function of Y denoted by $f_Y(y)$ and $f_{Xi}(x_i)$ using the Dirac delta function:

$$f_Y(y) = \int_{-\infty}^{\infty} \int_{-\infty}^{\infty} \cdots \int_{-\infty}^{\infty} f_{X_1}(x_1) f_{X_2}(x_2) \cdots f_{X_n}(x_n) \delta(y - G(x_1, x_2, \cdots, x_n)) dx_1 dx_2 \cdots dx_n$$

Sums of Independent Random Variables

The probability density function of the sum of two independent random variables U and V, each of which has a probability density function, is the convolution of their separate density functions:

$$f_{U+V}(x) = \int_{-\infty}^{\infty} f_U(y) f_V(x-y) dy = (f_U * f_V)(x)$$

It is possible to generalize the previous relation to a sum of N independent random variables, with densities $U_1, ..., U_N$:

$$f_{U_1 + \cdots + U_N}(x) = (f_{U_1} * \cdots * f_{U_N})(x)$$

This can be derived from a two-way change of variables involving $Y=U+V$ and $Z=V$, similarly to the example below for the quotient of independent random variables.

Products and Quotients of Independent Random Variables

Given two independent random variables U and V, each of which has a probability density function, the density of the product $Y=UV$ and quotient $Y=U/V$ can be computed by a change of variables.

Example: Quotient Distribution

To compute the quotient $Y=U/V$ of two independent random variables U and V, define the following transformation:

$Y = U/V$

$Z = V$

Then, the joint density $p(Y,Z)$ can be computed by a change of variables from U,V to Y,Z, and Y can be derived by marginalizing out Z from the joint density.

The inverse transformation is

$U = YZ$

$V = Z$

The Jacobian matrix $J(U,V \mid Y,Z)$ of this transformation is

$$\begin{vmatrix} \dfrac{\partial U}{\partial Y} & \dfrac{\partial U}{\partial Z} \\ \dfrac{\partial V}{\partial Y} & \dfrac{\partial V}{\partial Z} \end{vmatrix} = \begin{vmatrix} Z & Y \\ 0 & 1 \end{vmatrix} = |Z|.$$

Thus:

$$p(Y,Z)=p(U,V)J(U,V\mid Y,Z)=p(U)p(V)J(U,V\mid Y,Z)=p_U(YZ)p_V(Z)|Z|.$$

And the distribution of Y can be computed by marginalizing out Z:

$$p(Y)=\int_{-\infty}^{\infty}p_U(YZ)p_V(Z)|Z|dZ$$

Note that this method crucially requires that the transformation from U,V to Y,Z be bijective. The above transformation meets this because Z can be mapped directly back to V, and for a given V the quotient U/V is monotonic. This is similarly the case for the sum $U+V$, difference $U-V$ and product UV.

Exactly the same method can be used to compute the distribution of other functions of multiple independent random variables.

Example: Quotient of Two Standard Normals

Given two standard normal variables U and V, the quotient can be computed as follows. First, the variables have the following density functions:

$$p(U)=\frac{1}{\sqrt{2\pi}}e^{-\frac{U^2}{2}}$$

$$p(V)=\frac{1}{\sqrt{2\pi}}e^{-\frac{V^2}{2}}$$

We transform as described above:

$$Y=U/V$$

$$Z=V$$

This leads to:

$$p(Y)=\int_{-\infty}^{\infty}p_U(YZ)p_V(Z)|Z|dZ$$

$$=\int_{-\infty}^{\infty}\frac{1}{\sqrt{2\pi}}e^{-\frac{1}{2}Y^2Z^2}\frac{1}{\sqrt{2\pi}}e^{-\frac{1}{2}Z^2}|Z|dZ$$

$$=\int_{-\infty}^{\infty}\frac{1}{2\pi}e^{-\frac{1}{2}(Y^2+1)Z^2}|Z|dZ$$

$$=2\int_{0}^{\infty}\frac{1}{2\pi}e^{-\frac{1}{2}(Y^2+1)Z^2}ZdZ$$

$$=\int_{0}^{\infty}\frac{1}{\pi}e^{-(Y^2+1)u}du=\left(u=\tfrac{1}{2}Z^2\right)$$

$$=-\frac{1}{\pi(Y^2+1)}e^{-(Y^2+1)u}\Big]_{u=0}^{\infty}$$

$$=\frac{1}{\pi(Y^2+1)}$$

This is a standard Cauchy distribution.

Cumulative Distribution Function

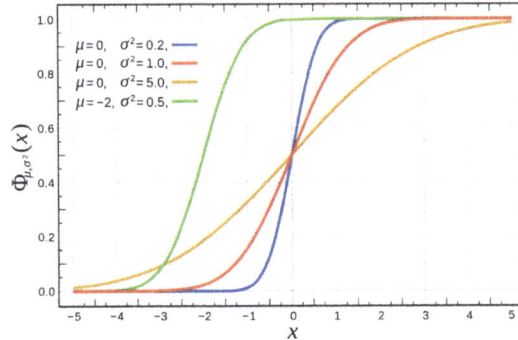

Cumulative Distribution Function for the normal distribution

In probability theory and statistics, the cumulative distribution function (CDF) of a real-valued random variable X, or just distribution function of X, evaluated at x, is the probability that X will take a value less than or equal to x.

In the case of a continuous distribution, it gives the area under the probability density function from minus infinity to x. Cumulative distribution functions are also used to specify the distribution of multivariate random variables.

Definition

The cumulative distribution function of a real-valued random variable X is the function given by

$$F_X(x) = P(X \leq x),$$

where the right-hand side represents the probability that the random variable X takes on a value less than or equal to x. The probability that X lies in the semi-closed interval $(a, b]$, where $a < b$, is therefore

$$P(a < X \leq b) = F_X(b) - F_X(a).$$

In the definition above, the "less than or equal to" sign, "$\leq$", is a convention, not a universally used one (e.g. Hungarian literature uses "$<$"), but is important for discrete distributions. The proper use of tables of the binomial and Poisson distributions depends upon this convention. Moreover, important formulas like Paul Lévy's inversion formula for the characteristic function also rely on the "less than or equal" formulation.

If treating several random variables X, Y, ... etc. the corresponding letters are used as subscripts while, if treating only one, the subscript is usually omitted. It is conventional to use a capital F for a cumulative distribution function, in contrast to the lower-case f used for probability density functions and probability mass functions. This applies when discussing general distributions: some specific distributions have their own conventional notation, for example the normal distribution.

The CDF of a continuous random variable X can be expressed as the integral of its probability density function f_X as follows:

$$F_X(x) = \int_{-\infty}^{x} f_X(t)dt.$$

In the case of a random variable X which has distribution having a discrete component at a value b,

$$P(X = b) = F_X(b) - \lim_{x \to b^-} F_X(x).$$

If F_X is continuous at b, this equals zero and there is no discrete component at b.

Properties

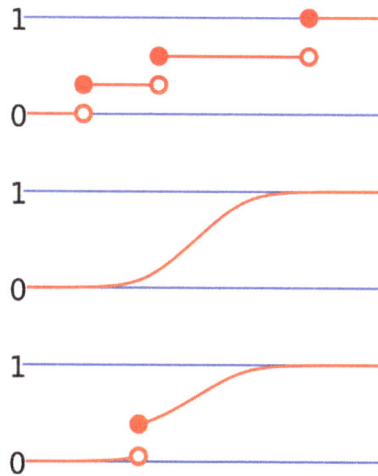

From top to bottom, the cumulative distribution function of a discrete probability distribution, continuous probability distribution, and a distribution which has both a continuous part and a discrete part.

Every cumulative distribution function F is non-decreasing and right-continuous, which makes it a càdlàg function. Furthermore,

$$\lim_{x \to -\infty} F(x) = 0, \quad \lim_{x \to +\infty} F(x) = 1.$$

Every function with these four properties is a CDF, i.e., for every such function, a random variable can be defined such that the function is the cumulative distribution function of that random variable.

If X is a purely discrete random variable, then it attains values $x_1, x_2, \ldots$ with probability $p_i = P(x_i)$, and the CDF of X will be discontinuous at the points x_i and constant in between:

$$F(x) = P(X \le x) = \sum_{x_i \le x} P(X = x_i) = \sum_{x_i \le x} p(x_i).$$

If the CDF F of a real valued random variable X is continuous, then X is a continuous random variable; if furthermore F is absolutely continuous, then there exists a Lebesgue-integrable function $f(x)$ such that

$$F(b) - F(a) = P(a < X \le b) = \int_{a}^{b} f(x)dx$$

for all real numbers a and b. The function f is equal to the derivative of F almost everywhere, and it is called the probability density function of the distribution of X.

Examples

As an example, suppose   is uniformly distributed on the unit interval [0, 1]. Then the CDF of X is given by

$$F(x) = \begin{cases} 0 & : x < 0 \\ x & : 0 \le x < 1 \\ 1 & : x \ge 1. \end{cases}$$

Suppose instead that X takes only the discrete values 0 and 1, with equal probability. Then the CDF of X is given by

$$F(x) = \begin{cases} 0 & : x < 0 \\ 1/2 & : 0 \le x < 1 \\ 1 & : x \ge 1. \end{cases}$$

Derived Functions

Complementary Cumulative Distribution Function (Tail Distribution)

Sometimes, it is useful to study the opposite question and ask how often the random variable is *above* a particular level. This is called the complementary cumulative distribution function (ccdf) or simply the tail distribution or exceedance, and is defined as

$$\bar{F}(x) = P(X > x) = 1 - F(x).$$

This has applications in statistical hypothesis testing, for example, because the one-sided p-value is the probability of observing a test statistic *at least* as extreme as the one observed. Thus, provided that the test statistic, T, has a continuous distribution, the one-sided p-value is simply given by the ccdf: for an observed value t of the test statistic

$$p = P(T \ge t) = P(T > t) = 1 - F_T(t).$$

In survival analysis, $\bar{F}(x)$ is called the survival function and denoted $S(x)$, while the term *reliability function* is common in engineering.

Properties

- For a non-negative continuous random variable having an expectation, Markov's inequality states that

$$\bar{F}(x) \le \frac{\mathbb{E}(X)}{x}.$$

- As $x \to \infty, \bar{F}(x) \to 0$, and in fact $\bar{F}(x) = o(1/x)$ provided that $\mathbb{E}(X)$ is finite.

Proof: Assuming X has a density function f, for any $c > 0$

$$\mathbb{E}(X) = \int_0^\infty x f(x) dx \geq \int_0^c x f(x) dx + c \int_c^\infty f(x) dx$$

Then, on recognizing $\bar{F}(c) = \int_c^\infty f(x) dx$ and rearranging terms,

$$0 \leq c\bar{F}(c) \leq \mathbb{E}(X) - \int_0^c x f(x) dx \to 0 \ \text{ as } \ c \to \infty$$

as claimed.

Folded Cumulative Distribution

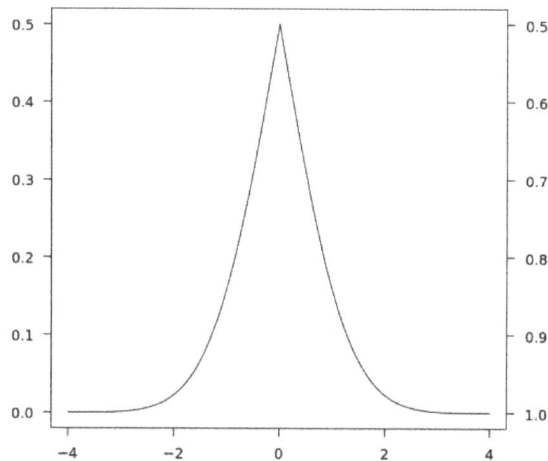

Example of the folded cumulative distribution for a normal distribution function with an expected value of 0 and a standard deviation of 1.

While the plot of a cumulative distribution often has an S-like shape, an alternative illustration is the folded cumulative distribution or mountain plot, which folds the top half of the graph over, thus using two scales, one for the upslope and another for the downslope. This form of illustration emphasises the median and dispersion (the mean absolute deviation from the median) of the distribution or of the empirical results.

Inverse Distribution Function (Quantile Function)

If the CDF F is strictly increasing and continuous then $F^{-1}(p), p \in [0,1]$, is the unique real number x such that $F(x) = p$. In such a case, this defines the inverse distribution function or quantile function.

Some distributions do not have a unique inverse (for example in the case where $f_X(x) = 0$ for all $a < x < b$, causing F_X to be constant). This problem can be solved by defining, for $p \in [0,1]$, the generalized inverse distribution function:

$$F^{-1}(p) = \inf\{x \in \mathbb{R} : F(x) \geq p\}.$$

- Example 1: The median is $F^{-1}(0.5)$.

- Example 2: Put $\tau = F^{-1}(0.95)$. Then we call τ the 95th percentile.

Some useful properties of the inverse cdf (which are also preserved in the definition of the generalized inverse distribution function) are:

1. F^{-1} is nondecreasing

2. $F^{-1}(F(x)) \leq x$

3. $F(F^{-1}(p)) \geq p$

4. $F^{-1}(p) \leq x$ if and only if $y \leq F(p)$

5. If Y has a $U[0,1]$ distribution then $F^{-1}(Y)$ is distributed as F. This is used in random number generation using the inverse transform sampling-method.

6. If $\{X_\alpha\}$ is a collection of independent F-distributed random variables defined on the same sample space, then there exist random variables Y_α such that Y_α is distributed as $U[0,1]$ and $F^{-1}(Y_\alpha) = X_\alpha$ with probability 1 for all α.

The inverse of the cdf can be used to translate results obtained for the uniform distribution to other distributions.

Multivariate Case

When dealing simultaneously with more than one random variable the *joint* cumulative distribution function can also be defined. For example, for a pair of random variables X,Y, the joint CDF F is given by

$$F(x,y) = P(X \leq x, Y \leq y),$$

where the right-hand side represents the probability that the random variable X takes on a value less than or equal to x and that Y takes on a value less than or equal to y.

Every multivariate CDF is:

1. Monotonically non-decreasing for each of its variables

2. Right-continuous for each of its variables.

3. $0 \leq F(x_1,...,x_n) \leq 1$

4. $\lim_{x_1,...,x_n \to +\infty} F(x_1,...,x_n) = 1$ and $\lim_{x_i \to -\infty} F(x_1,...,x_n) = 0$, for all i

Use in Statistical Analysis

The concept of the cumulative distribution function makes an explicit appearance in statistical analysis in two (similar) ways. Cumulative frequency analysis is the analysis of the frequency of occurrence of values of a phenomenon less than a reference value. The empirical distribution function is a formal direct estimate of the cumulative distribution function for which simple statistical properties can be derived and which can form the basis of various statistical hypothesis tests. Such tests can assess whether there is evidence against a sample of data having arisen from a given distribution, or evidence against two samples of data having arisen from the same (unknown) population distribution.

Kolmogorov–Smirnov and Kuiper's tests

The Kolmogorov–Smirnov test is based on cumulative distribution functions and can be used to test to see whether two empirical distributions are different or whether an empirical distribution is different from an ideal distribution. The closely related Kuiper's test is useful if the domain of the distribution is cyclic as in day of the week. For instance Kuiper's test might be used to see if the number of tornadoes varies during the year or if sales of a product vary by day of the week or day of the month.

Quantile Function

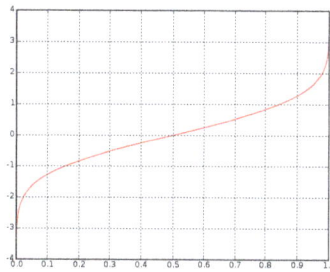

The probit is the quantile function of the normal distribution.

In probability and statistics, the quantile function specifies, for a given probability in the probability distribution of a random variable, the value at which the probability of the random variable is less than or equal to the given probability. It is also called the percent point function or inverse cumulative distribution function.

Definition

With reference to a continuous and strictly monotonic distribution function, for example the cumulative distribution function $F_X : R \to [0,1]$ of a random variable X, the quantile function Q returns a threshold value x below which random draws from the given c.d.f would fall p percent of the time.

In terms of the distribution function F, the quantile function Q returns the value x such that

$$F_X(x) := \Pr(X \leq x) = p.$$

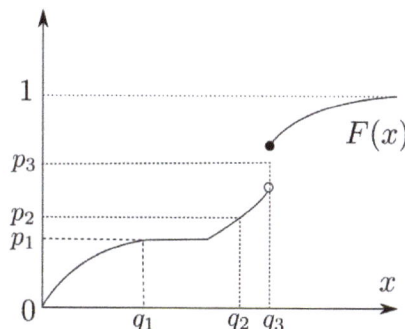

The cumulative distribution function (shown as $F(x)$) gives the p values as a function of the q values. The quantile function does the opposite: it gives the q values as a function of the p values.

Another way to express the quantile function is

$$Q(p)=\inf\left\{x\in\mathbb{R}:p\leq F(x)\right\}$$

for a probability $0<p<1$. Here we capture the fact that the quantile function returns the minimum value of x from amongst all those values whose c.d.f value exceeds p, which is equivalent to the previous probability statement.

If the function F is continuous and strictly monotonically increasing, then the infimum function can be replaced by the minimum function and

$$Q=F^{-1}$$

However, if F has flat regions or is discontinuous, then the inverse of F is not well-defined. In this case we need to use the more complicated formula above.

Simple Example

For example, the cumulative distribution function of Exponential(λ) (i.e. intensity λ and expected value (mean) $1/\lambda$) is

$$F(x;\lambda)=\begin{cases}1-e^{-\lambda x} & x\geq 0,\\ 0 & x<0.\end{cases}$$

The quantile function for Exponential(λ) is derived by finding the value of Q for which $1-e^{-\lambda Q}=p$:

$$Q(p;\lambda)=\frac{-\ln(1-p)}{\lambda},$$

for $0\leq p<1$. The quartiles are therefore:

first quartile (p = 1/4)

$$\ln(4/3)/\lambda$$

median (p = 2/4)

$$\ln(2)/\lambda$$

third quartile (p = 3/4)

$$\ln(4)/\lambda.$$

Applications

Quantile functions are used in both statistical applications and Monte Carlo methods.

The quantile function is one way of prescribing a probability distribution, and it is an alternative to the probability density function (pdf) or probability mass function, the cumulative distribution function (cdf) and the characteristic function. The quantile function, Q, of a probability distribution is the inverse of its cumulative distribution function F. The derivative of the quantile function, namely the quantile density function, is yet another way of prescribing a probability distribution. It is the reciprocal of the pdf composed with the quantile function.

For statistical applications, users need to know key percentage points of a given distribution. For example, they require the median and 25% and 75% quartiles as in the example above or 5%, 95%, 2.5%, 97.5% levels for other applications such as assessing the statistical significance of an observation whose distribution is known. Before the popularization of computers, it was not uncommon for books to have appendices with statistical tables sampling the quantile function. Statistical applications of quantile functions are discussed extensively by Gilchrist.

Monte-Carlo simulations employ quantile functions to produce non-uniform random or pseudo-random numbers for use in diverse types of simulation calculations. A sample from a given distribution may be obtained in principle by applying its quantile function to a sample from a uniform distribution. The demands, for example, of simulation methods in modern computational finance are focusing increasing attention on methods based on quantile functions, as they work well with multivariate techniques based on either copula or quasi-Monte-Carlo methods and Monte Carlo methods in finance.

Calculation

The evaluation of quantile functions often involves numerical methods, as the example of the exponential distribution above is one of the few distributions where a closed-form expression can be found (others include the uniform, the Weibull, the Tukey lambda (which includes the logistic) and the log-logistic). When the cdf itself has a closed-form expression, one can always use a numerical root-finding algorithm such as the bisection method to invert the cdf. Other algorithms to evaluate quantile functions are given in the Numerical Recipes series of books. Algorithms for common distributions are built into many statistical software packages.

Quantile functions may also be characterized as solutions of non-linear ordinary and partial differential equations. The ordinary differential equations for the cases of the normal, Student, beta and gamma distributions have been given and solved.

Normal Distribution

The normal distribution is perhaps the most important case. Because the normal distribution is a location-scale family, its quantile function for arbitrary parameters can be derived from a simple transformation of the quantile function of the standard normal distribution, known as the probit function. Unfortunately, this function has no closed-form representation using basic algebraic functions; as a result, approximate representations are usually used. Thorough composite rational and polynomial approximations have been given by Wichura and Acklam. Non-composite rational approximations have been developed by Shaw.

Ordinary Differential Equation for the Normal Quantile

A non-linear ordinary differential equation for the normal quantile, $w(p)$, may be given. It is

$$\frac{d^2w}{dp^2} = w\left(\frac{dw}{dp}\right)^2$$

with the centre (initial) conditions

$$w(1/2) = 0$$

$$w'(1/2) = \sqrt{2\pi}.$$

This equation may be solved by several methods, including the classical power series approach. From this solutions of arbitrarily high accuracy may be developed.

Student's t-distribution

This has historically been one of the more intractable cases, as the presence of a parameter, v, the degrees of freedom, makes the use of rational and other approximations awkward. Simple formulas exist when the v = 1, 2, 4 and the problem may be reduced to the solution of a polynomial when v is even. In other cases the quantile functions may be developed as power series. The simple cases are as follows:

v = 1 (Cauchy distribution)

$$Q(p) = \tan(\pi(p-1/2))$$

v = 2

$$Q(p) = 2(p-1/2)\sqrt{\frac{2}{\alpha}}$$

v = 4

$$Q(p) = \text{sign}(p-1/2)2\sqrt{q-1}$$

where

$$q = \frac{\cos\left(\frac{1}{3}\arccos\left(\sqrt{\alpha}\right)\right)}{\sqrt{\alpha}}$$

and

$$\alpha = 4p(1-p).$$

In the above the "sign" function is +1 for positive arguments, -1 for negative arguments and zero at zero. It should not be confused with the trigonometric sine function.

Quantile Mixtures

Analogously to the mixtures of densities, distributions can be defined as quantile mixtures

$$Q(p) = \sum_{i=1}^{m} a_i Q_i(p),$$

where $Q_i(p)$, $i = 1, \ldots, m$ are quantile functions and a_i, $i = 1, \ldots, m$ are the model parameters. The parameters a_i must be selected so that $Q(p)$ is a quantile function. Two four-parametric quantile mixtures, the normal-polynomial quantile mixture and the Cauchy-polynomial quantile mixture, are presented by Karvanen.

Non-linear Differential Equations for Quantile Functions

The non-linear ordinary differential equation given for normal distribution is a special case of that available for any quantile function whose second derivative exists. In general the equation for a quantile, $Q(p)$, may be given. It is

$$\frac{d^2Q}{dp^2} = H(Q)\left(\frac{dQ}{dp}\right)^2$$

augmented by suitable boundary conditions, where

$$H(x) = -\frac{d\log[f(x)]}{dx}$$

and $f(x)$ is the probability density function. The forms of this equation, and its classical analysis by series and asymptotic solutions, for the cases of the normal, Student, gamma and beta distributions has been elucidated by Steinbrecher and Shaw (2008). Such solutions provide accurate benchmarks, and in the case of the Student, suitable series for live Monte Carlo use.

Expected Value

In probability theory, the expected value of a random variable, intuitively, is the long-run average value of repetitions of the experiment it represents. For example, the expected value in rolling a six-sided die is 3.5 because, roughly speaking, the average of all the numbers that come up in an extremely large number of rolls is very nearly always quite close to three and a half. Less roughly, the law of large numbers states that the arithmetic mean of the values almost surely converges to the expected value as the number of repetitions approaches infinity. The expected value is also known as the expectation, mathematical expectation, EV, average, mean value, mean, or first moment.

More practically, the expected value of a discrete random variable is the probability-weighted average of all possible values. In other words, each possible value the random variable can assume is multiplied by its probability of occurring, and the resulting products are summed to produce the expected value. The same principle applies to a continuous random variable, except that an integral of the variable with respect to its probability density replaces the sum. The formal definition subsumes both of these and also works for distributions which are neither discrete nor continuous: the expected value of a random variable is the integral of the random variable with respect to its probability measure.

The expected value does not exist for random variables having some distributions with large "tails", such as the Cauchy distribution. For random variables such as these, the long-tails of the distribution prevent the sum/integral from converging.

The expected value is a key aspect of how one characterizes a probability distribution; it is one type of location parameter. By contrast, the variance is a measure of dispersion of the possible values of the random variable around the expected value. The variance itself is defined in terms of two expectations: it is the expected value of the squared deviation of the variable's value from the variable's expected value.

The expected value plays important roles in a variety of contexts. In regression analysis, one desires a formula in terms of observed data that will give a "good" estimate of the parameter giving the effect of some explanatory variable upon a dependent variable. The formula will give different estimates using different samples of data, so the estimate it gives is itself a random variable. A formula is typically considered good in this context if it is an unbiased estimator—that is, if the expected value of the estimate (the average value it would give over an arbitrarily large number of separate samples) can be shown to equal the true value of the desired parameter.

In decision theory, and in particular in choice under uncertainty, an agent is described as making an optimal choice in the context of incomplete information. For risk neutral agents, the choice involves using the expected values of uncertain quantities, while for risk averse agents it involves maximizing the expected value of some objective function such as a von Neumann–Morgenstern utility function. One example of using expected value in reaching optimal decisions is the Gordon–Loeb model of information security investment. According to the model, one can conclude that the amount a firm spends to protect information should generally be only a small fraction of the expected loss (i.e., the expected value of the loss resulting from a cyber/information security breach).

Definition

Univariate Discrete Random Variable, Finite Case

Suppose random variable X can take value x_1 with probability p_1, value x_2 with probability p_2, and so on, up to value x_k with probability p_k. Then the expectation of this random variable X is defined as

$$E[X] = x_1 p_1 + x_2 p_2 + \cdots + x_k p_k .$$

Since all probabilities p_i add up to one ($p_1 + p_2 + \ldots + p_k = 1$), the expected value can be viewed as the weighted average, with p_i's being the weights:

$$E[X] = \frac{x_1 p_1 + x_2 p_2 + \cdots + x_k p_k}{1} = \frac{x_1 p_1 + x_2 p_2 + \cdots + x_k p_k}{p_1 + p_2 + \cdots + p_k} .$$

If all outcomes x_i are equally likely (that is, $p_1 = p_2 = \ldots = p_k$), then the weighted average turns into the simple average. This is intuitive: the expected value of a random variable is the average of all values it can take; thus the expected value is what one expects to happen *on average*. If the outcomes x_i are not equally probable, then the simple average must be replaced with the weighted average, which takes into account the fact that some outcomes are more likely than the others. The intuition however remains the same: the expected value of X is what one expects to happen *on average*.

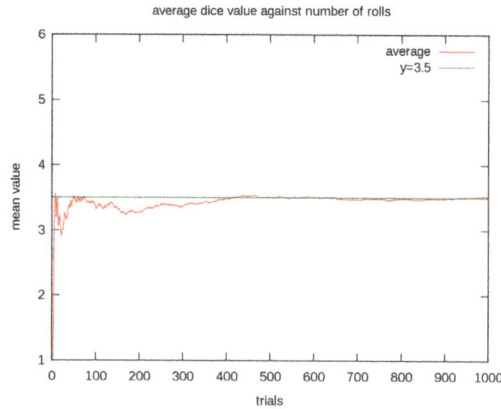

An illustration of the convergence of sequence averages of rolls of a die to the expected value of 3.5 as the number of rolls (trials) grows.

Example 1. Let X represent the outcome of a roll of a fair six-sided die. More specifically, X will be the number of pips showing on the top face of the die after the toss. The possible values for X are 1, 2, 3, 4, 5, and 6, all equally likely (each having the probability of 1/6). The expectation of X is

$$E[X] = 1 \cdot \frac{1}{6} + 2 \cdot \frac{1}{6} + 3 \cdot \frac{1}{6} + 4 \cdot \frac{1}{6} + 5 \cdot \frac{1}{6} + 6 \cdot \frac{1}{6} = 3.5.$$

If one rolls the die n times and computes the average (arithmetic mean) of the results, then as n grows, the average will almost surely converge to the expected value, a fact known as the strong law of large numbers. One example sequence of ten rolls of the die is 2, 3, 1, 2, 5, 6, 2, 2, 2, 6, which has the average of 3.1, with the distance of 0.4 from the expected value of 3.5. The convergence is relatively slow: the probability that the average falls within the range 3.5 ± 0.1 is 21.6% for ten rolls, 46.1% for a hundred rolls and 93.7% for a thousand rolls. For an illustration of the averages of longer sequences of rolls of the die and how they converge to the expected value of 3.5. More generally, the rate of convergence can be roughly quantified by e.g. Chebyshev's inequality and the Berry–Esseen theorem.

Example 2. The roulette game consists of a small ball and a wheel with 38 numbered pockets around the edge. As the wheel is spun, the ball bounces around randomly until it settles down in one of the pockets. Suppose random variable X represents the (monetary) outcome of a $1 bet on a single number ("straight up" bet). If the bet wins (which happens with probability 1/38 in American roulette), the payoff is $35; otherwise the player loses the bet. The expected profit from such a bet will be

$$E[\text{gain from } \$1 \text{ bet}] = -\$1 \cdot \frac{37}{38} + \$35 \cdot \frac{1}{38} = -\$0.0526.$$

i.e. the bet of $1 stands to lose $0.0526, so its expected value is -$0.0526.

Univariate Discrete Random Variable, Countable Case

Let X be a discrete random variable taking values $x_1, x_2, \ldots$ with probabilities $p_1, p_2, \ldots$ respectively. Then the expected value of this random variable is the infinite sum

$$E[X] = \sum_{i=1}^{\infty} x_i \, p_i,$$

provided that this series converges absolutely (that is, the sum must remain finite if we were to replace all x_i's with their absolute values). If this series does not converge absolutely, we say that the expected value of X does not exist.

For example, suppose random variable X takes values $1, -2, 3, -4, \ldots$, with respective probabilities $c/1^2, c/2^2, c/3^2, c/4^2, \ldots$, where $c = 6/\pi^2$ is a normalizing constant that ensures the probabilities sum up to one. Then the infinite sum

$$\sum_{i=1}^{\infty} x_i \, p_i = c \left(1 - \frac{1}{2} + \frac{1}{3} - \frac{1}{4} + \cdots \right)$$

converges and its sum is equal to $\ln(2)6/\pi^2 \simeq 0.421383$. However it would be incorrect to claim that the expected value of X is equal to this number—in fact $E[X]$ does not exist, as this series does not converge absolutely.

Univariate Continuous Random Variable

If the probability distribution of X admits a probability density function $f(x)$, then the expected value can be computed as

$$E[X] = \int_{-\infty}^{\infty} x f(x) \mathrm{d}x.$$

General Definition

In general, if X is a random variable defined on a probability space (Ω, Σ, P), then the expected value of X, denoted by $E[X]$, $\langle X \rangle$, $\overline{X}$ or $E[X]$, is defined as the Lebesgue integral

$$E[X] = \int_{\Omega} X \, \mathrm{d}P = \int_{\Omega} X(\omega) P(\mathrm{d}\omega)$$

When this integral exists, it is defined as the expectation of X. Not all random variables have a finite expected value, since the integral may not converge absolutely; furthermore, for some it is not defined at all (e.g., Cauchy distribution). Two variables with the same probability distribution will have the same expected value, if it is defined.

It follows directly from the discrete case definition that if X is a constant random variable, i.e. $X = b$ for some fixed real number b, then the expected value of X is also b.

The expected value of a measurable function of X, $g(X)$, given that X has a probability density function $f(x)$, is given by the inner product of f and g:

$$E[g(X)] = \int_{-\infty}^{\infty} g(x) f(x) \mathrm{d}x.$$

This is sometimes called the law of the unconscious statistician. Using representations as Riemann–Stieltjes integral and integration by parts the formula can be restated as

$$E[g(X)] = \int_{-\infty}^{\infty} g(x) \, \mathrm{d}P(X \le x) = \begin{cases} g(a) + \int_{a}^{\infty} g'(x) P(X > x) \, \mathrm{d}x & \text{if } P(X \ge a) = 1 \\ g(b) - \int_{-\infty}^{b} g'(x) P(X \le x) \, \mathrm{d}x & \text{if } P(X \le b) = 1. \end{cases}$$

As a special case let α denote a positive real number. Then

$$E\left[|X|^{\alpha}\right] = \alpha \int_{0}^{\infty} t^{\alpha-1} P(|X| > t) \, \mathrm{d}t.$$

In particular, if $\alpha = 1$ and $\Pr[X \ge 0] = 1$, then this reduces to

$$E[|X|] = E[X] = \int_{0}^{\infty} \{1 - F(t)\} \, \mathrm{d}t,$$

where F is the cumulative distribution function of X. This last identity is an instance of what, in a non-probabilistic setting, has been called the layer cake representation.

The law of the unconscious statistician applies also to a measurable function g of several random variables $X_1, \dots X_n$ having a joint density f:

$$E[g(X_1, \dots, X_n)] = \int_{-\infty}^{\infty} \cdots \int_{-\infty}^{\infty} g(x_1, \dots, x_n) f(x_1, \dots, x_n) \, \mathrm{d}x_1 \cdots \mathrm{d}x_n.$$

Properties

Constants

The expected value of a constant is equal to the constant itself; i.e., if c is a constant, then $E[c] = c$.

Monotonicity

If X and Y are random variables such that $X \le Y$ almost surely, then $E[X] \le E[Y]$.

Linearity

The expected value operator (or expectation operator) $E[\cdot]$ is linear in the sense that

$$E[X + c] = E[X] + c$$
$$E[X + Y] = E[X] + E[Y]$$
$$E[aX] = a E[X]$$

The second result is valid even if X is not statistically independent of Y. Combining the results from the previous three equations, we can see that

$$E[aX + bY + c] = a E[X] + b E[Y] + c$$

for any two random variables X and Y (which need to be defined on the same probability space) and any real numbers a, b and c.

Iterated Expectation

Iterated Expectation for Discrete Random Variables

For any two discrete random variables X, Y one may define the conditional expectation:

$$E[X \mid Y = y] = \sum_x x \cdot P(X = x \mid Y = y),$$

which means that $E[X \mid Y = y]$ is a function of y. Let $g(y)$ be that function of y; then the notation $E[X \mid Y]$ is a random variable in its own right, equal to $g(Y)$.

Lemma. Then the expectation of X satisfies:

$$E[X] = E\big[E[X \mid Y]\big].$$

Proof

$$E\big[E[X \mid Y]\big] = \sum_y E[X \mid Y = y] \cdot P(Y = y)$$

$$= \sum_y \left(\sum_x x \cdot P(X = x \mid Y = y) \right) \cdot P(Y = y)$$

$$= \sum_y \sum_x x \cdot P(X = x \mid Y = y) \cdot P(Y = y)$$

$$= \sum_y \sum_x x \cdot P(Y = y \mid X = x) \cdot P(X = x)$$

$$= \sum_x x \cdot P(X = x) \cdot \left(\sum_y P(Y = y \mid X = x) \right)$$

$$= \sum_x x \cdot P(X = x)$$

$$= E[X]$$

The left-hand side of this equation is referred to as the *iterated expectation*. The equation is sometimes called the *tower rule* or the *tower property*; it is treated under law of total expectation.

Iterated Expectation for Continuous Random Variables

In the continuous case, the results are completely analogous. The definition of conditional expectation would use inequalities, density functions, and integrals to replace equalities, mass functions, and summations, respectively. However, the main result still holds:

$$E[X] = E[E[X \mid Y]]$$

Inequality

If a random variable X is always less than or equal to another random variable Y, the expectation of X is less than or equal to that of Y:

If $X \leq Y$, then $E[X] \leq E[Y]$.

In particular, if we set Y to $|X|$ we know $X \leq Y$ and $-X \leq Y$. Therefore, we know $E[X] \leq E[Y]$ and $E[-X] \leq E[Y]$. From the linearity of expectation we know $-E[X] \leq E[Y]$. Therefore, the absolute value of expectation of a random variable is less than or equal to the expectation of its absolute value:

$$|E[X]| \leq E[|X|]$$

This is a special case of Jensen's inequality.

Non-multiplicativity

If one considers the joint probability density function of X and Y, say $j(x,y)$, then the expectation of XY is

$$E[XY] = \iint xy\, j(x,y)\mathrm{d}x\mathrm{d}y.$$

In general, the expected value operator is not multiplicative, i.e. $E[XY]$ is not necessarily equal to $E[X] \cdot E[Y]$. In fact, the amount by which multiplicativity fails is called the covariance:

$$\mathrm{Cov}(X,Y) = E[XY] - E[X]E[Y].$$

Thus multiplicativity holds precisely when $\mathrm{Cov}(X, Y) = 0$, in which case X and Y are said to be uncorrelated (independent variables are a notable case of uncorrelated variables).

Now if X and Y are independent, then by definition $j(x,y) = f(x)g(y)$ where f and g are the marginal PDFs for X and Y. Then

$$E[XY] = \iint xy\, j(x,y)\mathrm{d}x\mathrm{d}y = \iint xyf(x)g(y)\mathrm{d}y\mathrm{d}x$$
$$= \left[\int xf(x)\mathrm{d}x\right]\left[\int yg(y)\mathrm{d}y\right] = E[X]E[Y]$$

and $\mathrm{Cov}(X, Y) = 0$.

Observe that independence of X and Y is required only to write $j(x, y) = f(x)g(y)$, and this is required to establish the second equality above. The third equality follows from a basic application of the Fubini–Tonelli theorem.

Functional Non-invariance

In general, with the exception of linear functions, the expectation operator and functions of random variables do not commute; that is

$$E[g(X)] = \int_{\Omega} g(X)\mathrm{d}P \neq g(E[X]).$$

A notable inequality concerning this topic is Jensen's inequality, involving expected values of convex (or concave) functions.

Relation to Characteristic Function

The probability density function of a scalar random variable X is related to its characteristic function φ_X by the inversion formula:

$$f_X(x) = \frac{1}{2\pi} \int_{\mathbf{R}} e^{-itx} \varphi_X(t)\,dt.$$

We can use this inversion formula in expected value of a function $g(X)$ to obtain

$$\mathrm{E}[g(X)] = \frac{1}{2\pi} \int\int g(x) e^{-itx} \varphi_X(t)\,dt\,dx$$

Changing the order of integration, we get

$$\mathrm{E}[g(X)] = \frac{1}{2\pi} \int G(t) \varphi_X(t)\,dt,$$

where

$$G(t) = \int g(x) e^{-itx}\,dx$$

is the Fourier transform of $g(x)$. This can also be seen as a direct consequence of Plancherel theorem.

Uses and Applications

It is possible to construct an expected value equal to the probability of an event by taking the expectation of an indicator function that is one if the event has occurred and zero otherwise. This relationship can be used to translate properties of expected values into properties of probabilities, e.g. using the law of large numbers to justify estimating probabilities by frequencies.

The expected values of the powers of X are called the moments of X; the moments about the mean of X are expected values of powers of $X - \mathrm{E}[X]$. The moments of some random variables can be used to specify their distributions, via their moment generating functions.

To empirically estimate the expected value of a random variable, one repeatedly measures observations of the variable and computes the arithmetic mean of the results. If the expected value exists, this procedure estimates the true expected value in an unbiased manner and has the property of minimizing the sum of the squares of the residuals (the sum of the squared differences between the observations and the estimate). The law of large numbers demonstrates (under fairly mild conditions) that, as the size of the sample gets larger, the variance of this estimate gets smaller.

This property is often exploited in a wide variety of applications, including general problems of statistical estimation and machine learning, to estimate (probabilistic) quantities of interest via Monte Carlo methods, since most quantities of interest can be written in terms of expectation, e.g. $P(X \in \mathcal{A}) = E[I_A(X)]$ where $I_A(X)$ is the indicator function for set $\mathcal{A}$, i.e. $X \in \mathcal{A} \rightarrow I_A(X) = 1, X \notin \mathcal{A} \rightarrow I_A(X) = 0.$ ·

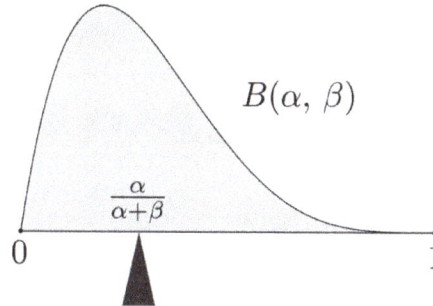

The mass of probability distribution is balanced at the expected value, here a Beta(α,β) distribution with expected value α/(α+β).

In classical mechanics, the center of mass is an analogous concept to expectation. For example, suppose X is a discrete random variable with values x_i and corresponding probabilities p_i. Now consider a weightless rod on which are placed weights, at locations x_i along the rod and having masses p_i (whose sum is one). The point at which the rod balances is E[X].

Expected values can also be used to compute the variance, by means of the computational formula for the variance

$$\mathrm{Var}(X) = E[X^2] - (E[X])^2.$$

A very important application of the expectation value is in the field of quantum mechanics. The expectation value of a quantum mechanical operator $\hat{A}$ operating on a quantum state vector $|\psi\rangle$ is written as $\langle \hat{A} \rangle = \langle \psi | A | \psi \rangle$. The uncertainty in $\hat{A}$ can be calculated using the formula $(\Delta A)^2 = \langle \hat{A}^2 \rangle - \langle \hat{A} \rangle^2$.

Expectation of Matrices

If X is an $m \times n$ matrix, then the expected value of the matrix is defined as the matrix of expected values:

$$E[X] = E\left[\begin{pmatrix} x_{1,1} & x_{1,2} & \cdots & x_{1,n} \\ x_{2,1} & x_{2,2} & \cdots & x_{2,n} \\ \vdots & \vdots & \ddots & \vdots \\ x_{m,1} & x_{m,2} & \cdots & x_{m,n} \end{pmatrix}\right] = \begin{pmatrix} E[x_{1,1}] & E[x_{1,2}] & \cdots & E[x_{1,n}] \\ E[x_{2,1}] & E[x_{2,2}] & \cdots & E[x_{2,n}] \\ \vdots & \vdots & \ddots & \vdots \\ E[x_{m,1}] & E[x_{m,2}] & \cdots & E[x_{m,n}] \end{pmatrix}.$$

This is utilized in covariance matrices.

Formulas for Special Cases

Discrete Distribution Taking only Non-negative Integer Values

When a random variable takes only values in {0, 1, 2, 3, ...} we can use the following formula for computing its expectation (even when the expectation is infinite):

$$E[X] = \sum_{i=1}^{\infty} P(X \geq i).$$

Proof.

$$\sum_{i=1}^{\infty} P(X \geq i) = \sum_{i=1}^{\infty} \sum_{j=i}^{\infty} P(X = j).$$

Interchanging the order of summation, we have

$$\sum_{i=1}^{\infty} \sum_{j=i}^{\infty} P(X = j) = \sum_{j=1}^{\infty} \sum_{i=1}^{\infty} P(X = j)$$

$$= \sum_{j=1}^{\infty} j P(X = j)$$

$$E[\quad].$$

This result can be a useful computational shortcut. For example, suppose we toss a coin where the probability of heads is p. How many tosses can we expect until the first heads (not including the heads itself)? Let X be this number. We are counting only the tails and not the heads which ends the experiment; in particular, we can have $X = 0$. The expectation of X may be computed by $\sum_{i=1}^{\infty} (1-p)^i = \frac{1}{p} - 1$. This is because, when the first i tosses yield tails, the number of tosses is at least i. The last equality used the formula for a geometric progression, $\sum_{i=1}^{\infty} r^i = \sum_{i=0}^{\infty} r^i - 1 = \frac{1}{1-r} - 1$, where $r = 1 - p < 1$.

Continuous Distribution Taking Non-negative Values

Analogously with the discrete case above, when a continuous random variable X takes only non-negative values, we can use the following formula for computing its expectation (even when the expectation is infinite):

$$E[X] = \int_0^{\infty} P(X \geq x)\, dx$$

Proof: It is first assumed that X has a density $f_X(x)$. We present two techniques:

- Using integration by parts (a special case of Section 1.4 above):

$$E[X] = \int_0^{\infty} (-x)(-f_X(x))\, dx = [-x(1-F(x))]_0^{\infty} + \int_0^{\infty} (1-F(x))\, dx$$

    and the bracket vanishes because

$1 - F(x) = o\left(\dfrac{1}{x}\right)$ as $x \to \infty$.

- Using an interchange in order of integration:

$$\int_0^\infty \Pr(X \geq x)dx = \int_0^\infty \int_x^\infty f_X(t) \, dt \, dx$$

$$= \int_0^\infty \int_0^t f_X(t) \, dx \, dt = \int_0^\infty t f_X(t) \, dt = E[X]$$

In case no density exists, it is seen that

$$E[X] = \int_0^\infty \int_0^x dt \, dF(x) = \int_0^\infty \int_t^\infty dF(x)dt = \int_0^\infty (1 - F(t))dt.$$

If both negative and non-negative values are taken for X, then its expectation is

$$E[X] = \int_0^\infty \Pr(X \geq x) \, dx - \int_{-\infty}^0 \Pr(X \leq x) \, dx$$

History

The idea of the expected value originated in the middle of the 17th century from the study of the so-called problem of points, which seeks to divide the stakes *in a fair way* between two players who have to end their game before it's properly finished. This problem had been debated for centuries, and many conflicting proposals and solutions had been suggested over the years, when it was posed in 1654 to Blaise Pascal by French writer and amateur mathematician Chevalier de Méré. de Méré claimed that this problem couldn't be solved and that it showed just how flawed mathematics was when it came to its application to the real world. Pascal, being a mathematician, was provoked and determined to solve the problem once and for all. He began to discuss the problem in a now famous series of letters to Pierre de Fermat. Soon enough they both independently came up with a solution. They solved the problem in different computational ways but their results were identical because their computations were based on the same fundamental principle. The principle is that the value of a future gain should be directly proportional to the chance of getting it. This principle seemed to have come naturally to both of them. They were very pleased by the fact that they had found essentially the same solution and this in turn made them absolutely convinced they had solved the problem conclusively. However, they did not publish their findings. They only informed a small circle of mutual scientific friends in Paris about it.

Three years later, in 1657, a Dutch mathematician Christiaan Huygens, who had just visited Paris, published a treatise *"De ratiociniis in ludo aleæ"* on probability theory. In this book he considered the problem of points and presented a solution based on the same principle as the solutions of Pascal and Fermat. Huygens also extended the concept of expectation by adding rules for how to calculate expectations in more complicated situations than the original problem (e.g., for three or more players). In this sense this book can be seen as the first successful attempt of laying down the foundations of the theory of probability.

In the foreword to his book, Huygens wrote: "It should be said, also, that for some time some of the best mathematicians of France have occupied themselves with this kind of calculus so that no one should attribute to me the honour of the first invention. This does not belong to me. But these

savants, although they put each other to the test by proposing to each other many questions difficult to solve, have hidden their methods. I have had therefore to examine and go deeply for myself
into this matter by beginning with the elements, and it is impossible for me for this reason to af-
firm that I have even started from the same principle. But finally I have found that my answers in
many cases do not differ from theirs." (cited by Edwards (2002)). Thus, Huygens learned about de
Méré's Problem in 1655 during his visit to France; later on in 1656 from his correspondence with
Carcavi he learned that his method was essentially the same as Pascal's; so that before his book
went to press in 1657 he knew about Pascal's priority in this subject.

Neither Pascal nor Huygens used the term "expectation" in its modern sense. In particular, Huygens writes: "That my Chance or Expectation to win any thing is worth just such a Sum, as wou'd
procure me in the same Chance and Expectation at a fair Lay. ... If I expect a or b, and have an
equal Chance of gaining them, my Expectation is worth a+b/2." More than a hundred years later,
in 1814, Pierre-Simon Laplace published his tract "*Théorie analytique des probabilités*", where the
concept of expected value was defined explicitly:

... this advantage in the theory of chance is the product of the sum hoped for by the probability of obtaining it; it is the partial sum which ought to result when we do not wish to run the risks of the event
in supposing that the division is made proportional to the probabilities. This division is the only
equitable one when all strange circumstances are eliminated; because an equal degree of probability
gives an equal right for the sum hoped for. We will call this advantage *mathematical hope*.

The use of the letter E to denote expected value goes back to W.A. Whitworth in 1901, who used a script
E. The symbol has become popular since for English writers it meant "Expectation", for Germans "Er-
wartungswert", for Spanish "Esperanza matemática" and for French "Espérance mathématique".

Variance

In probability theory and statistics, variance is the expectation of the squared deviation of a random variable from its mean, and it informally measures how far a set of (random) numbers are
spread out from their mean. The variance has a central role in statistics. It is used in descriptive
statistics, statistical inference, hypothesis testing, goodness of fit, Monte Carlo sampling, amongst
many others. This makes it a central quantity in numerous fields such as physics, biology, chemistry, economics, and finance. The variance is the square of the standard deviation, the second
central moment of a distribution, and the covariance of the random variable with itself, and it is
often represented by σ^2, s^2, or $\mathrm{Var}(X)$.

Definition

The variance of a random variable X is the expected value of the squared deviation from the mean
of X, $\mu = \mathrm{E}[X]$:

$$\mathrm{Var}(X) = \mathrm{E}\left[(X - \mu)^2\right].$$

This definition encompasses random variables that are generated by processes that are discrete, continuous, neither, or mixed. The variance can also be thought of as the covariance of a random variable with itself:

$$\mathrm{Var}(X) = \mathrm{Cov}(X, X).$$

The variance is also equivalent to the second cumulant of a probability distribution that generates X. The variance is typically designated as $\mathrm{Var}(X)$, σ_X^2, or simply σ^2 (pronounced "sigma squared"). The expression for the variance can be expanded:

$$\begin{aligned}
\mathrm{Var}(X) &= \mathrm{E}\left[(X - \mathrm{E}[X])^2\right] \\
&= \mathrm{E}\left[X^2 - 2X\,\mathrm{E}[X] + (\mathrm{E}[X])^2\right] \\
&= \mathrm{E}\left[X^2\right] - 2\,\mathrm{E}[X]\mathrm{E}[X] + (\mathrm{E}[X])^2 \\
&= \mathrm{E}\left[X^2\right] - (\mathrm{E}[X])^2
\end{aligned}$$

A mnemonic for the above expression is "mean of square minus square of mean". On computational floating point arithmetic, this equation should not be used, because it suffers from catastrophic cancellation if the two components of the equation are similar in magnitude. There exist numerically stable alternatives.

Continuous Random Variable

If the random variable X represents samples generated by a continuous distribution with probability density function $f(x)$, then the population variance is given by

$$\mathrm{Var}(X) = \sigma^2 = \int (x-\mu)^2 f(x)dx = \int x^2 f(x)dx - 2\mu\int xf(x)dx + \int \mu^2 f(x)dx = \int x^2 f(x)dx - \mu^2$$

where μ is the expected value

$$\mu = \int x f(x)dx$$

and where the integrals are definite integrals taken for x ranging over the range of X.

If a continuous distribution does not have an expected value, as is the case for the Cauchy distribution, it does not have a variance either. Many other distributions for which the expected value does exist also do not have a finite variance because the integral in the variance definition diverges. An example is a Pareto distribution whose index k satisfies $1 < k \le 2$.

Discrete Random Variable

If the generator of random variable X is discrete with probability mass function $x_1 \mapsto p_1, x_2 \mapsto p_2, \ldots, x_n \mapsto p_n$ then

$$\mathrm{Var}(X) = \sum_{i=1}^{n} p_i \cdot (x_i - \mu)^2,$$

or equivalently

$$\text{Var}(X) = \sum_{i=1}^{n} p_i x_i^2 - \mu^2,$$

where μ is the expected value, i.e.

$$\mu = \sum_{i=1}^{n} p_i \cdot x_i.$$

(When such a discrete weighted variance is specified by weights whose sum is not 1, then one divides by the sum of the weights.)

The variance of a set of n equally likely values can be written as

$$\text{Var}(X) = \frac{1}{n} \sum_{i=1}^{n} (x_i - \mu)^2.$$

where μ is the expected value, i.e.,

$$\mu = \frac{1}{n} \sum_{i=1}^{n} x_i$$

The variance of a set of n equally likely values can be equivalently expressed, without directly referring to the mean, in terms of squared deviations of all points from each other:

$$\text{Var}(X) = \frac{1}{n^2} \sum_{i=1}^{n} \sum_{j=1}^{n} \frac{1}{2} (x_i - x_j)^2 = \frac{1}{n^2} \sum_{i} \sum_{j>i} (x_i - x_j)^2.$$

Examples

Normal Distribution

The normal distribution with parameters μ and σ is a continuous distribution whose probability density function is given by

$$f(x) = \frac{1}{\sqrt{2\pi\sigma^2}} e^{-\frac{(x-\mu)^2}{2\sigma^2}}.$$

In this distribution, $E[X] = \mu$ and the variance $\text{Var}(X)$ is related with σ via

$$\text{Var}(X) = \int_{-\infty}^{\infty} \frac{(x-\mu)^2}{\sqrt{2\pi\sigma^2}} e^{-\frac{(x-\mu)^2}{2\sigma^2}} dx = \sigma^2.$$

The role of the normal distribution in the central limit theorem is in part responsible for the prevalence of the variance in probability and statistics.

Exponential Distribution

The exponential distribution with parameter λ is a continuous distribution whose support is the semi-infinite interval $[0, \infty]$. Its probability density function is given by

$$f(x) = \lambda e^{-\lambda x},$$

and it has expected value $\mu = \lambda^{-1}$. The variance is equal to

$$\text{Var}(X) = \int_0^\infty (x - \lambda^{-1})^2 \lambda e^{-\lambda x} dx = \lambda^{-2}.$$

So for an exponentially distributed random variable, $\sigma^2 = \mu^2$.

Poisson Distribution

The Poisson distribution with parameter λ is a discrete distribution for $k = 0, 1, 2, \ldots$. Its probability mass function is given by

$$p(k) = \frac{\lambda^k}{k!} e^{-\lambda},$$

and it has expected value $\mu = \lambda$. The variance is equal to

$$\text{Var}(X) = \sum_{k=0}^\infty \frac{\lambda^k}{k!} e^{-\lambda} (k - \lambda)^2 = \lambda,$$

So for a Poisson-distributed random variable, $\sigma^2 = \mu$.

Binomial Distribution

The binomial distribution with parameters n and p is a discrete distribution for $k = 0, 1, 2, \ldots, n$. Its probability mass function is given by

$$p(k) = \binom{n}{k} p^k (1 - p)^{n-k},$$

and it has expected value $\mu = np$. The variance is equal to

$$\text{Var}(X) = \sum_{k=0}^n \binom{n}{k} p^k (1 - p)^{n-k} (k - np)^2 = np(1 - p).$$

Coin Toss

The binomial distribution with $p = 0.5$ describes the probability of getting k heads in n tosses.

Thus the expected value of the number of heads is $\frac{n}{2}$, and the variance is $\frac{n}{4}$.

Fair Die

A six-sided fair die can be modelled with a discrete random variable with outcomes 1 through 6, each with equal probability $\frac{1}{6}$. The expected value is $\frac{1+2+3+4+5+6}{6} = 3.5$. Therefore, the variance can be computed to be

$$\sum_{i=1}^{6} \tfrac{1}{6}(i-3.5)^2 = \tfrac{1}{6}\sum_{i=1}^{6}(i-3.5)^2 = \tfrac{1}{6}\left((-2.5)^2 +(-1.5)^2 +(-0.5)^2 +0.5^2 +1.5^2 +2.5^2\right)$$

$$= \tfrac{1}{6}\cdot 17.50 = \tfrac{35}{12} \approx 2.92.$$

The general formula for the variance of the outcome X of a die of n sides is

$$\sigma^2 = E(X^2)-(E(X))^2 = \frac{1}{n}\sum_{i=1}^{n}i^2 -\left(\frac{1}{n}\sum_{i=1}^{n}i\right)^2$$

$$= \tfrac{1}{6}(n+1)(2n+1)-\tfrac{1}{4}(n+1)^2$$

$$= \frac{n^2-1}{12}.$$

Properties

Basic Properties

Variance is non-negative because the squares are positive or zero.

$$\mathrm{Var}(X) \geq 0.$$

The variance of a constant random variable is zero, and if the variance of a variable in a data set is 0, then all the entries have the same value.

$$P(X = a) = 1 \Leftrightarrow \mathrm{Var}(X) = 0.$$

Variance is invariant with respect to changes in a location parameter. That is, if a constant is added to all values of the variable, the variance is unchanged.

$$\mathrm{Var}(X + a) = \mathrm{Var}(X).$$

If all values are scaled by a constant, the variance is scaled by the square of that constant.

$$\mathrm{Var}(aX) = a^2 \,\mathrm{Var}(X).$$

The variance of a sum of two random variables is given by:

$$\mathrm{Var}(aX + bY) = a^2 \,\mathrm{Var}(X)+b^2 \,\mathrm{Var}(Y)+2ab\mathrm{Cov}(X,Y),$$

$$\mathrm{Var}(aX - bY) = a^2 \,\mathrm{Var}(X)+b^2 \,\mathrm{Var}(Y)-2ab\mathrm{Cov}(X,Y),$$

where $\mathrm{Cov}(\cdot,\cdot)$ is the covariance. In general we have for the sum of N random variables $\{X_1,\ldots,X_N\}$:

$$\mathrm{Var}\left(\sum_{i=1}^{N} X_i\right) = \sum_{i,j=1}^{N} \mathrm{Cov}(X_i, X_j) = \sum_{i=1}^{N} \mathrm{Var}(X_i) + \sum_{i \neq j} \mathrm{Cov}(X_i, X_j).$$

These results lead to the variance of a linear combination as:

$$\mathrm{Var}\left(\sum_{i=1}^{N} a_i X_i\right) = \sum_{i,j=1}^{N} a_i a_j \, \mathrm{Cov}(X_i, X_j)$$

$$= \sum_{i=1}^{N} a_i^2 \, \mathrm{Var}(X_i) + \sum_{i \neq j} a_i a_j \, \mathrm{Cov}(X_i, X_j)$$

$$= \sum_{i=1}^{N} a_i^2 \, \mathrm{Var}(X_i) + 2 \sum_{1 \leq i < j \leq N} a_i a_j \, \mathrm{Cov}(X_i, X_j).$$

If the random variables $X_1, \ldots, X_N$ are such that

$$\mathrm{Cov}(X_i, X_j) = 0, \, \forall \, (i \neq j),$$

they are said to be uncorrelated. It follows immediately from the expression given earlier that if the random variables $X_1, \ldots, X_N$ are uncorrelated, then the variance of their sum is equal to the sum of their variances, or, expressed symbolically:

$$\mathrm{Var}\left(\sum_{i=1}^{N} X_i\right) = \sum_{i=1}^{N} \mathrm{Var}(X_i).$$

Since independent random variables are always uncorrelated, the equation above holds in particular when the random variables $X_1, \ldots, X_n$ are independent. Thus independence is sufficient but not necessary for the variance of the sum to equal the sum of the variances.

Sum of Uncorrelated Variables (Bienaymé Formula)

One reason for the use of the variance in preference to other measures of dispersion is that the variance of the sum (or the difference) of uncorrelated random variables is the sum of their variances:

$$\mathrm{Var}\left(\sum_{i=1}^{n} X_i\right) = \sum_{i=1}^{n} \mathrm{Var}(X_i).$$

This statement is called the Bienaymé formula and was discovered in 1853. It is often made with the stronger condition that the variables are independent, but being uncorrelated suffices. So if all the variables have the same variance σ^2, then, since division by n is a linear transformation, this formula immediately implies that the variance of their mean is

$$\mathrm{Var}\left(\overline{X}\right) = \mathrm{Var}\left(\frac{1}{n}\sum_{i=1}^{n} X_i\right) = \frac{1}{n^2}\sum_{i=1}^{n} \mathrm{Var}\left(X_i\right) = \frac{\sigma^2}{n}.$$

That is, the variance of the mean decreases when n increases. This formula for the variance of the mean is used in the definition of the standard error of the sample mean, which is used in the central limit theorem.

Sum of Correlated Variables

In general, if the variables are correlated, then the variance of their sum is the sum of their covariances:

$$\mathrm{Var}\left(\sum_{i=1}^{n} X_i\right) = \sum_{i=1}^{n}\sum_{j=1}^{n} \mathrm{Cov}(X_i, X_j) = \sum_{i=1}^{n} \mathrm{Var}(X_i) + 2\sum_{1\le i<j\le n} \mathrm{Cov}(X_i, X_j).$$

(Note: The second equality comes from the fact that $\mathrm{Cov}(X_i, X_i) = \mathrm{Var}(X_i)$.)

Here $\mathrm{Cov}(\cdot, \cdot)$ is the covariance, which is zero for independent random variables (if it exists). The formula states that the variance of a sum is equal to the sum of all elements in the covariance matrix of the components. This formula is used in the theory of Cronbach's alpha in classical test theory.

So if the variables have equal variance σ^2 and the average correlation of distinct variables is ρ, then the variance of their mean is

$$\mathrm{Var}(\bar{X}) = \frac{\sigma^2}{n} + \frac{n-1}{n}\rho\sigma^2.$$

This implies that the variance of the mean increases with the average of the correlations. In other words, additional correlated observations are not as effective as additional independent observations at reducing the uncertainty of the mean. Moreover, if the variables have unit variance, for example if they are standardized, then this simplifies to

$$\mathrm{Var}(\bar{X}) = \frac{1}{n} + \frac{n-1}{n}\rho.$$

This formula is used in the Spearman–Brown prediction formula of classical test theory. This converges to ρ if n goes to infinity, provided that the average correlation remains constant or converges too. So for the variance of the mean of standardized variables with equal correlations or converging average correlation we have

$$\lim_{n\to\infty} \mathrm{Var}(\bar{X}) = \rho.$$

Therefore, the variance of the mean of a large number of standardized variables is approximately equal to their average correlation. This makes clear that the sample mean of correlated variables does not generally converge to the population mean, even though the Law of large numbers states that the sample mean will converge for independent variables.

Matrix Notation for the Variance of a Linear Combination

Define X as a column vector of n random variables $X_1,\ldots,X_n$, and c as a column vector of n scalars $c_1,\ldots,c_n$. Therefore, $c^T X$ is a linear combination of these random variables, where c^T denotes the transpose of c. Also let Σ be the covariance matrix of X. The variance of $c^T X$ is then given by:

$$\mathrm{Var}(c^T X) = c^T \Sigma c.$$

Weighted Sum of Variables

The scaling property and the Bienaymé formula, along with the property of the covariance $\mathrm{Cov}(aX, bY) = ab\,\mathrm{Cov}(X, Y)$ jointly imply that

$$\mathrm{Var}(aX \pm bY) = a^2\,\mathrm{Var}(X) + b^2\,\mathrm{Var}(Y) \pm 2ab\,\mathrm{Cov}(X,Y).$$

This implies that in a weighted sum of variables, the variable with the largest weight will have a disproportionally large weight in the variance of the total. For example, if X and Y are uncorrelated and the weight of X is two times the weight of Y, then the weight of the variance of X will be four times the weight of the variance of Y.

The expression above can be extended to a weighted sum of multiple variables:

$$\mathrm{Var}\left(\sum_i^n a_i X_i\right) = \sum_{i=1}^n a_i^2\,\mathrm{Var}(X_i) + 2\sum_{1\le i\,<j\le n}\sum a_i a_j\,\mathrm{Cov}(X_i, X_j)$$

Product of Independent Variables

If two variables X and Y are independent, the variance of their product is given by

$$\mathrm{Var}(XY) = [E(X)]^2\,\mathrm{Var}(Y) + [E(Y)]^2\,\mathrm{Var}(X) + \mathrm{Var}(X)\,\mathrm{Var}(Y).$$

Equivalently, using the basic properties of expectation, it is given by

$$\mathrm{Var}(XY) = E(X^2)E(Y^2) - [E(X)]^2[E(Y)]^2.$$

Product of Correlated Variables

In general, if two variables are correlated, the variance of their product is given by:

$$\begin{aligned}\mathrm{Var}(XY) &= E[X^2Y^2] - [E(XY)]^2\\ &= \mathrm{Cov}(X^2,Y^2) + E(X^2)E(Y^2) - [E(XY)]^2\\ &= \mathrm{Cov}(X^2,Y^2) + (\mathrm{Var}(X)+[E(X)]^2)(\mathrm{Var}(Y)+[E(Y)]^2) - [\mathrm{Cov}(X,Y)+E(X)E(Y)]^2\end{aligned}$$

Decomposition

The general formula for variance decomposition or the law of total variance is: If X and Y are two random variables, and the variance of X exists, then

$$\mathrm{Var}[X] = E_Y(\mathrm{Var}[X\,|\,Y]) + \mathrm{Var}_Y(E[X\,|\,Y]).$$

where $E(X\,|\,Y)$ is the conditional expectation of X given Y, and $\mathrm{Var}(X\,|\,Y)$ is the conditional variance of X given Y. (A more intuitive explanation is that given a particular value of Y, then X follows a distribution with mean $E(X\,|\,Y)$ and variance $\mathrm{Var}(X\,|\,Y)$). As $E(X\,|\,Y)$ is a function of the variable Y, the outer expectation or variance is taken with respect to Y. The above formula tells how to find $\mathrm{Var}(X)$ based on the distributions of these two quantities when Y is allowed to vary.

In particular, if Y is a discrete random variable assuming $y_1, y_2, \ldots, y_n$ with corresponding probability masses $p_1, p_2, \ldots, p_n$, then in the formula for total variance, the first term on the right-hand side becomes

$$E_Y(\mathrm{Var}[X\,|\,Y]) = \sum_{i=1}^{n} p_i \sigma_i^2,$$

where $\acute{o}_i^2 = \mathrm{Var}[X\,|\,y_i]$ Similarly, the second term on the right-hand side becomes

$$\mathrm{Var}_Y(E[X\,|\,Y]) = \sum_{i=1}^{n} p_i \mu_i^2 - \left(\sum_{i=1}^{n} p_i \mu_i\right)^2 = \sum_{i=1}^{n} p_i \mu_i^2 - \mu^2,$$

where $\mu_i = E[X\,|\,y_i]$ and $\mu = \sum_{i=1}^{n} p_i \mu_i$. Thus the total variance is given by

$$\mathrm{Var}[X] = \sum_{i=1}^{n} p_i \sigma_i^2 + \left(\sum_{i=1}^{n} p_i \mu_i^2 - \mu^2\right).$$

A similar formula is applied in analysis of variance, where the corresponding formula is

$$MS_{\mathrm{total}} = MS_{\mathrm{between}} + MS_{\mathrm{within}};$$

here MS refers to the Mean of the Squares. In linear regression analysis the corresponding formula is

$$MS_{\mathrm{total}} = MS_{\mathrm{regression}} + MS_{\mathrm{residual}}.$$

This can also be derived from the additivity of variances, since the total (observed) score is the sum of the predicted score and the error score, where the latter two are uncorrelated.

Similar decompositions are possible for the sum of squared deviations (sum of squares, SS):

$$SS_{\mathrm{total}} = SS_{\mathrm{between}} + SS_{\mathrm{within}},$$

$$SS_{\mathrm{total}} = SS_{\mathrm{regression}} + SS_{\mathrm{residual}}.$$

Formulae for the Variance

A formula often used for deriving the variance of a theoretical distribution is as follows:

$$\mathrm{Var}(X) = E(X^2) - (E(X))^2.$$

This will be useful when it is possible to derive formulae for the expected value and for the expected value of the square.

This formula is also sometimes used in connection with the sample variance. While useful for hand calculations, it is not advised for computer calculations as it suffers from catastrophic cancellation if the two components of the equation are similar in magnitude and floating point arithmetic is used.

Calculation from the CDF

The population variance for a non-negative random variable can be expressed in terms of the cumulative distribution function F using

$$2\int_0^\infty u(1-F(u))du - \left(\int_0^\infty 1-F(u)du\right)^2 .$$

This expression can be used to calculate the variance in situations where the CDF, but not the density, can be conveniently expressed.

Characteristic Property

The second moment of a random variable attains the minimum value when taken around the first moment (i.e., mean) of the random variable, i.e. $\operatorname{argmin}_m \operatorname{E}((X-m)^2) = \operatorname{E}(X)$. Conversely, if a continuous function φ satisfies $\operatorname{argmin}_m \operatorname{E}(\varphi(X-m)) = \operatorname{E}(X)$ for all random variables X, then it is necessarily of the form $\varphi(x) = ax^2 + b$, , where $a > 0$. This also holds in the multidimensional case.

Units of Measurement

Unlike expected absolute deviation, the variance of a variable has units that are the square of the units of the variable itself. For example, a variable measured in meters will have a variance measured in meters squared. For this reason, describing data sets via their standard deviation or root mean square deviation is often preferred over using the variance. In the dice example the standard deviation is $\sqrt{2.9} \approx 1.7$, slightly larger than the expected absolute deviation of 1.5.

The standard deviation and the expected absolute deviation can both be used as an indicator of the "spread" of a distribution. The standard deviation is more amenable to algebraic manipulation than the expected absolute deviation, and, together with variance and its generalization covariance, is used frequently in theoretical statistics; however the expected absolute deviation tends to be more robust as it is less sensitive to outliers arising from measurement anomalies or an unduly heavy-tailed distribution.

Approximating the Variance of a Function

The delta method uses second-order Taylor expansions to approximate the variance of a function of one or more random variables. For example, the approximate variance of a function of one variable is given by

$$\operatorname{Var}\left[f(X)\right] \approx \left(f'(\operatorname{E}\left[X\right])\right)^2 \operatorname{Var}\left[X\right]$$

provided that f is twice differentiable and that the mean and variance of X are finite.

Population Variance and Sample Variance

Real-world observations such as the measurements of yesterday's rain throughout the day typically cannot be complete sets of all possible observations that could be made. As such, the variance calculated from the finite set will in general not match the variance that would have been calculated from the full population of possible observations. This means that one estimates the mean and variance that would have been calculated from an omniscient set of observations by using an estimator equation. The estimator is a function of the sample of n observations drawn without observational bias from the whole population of potential observations. In this example that sample would be the set of actual measurements of yesterday's rainfall from available rain gauges within the geography of interest.

The simplest estimators for population mean and population variance are simply the mean and variance of the sample, the sample mean and (uncorrected) sample variance – these are consistent estimators (they converge to the correct value as the number of samples increases), but can be improved. Estimating the population variance by taking the sample's variance is close to optimal in general, but can be improved in two ways. Most simply, the sample variance is computed as an average of squared deviations about the (sample) mean, by dividing by n. However, using values other than n improves the estimator in various ways. Four common values for the denominator are n, $n - 1$, $n + 1$, and $n - 1.5$: n is the simplest (population variance of the sample), $n - 1$ eliminates bias, $n + 1$ minimizes mean squared error for the normal distribution, and $n - 1.5$ mostly eliminates bias in unbiased estimation of standard deviation for the normal distribution.

Firstly, if the omniscient mean is unknown (and is computed as the sample mean), then the sample variance is a biased estimator: it underestimates the variance by a factor of $(n - 1) / n$; correcting by this factor (dividing by $n - 1$ instead of n) is called Bessel's correction. The resulting estimator is unbiased, and is called the (corrected) sample variance or unbiased sample variance. For example, when $n = 1$ the variance of a single observation about the sample mean (itself) is obviously zero regardless of the population variance. If the mean is determined in some other way than from the same samples used to estimate the variance then this bias does not arise and the variance can safely be estimated as that of the samples about the (independently known) mean.

Secondly, the sample variance does not generally minimize mean squared error between sample variance and population variance. Correcting for bias often makes this worse: one can always choose a scale factor that performs better than the corrected sample variance, though the optimal scale factor depends on the excess kurtosis of the population, and introduces bias. This always consists of scaling down the unbiased estimator (dividing by a number larger than $n - 1$), and is a simple example of a shrinkage estimator: one "shrinks" the unbiased estimator towards zero. For the normal distribution, dividing by $n + 1$ (instead of $n - 1$ or n) minimizes mean squared error. The resulting estimator is biased, however, and is known as the biased sample variation.

Population Variance

In general, the *population variance* of a *finite* population of size N with values x_i is given by

$$\sigma^2 = \frac{1}{N}\sum_{i=1}^{N}(x_i - \mu)^2 = \frac{1}{N}\sum_{i=1}^{N}\left(x_i^2 - 2\mu x_i + \mu^2\right) = \left(\frac{1}{N}\sum_{i=1}^{N}x_i^2\right) - 2\mu\left(\frac{1}{N}\sum_{i=1}^{N}x_i\right) + \mu^2 = \left(\frac{1}{N}\sum_{i=1}^{N}x_i^2\right) - \mu^2$$

where the population mean is

$$\mu = \frac{1}{N}\sum_{i=1}^{N} x_i.$$

The population variance can also be computed using

$$\sigma^2 = \frac{1}{N^2}\sum_{i<j}\left(x_i - x_j\right)^2 = \frac{1}{2N^2}\sum_{i,j=1}^{N}\left(x_i - x_j\right)^2$$

This is true because

$$\frac{1}{2N^2}\sum_{i,j=1}^{N}\left(x_i - x_j\right)^2 = \frac{1}{2N^2}\sum_{i,j=1}^{N}\left(x_i^2 - 2x_i x_j + x_j^2\right)$$

$$= \frac{1}{2N}\sum_{j=1}^{N}\left(\frac{1}{N}\sum_{i=1}^{N}x_i^2\right) - \left(\frac{1}{N}\sum_{i=1}^{N}x_i\right)\left(\frac{1}{N}\sum_{j=1}^{N}x_j\right) + \frac{1}{2N}\sum_{i=1}^{N}\left(\frac{1}{N}\sum_{j=1}^{N}x_j^2\right)$$

$$= \frac{1}{2}\left(\sigma^2 + \mu^2\right) - \mu^2 + \frac{1}{2}\left(\sigma^2 + \mu^2\right) = \sigma^2$$

The population variance matches the variance of the generating probability distribution. In this sense, the concept of population can be extended to continuous random variables with infinite populations.

Sample Variance

In many practical situations, the true variance of a population is not known *a priori* and must be computed somehow. When dealing with extremely large populations, it is not possible to count every object in the population, so the computation must be performed on a sample of the population. Sample variance can also be applied to the estimation of the variance of a continuous distribution from a sample of that distribution.

We take a sample with replacement of n values $y_1, ..., y_n$ from the population, where $n < N$, and estimate the variance on the basis of this sample. Directly taking the variance of the sample data gives the average of the squared deviations:

$$\sigma_y^2 = \frac{1}{n}\sum_{i=1}^{n}\left(y_i - \overline{y}\right)^2 = \left(\frac{1}{n}\sum_{i=1}^{n}y_i^2\right) - \overline{y}^2 = \frac{1}{n^2}\sum_{i<j}\left(y_i - y_j\right)^2.$$

Here, $\overline{y}$ denotes the sample mean:

$$\overline{y} = \frac{1}{n}\sum_{i=1}^{n}y_i.$$

Since the y_i are selected randomly, both $\overline{y}$ and σ_y^2 are random variables. Their expected values can be evaluated by summing over the ensemble of all possible samples $\{y_i\}$ from the population. For σ_y^2 this gives:

$$E[\sigma_y^2] = E\left[\frac{1}{n}\sum_{i=1}^{n}\left(y_i - \frac{1}{n}\sum_{j=1}^{n}y_j\right)^2\right]$$

$$= \frac{1}{n}\sum_{i=1}^{n}E\left[y_i^2 - \frac{2}{n}y_i\sum_{j=1}^{n}y_j + \frac{1}{n^2}\sum_{j=1}^{n}y_j\sum_{k=1}^{n}y_k\right]$$

$$= \frac{1}{n}\sum_{i=1}^{n}\left[\frac{n-2}{n}E[y_i^2] - \frac{2}{n}\sum_{j\neq i}E[y_iy_j] + \frac{1}{n^2}\sum_{j=1}^{n}\sum_{k\neq j}E[y_jy_k] + \frac{1}{n^2}\sum_{j=1}^{n}E[y_j^2]\right]$$

$$= \frac{1}{n}\sum_{i=1}^{n}\left[\frac{n-2}{n}(\sigma^2 + \mu^2) - \frac{2}{n}(n-1)\mu^2 + \frac{1}{n^2}n(n-1)\mu^2 + \frac{1}{n}(\sigma^2 + \mu^2)\right]$$

$$= \frac{n-1}{n}\sigma^2.$$

Hence σ_y^2 gives an estimate of the population variance that is biased by a factor of $\frac{n-1}{n}$. For this reason, σ_y^2 is referred to as the *biased sample variance*. Correcting for this bias yields the *unbiased sample variance*:

$$s^2 = \frac{n}{n-1}\sigma_y^2 = \frac{n}{n-1}\left(\frac{1}{n}\sum_{i=1}^{n}(y_i - \bar{y})^2\right) = \frac{1}{n-1}\sum_{i=1}^{n}(y_i - \bar{y})^2$$

Either estimator may be simply referred to as the *sample variance* when the version can be determined by context. The same proof is also applicable for samples taken from a continuous probability distribution.

The use of the term $n - 1$ is called Bessel's correction, and it is also used in sample covariance and the sample standard deviation (the square root of variance). The square root is a concave function and thus introduces negative bias (by Jensen's inequality), which depends on the distribution, and thus the corrected sample standard deviation (using Bessel's correction) is biased. The unbiased estimation of standard deviation is a technically involved problem, though for the normal distribution using the term $n - 1.5$ yields an almost unbiased estimator.

The unbiased sample variance is a U-statistic for the function $f(y_1, y_2) = (y_1 - y_2)^2/2$, meaning that it is obtained by averaging a 2-sample statistic over 2-element subsets of the population.

Distribution of the Sample Variance

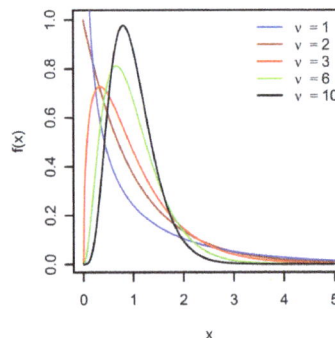

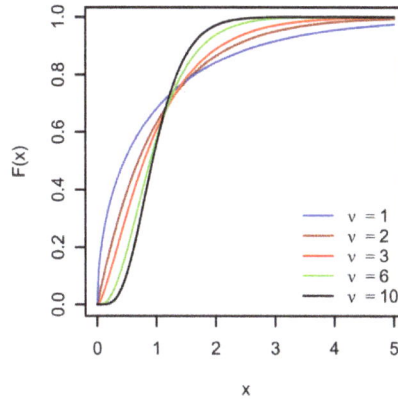

Distribution and cumulative distribution of s^2/σ^2, for various values of $v = n - 1$, when the y_i are independent normally distributed.

Being a function of random variables, the sample variance is itself a random variable, and it is natural to study its distribution. In the case that y_i are independent observations from a normal distribution, Cochran's theorem shows that s^2 follows a scaled chi-squared distribution:

$$(n-1)\frac{s^2}{\sigma^2} \sim \chi^2_{n-1}.$$

As a direct consequence, it follows that

$$E(s^2) = E\left(\frac{\sigma^2}{n-1}\chi^2_{n-1}\right) = \sigma^2,$$

and

$$\text{Var}[s^2] = \text{Var}\left(\frac{\sigma^2}{n-1}\chi^2_{n-1}\right) = \frac{\sigma^4}{(n-1)^2}\text{Var}\left(\chi^2_{n-1}\right) = \frac{2\sigma^4}{n-1}.$$

If the y_i are independent and identically distributed, but not necessarily normally distributed, then

$$E[s^2] = \sigma^2, \quad \text{Var}[s^2] = \frac{\sigma^4}{n}\left((\kappa-1)+\frac{2}{n-1}\right) = \frac{1}{n}\left(\mu_4 - \frac{n-3}{n-1}\sigma^4\right),$$

where κ is the kurtosis of the distribution and μ_4 is the fourth central moment.

If the conditions of the law of large numbers hold for the squared observations, s^2 is a consistent estimator of σ^2. One can see indeed that the variance of the estimator tends asymptotically to zero. An asymptotically equivalent formula was given in Kenney and Keeping (1951:164), Rose and Smith (2002:264), and Weisstein (n.d.).

Samuelson's Inequality

Samuelson's inequality is a result that states bounds on the values that individual observations in a sample can take, given that the sample mean and (biased) variance have been calculated. Values must lie within the limits $\bar{y} \pm \sigma_y (n-1)^{1/2}$.

Relations With the Harmonic and Arithmetic Means

It has been shown that for a sample $\{y_i\}$ of real numbers,

$$\sigma_y^2 \le 2 y_{max} (A - H),$$

where y_{max} is the maximum of the sample, A is the arithmetic mean, H is the harmonic mean of the sample and σ_y^2 is the (biased) variance of the sample.

This bound has been improved, and it is known that variance is bounded by

$$\sigma_y^2 \le \frac{y_{max} (A - H)(y_{max} - A)}{y_{max} - H},$$

$$\sigma_y^2 \ge \frac{y_{min} (A - H)(A - y_{min})}{H - y_{min}},$$

where y_{min} is the minimum of the sample.

Tests of Equality of Variances

Testing for the equality of two or more variances is difficult. The F test and chi square tests are both adversely affected by non-normality and are not recommended for this purpose.

Several non parametric tests have been proposed: these include the Barton–David–Ansari–Freund–Siegel–Tukey test, the Capon test, Mood test, the Klotz test and the Sukhatme test. The Sukhatme test applies to two variances and requires that both medians be known and equal to zero. The Mood, Klotz, Capon and Barton–David–Ansari–Freund–Siegel–Tukey tests also apply to two variances. They allow the median to be unknown but do require that the two medians are equal.

The Lehmann test is a parametric test of two variances. Of this test there are several variants known. Other tests of the equality of variances include the Box test, the Box–Anderson test and the Moses test.

Resampling methods, which include the bootstrap and the jackknife, may be used to test the equality of variances.

History

The term *variance* was first introduced by Ronald Fisher in his 1918 paper *The Correlation Between Relatives on the Supposition of Mendelian Inheritance*:

The great body of available statistics show us that the deviations of a human measurement from its mean follow very closely the Normal Law of Errors, and, therefore, that the variability may be uniformly measured by the standard deviation corresponding to the square root of the mean square error. When there are two independent causes of variability capable of producing in an otherwise uniform population distributions with standard deviations σ_1 and σ_2, it is found that the distribution, when both causes act together, has a standard deviation $\sqrt{\sigma_1^2 + \sigma_2^2}$. It is therefore desirable in analysing the causes of variability to deal with the square of the standard deviation as the measure of variability. We shall term this quantity the Variance...

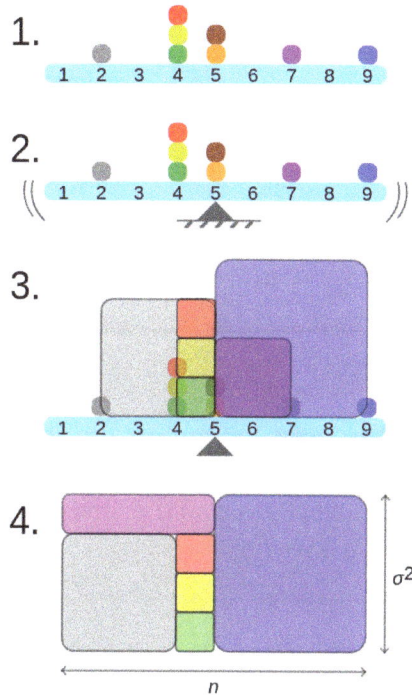

Geometric visualisation of the variance of an arbitrary distribution (2, 4, 4, 4, 5, 5, 7, 9):

1. A frequency distribution is constructed.

2. The centroid of the distribution gives its mean.

3. A square with sides equal to the difference of each value from the mean is formed for each value.

4. Arranging the squares into a rectangle with one side equal to the number of values, n, results in the other side being the distribution's variance, σ^2.

Moment of Inertia

The variance of a probability distribution is analogous to the moment of inertia in classical mechanics of a corresponding mass distribution along a line, with respect to rotation about its center of mass. It is because of this analogy that such things as the variance are called *moments* of probability distributions. The covariance matrix is related to the moment of inertia tensor for multivariate distributions. The moment of inertia of a cloud of n points with a covariance matrix of Σ is given by

$$I = n(\mathbf{1}_{3\times3}\operatorname{tr}(\Sigma) - \Sigma).$$

This difference between moment of inertia in physics and in statistics is clear for points that are gathered along a line. Suppose many points are close to the x axis and distributed along it. The covariance matrix might look like

$$\Sigma = \begin{bmatrix} 10 & 0 & 0 \\ 0 & 0.1 & 0 \\ 0 & 0 & 0.1 \end{bmatrix}.$$

That is, there is the most variance in the x direction. Physicists would consider this to have a low moment *about* the x axis so the moment-of-inertia tensor is

$$I = n \begin{bmatrix} 0.2 & 0 & 0 \\ 0 & 10.1 & 0 \\ 0 & 0 & 10.1 \end{bmatrix}.$$

Semivariance

The *semivariance* is calculated in the same manner as the variance but only those observations that fall below the mean are included in the calculation. It is sometimes described as a measure of downside risk in an investments context. For skewed distributions, the semivariance can provide additional information that a variance does not.

Generalizations

If x is a scalar complex-valued random variable, with values in $\mathbb{C}$, then its variance is $E((x - \mu)(x - \mu)^*)$, where x^* is the complex conjugate of x. This variance is a real scalar.

If X is a vector-valued random variable, with values in $\mathbb{R}^n$, and thought of as a column vector, then the natural generalization of variance is $E((X - \mu)(X - \mu)^T)$, where $\mu = E(X)$ and X^T is the transpose of X, and so is a row vector. The result is a positive semi-definite square matrix, commonly referred to as the variance-covariance matrix.

If X is a vector- and complex-valued random variable, with values in $\mathbb{C}^n$, then the generalization of its variance is $E((X - \mu)(X - \mu)^\dagger)$, where $X^\dagger$ is the conjugate transpose of X. This matrix is also positive semi-definite and square.

References

- B. S. Everitt: The Cambridge Dictionary of Statistics, Cambridge University Press, Cambridge (3rd edition, 2006). ISBN 0-521-69027-7

- Kolmogoroff (1933). Grundbegriffe der Wahrscheinlichkeitsrechnung. doi:10.1007/978-3-642-49888-6. ISBN 978-3-642-49888-6.

- Olav Kallenberg; Foundations of Modern Probability, 2nd ed. Springer Series in Statistics. (2002). 650 pp. ISBN 0-387-95313-2

- Olav Kallenberg; Probabilistic Symmetries and Invariance Principles. Springer -Verlag, New York (2005). 510 pp. ISBN 0-387-25115-4

- Stewart, William J. (2011). Probability, Markov Chains, Queues, and Simulation: The Mathematical Basis of Performance Modeling. Princeton University Press. p. 105. ISBN 978-1-4008-3281-1.

- Zwillinger, Daniel; Kokoska, Stephen (2010). CRC Standard Probability and Statistics Tables and Formulae. CRC Press. p. 49. ISBN 978-1-58488-059-2.

- Sheldon M Ross (2007). "§2.4 Expectation of a random variable". Introduction to probability models (9th ed.). Academic Press. p. 38 ff. ISBN 0-12-598062-0.

- Richard W Hamming (1991). "§2.5 Random variables, mean and the expected value". The art of probability for scientists and engineers. Addison–Wesley. p. 64 ff. ISBN 0-201-40686-1.

Conditional Probability: A Comprehensive Study

Measuring any event by the probability of another event occurring is known as conditional probability. It is the most fundamental theory of probability. Conditional probability is used in statistical inference as a revision of the probability of an event; the information on which this is based should be new. Conditional expectation, conditional probability distribution, regular conditional probability and Bayes theorem are some of the aspects of conditional probability explained in the following section.

Conditional Probability

In probability theory, conditional probability is a measure of the probability of an event given that (by assumption, presumption, assertion or evidence) another event has occurred. If the event of interest is A and the event B is known or assumed to have occurred, "the conditional probability of A given B", or "the probability of A under the condition B", is usually written as $P(A|B)$, or sometimes $P_B(A)$. For example, the probability that any given person has a cough on any given day may be only 5%. But if we know or assume that the person has a cold, then they are much more likely to be coughing. The conditional probability of coughing given that you have a cold might be a much higher 75%.

The concept of conditional probability is one of the most fundamental and one of the most important concepts in probability theory. But conditional probabilities can be quite slippery and require careful interpretation. For example, there need not be a causal or temporal relationship between A and B.

$P(A|B)$ may or may not be equal to $P(A)$ (the unconditional probability of A). If $P(A|B) = P(A)$ and $P(B|A) = P(B)$, then events A and B are said to be independent: in such a case, having knowledge about either event does not change our knowledge about the other event. Also, in general, $P(A|B)$ (the conditional probability of A given B) is not equal to $P(B|A)$. For example, if you have cancer you might have a 90% chance of testing positive for cancer. In this case what is being measured is that the if event B "having cancer" has occurred, the probability A - *test is positive* given that B *having cancer* occurred is 90%, $P(A|B)$ = 90%. Alternatively, you can test positive for cancer but you may have only a 10% chance of actually having cancer because cancer is very rare. In this case what is being measured is the probability of the event B - *having cancer* given that the event A - *test is positive* has occurred, $P(B|A)$ = 10%. Falsely equating the two probabilities causes various errors of reasoning such as the base rate fallacy. Conditional probabilities can be correctly reversed using Bayes' theorem.

Conditioning on an Event

Kolmogorov Definition

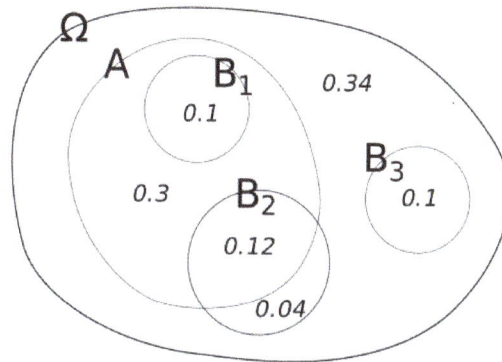

Illustration of conditional probabilities with an Euler diagram. The unconditional probability $P(A) = 0.52$. However, the conditional probability $P(A|B_1) = 1$, $P(A|B_2) = 0.75$, and $P(A|B_3) = 0$.

Given two events A and B from the sigma-field of a probability space with $P(B) > 0$, the conditional probability of A given B is defined as the quotient of the probability of the joint of events A and B, and the probability of B:

$$P(A|B) = \frac{P(A \cap B)}{P(B)}$$

This may be visualized as restricting the sample space to B. The logic behind this equation is that if the outcomes are restricted to B, this set serves as the new sample space.

Note that this is a definition but not a theoretical result. We just denote the quantity $P(A \cap B) / P(B)$ as $P(A|B)$ and call it the conditional probability of A given B.

As an Axiom of Probability

Some authors, such as De Finetti, prefer to introduce conditional probability as an axiom of probability:

$$P(A \cap B) = P(A|B)P(B)$$

Although mathematically equivalent, this may be preferred philosophically; under major probability interpretations such as the subjective theory, conditional probability is considered a primitive entity. Further, this "multiplication axiom" introduces a symmetry with the summation axiom for mutually exclusive events:

$$P(A \cup B) = P(A) + P(B) - P(A \cap B)^{0}$$

Measure-theoretic Definition

If $P(B) = 0$, then the simple definition of $P(A|B)$ is undefined. However, it is possible to define a conditional probability with respect to a σ-algebra of such events (such as those arising from a continuous random variable).

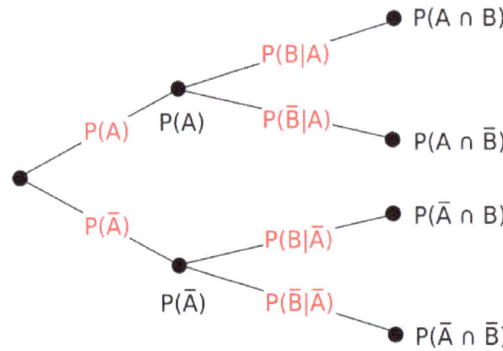

On a tree diagram, branch probabilities are conditional on the event associated with the parent node.

For example, if X and Y are non-degenerate and jointly continuous random variables with density $f_{X,Y}(x, y)$ then, if B has positive measure,

$$P(X \in A \mid Y \in B) = \frac{\displaystyle\int_{y \in B} \int_{x \in A} f_{X,Y}(x, y) dx dy}{\displaystyle\int_{y \in B} \int_{x \in \mathbb{R}} f_{X,Y}(x, y) dx dy}.$$

The case where B has zero measure is problematic. For the case that $B = \{y_0\}$, representing a single point, the conditional probability could be defined as

$$P(X \in A \mid Y = y_0) = \frac{\displaystyle\int_{x \in A} f_{X,Y}(x, y_0) dx}{\displaystyle\int_{x \in \mathbb{R}} f_{X,Y}(x, y_0) dx},$$

however this approach leads to the Borel–Kolmogorov paradox. The more general case of zero measure is even more problematic, as can be seen by noting that the limit, as all δy_i approach zero, of

$$P(X \in A \mid Y \in \cup_i [y_i, y_i + \delta y_i]) \approx \frac{\displaystyle\sum_i \int_{x \in A} f_{X,Y}(x, y_i) dx \delta y_i}{\displaystyle\sum_i \int_{x \in \mathbb{R}} f_{X,Y}(x, y_i) dx \delta y_i},$$

depends on their relationship as they approach zero.

Conditioning on a Random Variable

Conditioning on an event may be generalized to conditioning on a random variable. Let X be a random variable; we assume for the sake of presentation that X is discrete, that is, X takes on only finitely many values x. Let A be an event. The conditional probability of A given X is defined as the random variable, written $P(A|X)$, *that takes on the value*

$P(A \mid X = x)$

whenever

$X = x.$

More formally:

$P(A \mid X)(\omega) = P(A \mid X = X(\omega)).$

The conditional probability $P(A|X)$ is a function of X, e.g., if the function g is defined as

$g(x) = P(A \mid X = x),$

then

$P(A \mid X) = g \circ X$

Note that $P(A|X)$ and X are now both random variables. From the law of total probability, the expected value of $P(A|X)$ is equal to the unconditional probability of A.

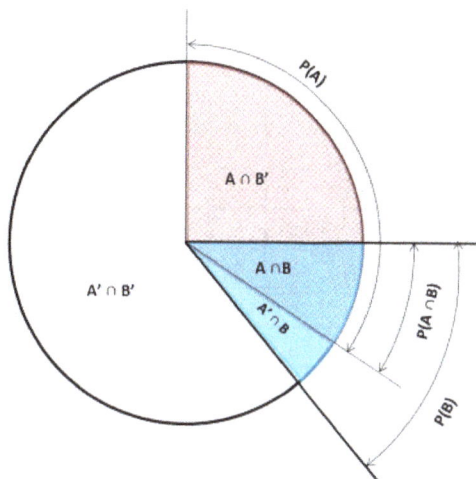

Venn Pie Chart describing conditional probabilities

Example

Suppose that somebody secretly rolls two fair six-sided dice, and we must predict the outcome (the sum of the two upward faces).

- Let A be the value rolled on die 1.

- Let B be the value rolled on die 2.

What is the probability that A = 2?

Table 1 shows the sample space of 36 outcomes

Clearly, $A = 2$ in exactly 6 of the 36 outcomes, thus $P(A=2) = \frac{6}{36} = \frac{1}{6}$.

Table 1							
+		**B**					
1		**2**	**3**	**4**	**5**	**6**	
	1	2	3	4	5	6	7
	2	3	4	5	6	7	8
	3	4	5	6	7	8	9
A	**4**	5	6	7	8	9	10
	5	6	7	8	9	10	11
	6	7	8	9	10	11	12

Suppose it is Revealed that A+B ≤ 5

What is the probability $A+B \leq 5$?

Table 2 shows that $A+B \leq 5$ for exactly 10 of the same 36 outcomes, thus $P(A+B \leq 5) = {}^{10}\!/_{36}$

Table 2							
+		**B**					
1		**2**	**3**	**4**	**5**	**6**	
	1	2	3	4	5	6	7
	2	3	4	5	6	7	8
	3	4	5	6	7	8	9
A	**4**	5	6	7	8	9	10
	5	6	7	8	9	10	11
	6	7	8	9	10	11	12

What is the Probability that A = 2 Given that A+B ≤ 5 ?

Table 3 shows that for 3 of these 10 outcomes, $A = 2$

Thus, the conditional probability $P(A=2 \mid A+B \leq 5) = {}^{3}\!/_{10} = 0.3$.

Table 3							
+		**B**					
1		**2**	**3**	**4**	**5**	**6**	
	1	2	3	4	5	6	7
	2	3	4	5	6	7	8
	3	4	5	6	7	8	9
A	**4**	5	6	7	8	9	10
	5	6	7	8	9	10	11
	6	7	8	9	10	11	12

Use in Inference

In statistical inference, the conditional probability is an update of the probability of an event based on new information. Incorporating the new information can be done as follows

- Let A the event of interest be in the sample space, say (X,P).

- The occurrence of the event A knowing that event B has or will have occurred, means the occurrence of A as it is restricted to B, i.e. $A \cap B$.

- Without the knowledge of the occurrence of B, the information about the occurrence of A would simply be $P(A)$

- The probability of A knowing that event B has or will have occurred, will be the probability of $A \cap B$ compared with $P(B)$, the probability B has occurred.

- This results in $P(A|B) = P(A \cap B)/P(B)$ whenever $P(B)>0$ and 0 otherwise.

Note: This approach results in a probability measure that is consistent with the original probability measure and satisfies all the Kolmogorov Axioms. This conditional probability measure also could have resulted by assuming that the relative magnitude of the probability of A with respect to X will be preserved with respect to B.

Note: The phraseology "evidence" or "information" is generally used in the Bayesian interpretation of probability. The conditioning event is interpreted as evidence for the conditioned event. That is, $P(A)$ is the probability of A before accounting for evidence E, and $P(A|E)$ is the probability of A after having accounted for evidence E or after having updated $P(A)$. This is consistent with the frequentist interpretation, which presumably is the first definition given above.

Statistical Independence

Events A and B are defined to be statistically independent if

$$P(A \cap B) = P(A)P(B).$$

If $P(B)$ is not zero, then this is equivalent to the statement that

$$P(A \mid B) = P(A).$$

Similarly, if $P(A)$ is not zero, then

$$P(B \mid A) = P(B).$$

is also equivalent. Although the derived forms may seem more intuitive, they are not the preferred definition as the conditional probabilities may be undefined, and the preferred definition is symmetrical in A and B.

Important Distinctions: Independent Events Vs Mutually Exclusive Events

The concepts of mutually independent events and mutually exclusive events are often confused.

As noted, statistical independence means:

P(A|B) = P(A), P(B|A) = P(B), provided that the probability for the conditioning event is not zero.

However, mutually exclusive events implies that:

$P(A|B)$ = 0, $P(B|A)$=0, and $P(A \cap B) = 0$.

Assuming Conditional Probability is of Similar Size to its Inverse

Relative size	Case B	Case $\bar{B}$	Total
Condition A	w	x	$w+x$
Condition $\bar{A}$	y	z	$y+z$
Total	$w+y$	$x+z$	$w+x+y+z$

$$P(A|B) \times P(B) = \frac{w}{w+y} \times \frac{w+y}{w+x+y+z} = \frac{w}{w+x+y+z}$$

$$P(B|A) \times P(A) = \frac{w}{w+x} \times \frac{w+x}{w+x+y+z} = \frac{w}{w+x+y+z}$$

A geometric visualisation of Bayes' theorem.

In the table, the values w, x, y and z give the relative weights of each corresponding condition and case. The figures denote the cells of the table involved in each metric, the probability being the fraction of each figure that is shaded. This shows that P(A|B) P(B) = P(B|A) P(A) i.e. P(A|B) = P(B|A) P(A)/P(B) . Similar reasoning can be used to show that P($\bar{A}$|B) = P(B|$\bar{A}$) P($\bar{A}$)/P(B) etc.

In general, it cannot be assumed that $P(A|B) \approx P(B|A)$. This can be an insidious error, even for those who are highly conversant with statistics. The relationship between $P(A|B)$ and $P(B|A)$ is given by Bayes' theorem:

$$P(B \mid A) = \frac{P(A \mid B)P(B)}{P(A)}$$

$$\Leftrightarrow \frac{P(B \mid A)}{P(A \mid B)} = \frac{P(B)}{P(A)}$$

That is, $P(A|B) \approx P(B|A)$ only if $P(B)/P(A) \approx 1$, or equivalently, $P(A) \approx P(B)$.

Alternatively, noting that $A \cap B = B \cap A$, and applying conditional probability:

$$P(A \mid B)P(B) = P(A \cap B) = P(B \cap A) = P(B \mid A)P(A)$$

Rearranging gives the result.

Assuming Marginal and Conditional Probabilities are of Similar Size

In general, it cannot be assumed that $P(A) \approx P(A|B)$. These probabilities are linked through the law of total probability:

$$P(A) = \sum_n P(A \cap B_n) = \sum_n P(A \mid B_n)P(B_n)$$

where the events (B_n) form a countable partition of A.

This fallacy may arise through selection bias. For example, in the context of a medical claim, let S_C be the event that a sequela (chronic disease) S occurs as a consequence of circumstance (acute condition) C. Let H be the event that an individual seeks medical help. Suppose that in most cases, C does not cause S so $P(S_C)$ is low. Suppose also that medical attention is only sought if S has occurred due to C. From experience of patients, a doctor may therefore erroneously conclude that $P(S_C)$ is high. The actual probability observed by the doctor is $P(S_C \mid H)$.

Over- or Under-weighting Priors

Not taking prior probability into account partially or completely is called *base rate neglect*. The reverse, insufficient adjustment from the prior probability is *conservatism*.

Formal Derivation

Formally, $P(A \mid B)$ is defined as the probability of A according to a new probability function on the sample space, such that outcomes not in B have probability 0 and that it is consistent with all original probability measures.

Let Ω be a sample space with elementary events $\{\omega\}$. Suppose we are told the event $B \subseteq \Omega$ has occurred. A new probability distribution (denoted by the conditional notation) is to be assigned on $\{\omega\}$ to reflect this. For events in B, it is reasonable to assume that the relative magnitudes of the probabilities will be preserved. For some constant scale factor α, the new distribution will therefore satisfy:

1. $\omega \in B : P(\omega \mid B) = \alpha P(\omega)$
2. $\omega \notin B : P(\omega \mid B) = 0$
3. $\sum_{\omega \in \Omega} P(\omega \mid B) = 1$.

Substituting 1 and 2 into 3 to select α:

$$1 = \sum_{\omega \in \Omega} P(\omega \mid B)$$
$$= \sum_{\omega \in B} \alpha P(\omega) + \sum_{\omega \notin B} 0$$
$$= \alpha \sum_{\omega \in B} P(\omega)$$
$$= \alpha \cdot P(B)$$
$$\Rightarrow \alpha = \frac{1}{P(B)}$$

So the new probability distribution is

$$1.\ \omega \in B : P(\omega \,|\, B) = \frac{P(\omega)}{P(B)}$$

$$2.\ \omega \notin B : P(\omega \,|\, B) = 0$$

Now for a general event A,

$$P(A|B) = \sum_{\omega \in A \cap B} P(\omega|B) + \cancel{\sum_{\omega \in A \cap B^c} P(\omega|B)}^{\,0}$$

$$= \sum_{\omega \in A \cap B} \frac{P(\omega)}{P(B)}$$

$$= \frac{P(A \cap B)}{P(B)}$$

Conditional Expectation

In probability theory, the conditional expectation of a random variable is another random variable equal to the average of the former over each possible "condition". In the case when the random variable is defined over a discrete probability space, the "conditions" are a partition of this probability space. This definition is then generalized to any probability space using measure theory.

Conditional expectation is also known as conditional expected value or conditional mean.

In modern probability theory the concept of conditional probability is defined in terms of conditional expectation.

Concept

The concept of conditional expectation can be nicely illustrated through the following example. Suppose we have daily rainfall data (mm of rain each day) collected by a weather station on every day of the ten year period from Jan 1, 1990 to Dec 31, 1999. The conditional expectation of daily rainfall knowing the month of the year is the average of daily rainfall over all days of the ten year period that fall in a given month. These data then may be considered either as a function of each day (so for example its value for Mar 3, 1992, would be the sum of daily rainfalls on all days that are in the month of March during the ten years, divided by the number of these days, which is 310) or as a function of just the month (so for example the value for March would be equal to the value of the previous example).

It is important to note the following.

- The conditional expectation of daily rainfall knowing that we are in a month of March of the given ten years is not a monthly rainfall data, that is it is not the average of the ten monthly total March rainfalls. That number would be 31 times higher.

- The average daily rainfall in March 1992 is not equal to the conditional expectation of daily rainfall knowing that we are in a month of March of the given ten years, because we have restricted ourselves to 1992, that is we have more conditions than just that of being in March. This shows that reasoning as "we are in March 1992, so I know we are in March, so the average daily rainfall is the March average daily rainfall" is incorrect. Stated differently, although we use the expression "conditional expectation knowing that we are in March" this really means "conditional expectation knowing nothing other than that we are in March".

History

The related concept of conditional probability dates back at least to Laplace who calculated conditional distributions. It was Andrey Kolmogorov who in 1933 formalized it using the Radon–Nikodym theorem. In works of Paul Halmos and Joseph L. Doob from 1953, conditional expectation was generalized to its modern definition using sub-sigma-algebras.

Classical Definition

Conditional Expectation With Respect to an Event

In classical probability theory the conditional expectation of X given an event H (which may be the event $Y=y$ for a random variable Y) is the average of X over all outcomes in H, that is

$$E(X \mid H) = \frac{\sum\limits_{\omega \in H} X(\omega)}{|H|}$$

The sum above can be grouped by different values of $X(\omega)$, to get a sum over the range x of X

$$E(X \mid H) = \sum_{x \in \mathcal{X}} x \frac{|\{\omega \in H \mid X(\omega) = x\}|}{|H|}$$

In modern probability theory, when H is an event with strictly positive probability, it is possible to give a similar formula. This is notably the case for a discrete random variable Y and for y in the range of Y if the event H is $Y=y$. Let $(\Omega, \mathcal{F}, P)$ be a probability space, X a random variable on that probability space, and $H \in \mathcal{F}$ an event with strictly positive probability $P(H) > 0$. Then the conditional expectation of X given the event H is

$$E(X \mid H) = \frac{E(1_H X)}{P(H)} = \int_{x \in \mathcal{X}} x \, P(dx \mid H),$$

where $\mathcal{X}$ is the range of X and $P(A \mid H) = \dfrac{P(A \cap H)}{P(H)}$ is the conditional probability of A knowing H.

When $P(H) = 0$ (for instance if Y is a continuous random variable and H is the event $Y=y$, this is in general the case), the Borel–Kolmogorov paradox demonstrates the ambiguity of attempting to define the conditional probability knowing the event H. The above formula shows that this problem transposes to the conditional expectation. So instead one only defines the conditional expectation with respect to a sigma-algebra or a random variable.

Conditional Expectation With Respect to a Random Variable

If Y is a discrete random variable with range $\mathcal{Y}$, then we can define on $\mathcal{Y}$ the function

$$g : y \mapsto E(X \mid Y = y).$$

Sometimes this function is called the conditional expectation of X with respect to Y. In fact, according to the modern definition, it is $g \circ Y$ that is called the conditional expectation of X with respect to Y, so that we have

$$E(X \mid Y) : \omega \mapsto E(X \mid Y = Y(\omega)).$$

which is a random variable.

As mentioned above, if Y is a continuous random variable, it is not possible to define $E(X \mid Y)$ by this method. As explained in the Borel–Kolmogorov paradox, we have to specify what limiting procedure produces the set $Y = y$. This can be naturally done by defining the set $H_y^\epsilon = \{\omega \mid \|Y(\omega) - y\| < \epsilon\}$, and taking the limit $\epsilon \to 0$, so that if $P(H_y^\epsilon) > 0$ for all $\epsilon > 0$, , then

$$g : y \mapsto \lim_{\epsilon \to 0} E(X \mid H_y^\epsilon).$$

The modern definition is analogous to the above except that the above limiting process is replaced by the Radon–Nikodym derivative, so the result holds more generally.

Formal Definition

Conditional Expectation With Respect to a σ-algebra

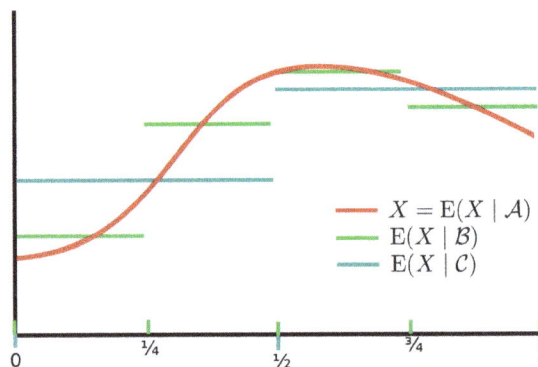

Conditional expectation with respect to a sigma-algebra: in this example the probability space $(\Omega, \mathcal{F}, P)$ is the [0,1] interval with the Lebesgue measure. We define the following σ-algebras: $\mathcal{A} = \mathcal{F}$ while $\mathcal{B}$ is the σ-algebra generated by the intervals with end-points 0, ¼, ½, ¾, 1 and $\mathcal{C}$ is the σ-algebra generated by the intervals with end-points 0, ½, 1. Here the conditional expectation is effectively the average over the minimal sets of the σ-algebra.

Consider the following

- $(\Omega, \mathcal{F}, \mathrm{P})$ is a probability space

- $\mathrm{X} : \Omega \to \mathbb{R}^n$ is a random variable on that probability space

- $\mathcal{H} \subseteq \mathcal{F}$ is a sub-σ-algebra of $\mathcal{F}$

Since $\mathcal{H}$ is a subalgebra of $\mathcal{F}$, the function $X : \Omega \to \mathbb{R}^n$ is usually not $\mathcal{H}$-measureable, thus the existence of the integrals of the form $\int_H X d \mathrm{P}|_{\mathcal{H}}$, where $H \in \mathcal{H}$ and $\mathrm{P}_{|\mathcal{H}}$ is the restriction of P to $\mathcal{H}$ cannot be stated in general. However, the local averages $\int_H X d \mathrm{P}$ can be recovered in $(\Omega, \mathcal{H}, \mathrm{P}|_{\mathcal{H}})$ with the help of the conditional expectation. A conditional expectation of X given $\mathcal{H}$, denoted as $\mathrm{E}(X \,|\, \mathcal{H})$, is any $\mathcal{H}$-measurable function $(\Omega \to \mathbb{R}^n)$ which satisfies:

$$\int_H \mathrm{E}(\mathrm{X} \,|\, \mathcal{H})\, d\mathrm{P} = \int_H \mathrm{X}\, d\mathrm{P} \qquad \textit{for each} \quad \mathrm{H} \in \mathcal{H}.$$

The existence of $\mathrm{E}(X \,|\, \mathcal{H})$ can be established by noting that $\mu^X : F \mapsto \int_F X$ for $F \in \mathcal{F}$ is a measure on $(\Omega, \mathcal{F})$ that is absolutely continuous with respect to P. Furthermore, if h is the natural injection from $\mathcal{H}$ to $\mathcal{F}$ then $\mu^X \circ h = \mu^X_{|\mathcal{H}}$ is the restriction of μ^X to $\mathcal{H}$ and $P \circ h = P_{|\mathcal{H}}$ is the restriction of P to $\mathcal{H}$ and $\mu^X \circ h$ is absolutely continuous with respect to $P \circ h$ since $P \circ h(H) = 0 \Leftrightarrow P(h(H)) = 0 \Rightarrow \mu^X(h(H)) = 0 \Leftrightarrow \mu^X \circ h(H) = 0.$. Thus, we have

$$\mathrm{E}(\mathrm{X} \,|\, \mathcal{H}) = \frac{d\mu^X_{|\mathcal{H}}}{d\mathrm{P}_{|\mathcal{H}}} = \frac{d(\mu^X \circ \mathrm{h})}{d(\mathrm{P} \circ \mathrm{h})}$$

where the derivatives are Radon–Nikodym derivatives of measures.

Conditional Expectation With Respect to a Random Variable

Consider further to the above

- (U, Σ) is a measurable space

- $Y : \Omega \to U$ is a random variable

Then for any Σ-measurable function $g : U \to \mathbb{R}^n$ which satisfies:

$$\int g(Y) f(Y) dP = \int X f(Y) dP \qquad \text{for every } \Sigma\text{-measurable function} \quad f : U \to \mathbb{R}^n.$$

the random variable $g(Y)$, denoted as $\mathrm{E}(X \,|\, Y)$ is a conditional expectation of X given Y.

This definition is equivalent to defining the conditional expectation using the pre-image of Σ with respect to Y. If we define

$$\mathcal{H} = \sigma(Y) := Y^{-1}(\Sigma) := \{Y^{-1}(B) : B \in \Sigma\}$$

then

$$\mathrm{E}(\mathrm{X} \,|\, \mathrm{Y}) = \mathrm{E}(\mathrm{X} \,|\, \mathcal{H}) = \frac{d(\mu^X \circ \mathrm{Y}^{-1})}{d(\mathrm{P} \circ \mathrm{Y}^{-1})} \circ \mathrm{Y}.$$

Discussion

A couple of points worth noting about the definition:

- This is not a constructive definition; we are merely given the required property that a conditional expectation must satisfy.

  o The definition of $E(X \mid \mathcal{H})$ may resemble that of $E(X \mid \mathcal{H})$ for an event H but these are very different objects, the former being a $\mathcal{H}$-measurable function $\Omega \to \mathbb{R}^n$, while the latter is an element of $\mathbb{R}^n$ for fixed H, or a function $\mathcal{F} \to \mathbb{R}^n$ if considered as the function $H \mapsto E(X \mid H)$. .

  o Existence of a conditional expectation function is determined by the Radon–Nikodym theorem, a sufficient condition is that the (unconditional) expected value for X exist.

  o Uniqueness can be shown to be almost sure: that is, versions of the same conditional expectation will only differ on a set of probability zero.

- The σ-algebra $\mathcal{H}$ controls the "granularity" of the conditioning. A conditional expectation $E(X \mid \mathcal{H})$ over a finer-grained σ-algebra $\mathcal{H}$ will allow us to condition on a wider variety of events.

Conditioning as Factorization

In the definition of conditional expectation that we provided above, the fact that Y is a *real* random element is irrelevant. Let (U, Σ) be a measurable space, where $\Sigma \subset \mathcal{P}(U)$ is a σ-algebra in U. A U-*valued random element* is a measurable function $Y : \Omega \to U$, i.e. $Y^{-1}(B) \in \mathcal{F}$ for all $B \in \Sigma$. . The *distribution* of Y is the probability measure $\mathbb{P}_Y : \Sigma \to \mathbb{R}$ such that $\mathbb{P}_Y(B) = \mathbb{P}(Y^{-1}(B))$.

Theorem. If $X : \Omega \to \mathbb{R}$ is an integrable random variable, then there exists a $\mathbb{P}_Y$ – unique integrable random element $E(X \mid Y) : U \to \mathbb{R}$, , such that

$$\int_{Y^{-1}(B)} X \, d\mathbb{P} = \int_B E(X \mid Y) \, d\mathbb{P}_Y,$$

for all $B \in \Sigma$..

Proof Sketch

Let $\mu : \Sigma \to \mathbb{R}$ be such that $\mu(B) = \int_{Y^{-1}(B)} X \, d\mathbb{P}$. Then μ is a signed measure which is absolutely continuous with respect to $\mathbb{P}_Y$. Indeed $\mathbb{P}_Y(B) = 0$ means exactly that $\mathbb{P}(Y^{-1}(B)) = 0$. Since the integral of an integrable function on a set of probability 0 is 0, this proves absolute continuity. The Radon–Nikodym theorem then proves the existence of a density of μ with respect to   , which we denote by $E(X \mid Y)$.

Comparing with conditional expectation with respect to sub-sigma algebras, it holds that

$$E(X \mid Y) \circ Y = E\left(X \mid Y^{-1}(\Sigma)\right).$$

We can further interpret this equality by considering the abstract change of variables formula to transport the integral on the right hand side to an integral over Ω:

$$\int_{Y^{-1}(B)} X \, d\mathbb{P} = \int_{Y^{-1}(B)} (E(X \mid Y) \circ Y) \, d\mathbb{P}.$$

The equation means that the integrals of X and the composition $E(X\,|\,Y)\circ Y$ over sets of the form $Y^{-1}(B)$, for $B\in\Sigma$, are identical.

This equation can be interpreted to say that the following diagram is commutative *in the average.*

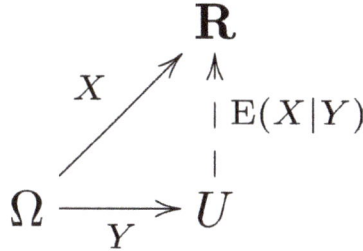

$$
\begin{array}{ccc}
 & & \mathbf{R} \\
 & \nearrow{\scriptstyle X} & \uparrow{\scriptstyle\,|\;E(X|Y)} \\
 & & | \\
\Omega & \xrightarrow[\;\;Y\;\;]{} & U
\end{array}
$$

Computation

When X and Y are both discrete random variables, then the conditional expectation of X given the event $Y=y$ can be considered as function of y for y in the range of Y

$$E(X\,|\,Y=y)=\sum_{x\in\mathcal{X}}x\,P(X=x\,|\,Y=y)=\sum_{x\in\mathcal{X}}x\,\frac{P(X=x,Y=y)}{P(Y=y)},$$

where $\mathcal{X}$ is the range of X.

If X is a continuous random variable, while Y remains a discrete variable, the conditional expectation is:

$$E(X\,|\,Y=y)=\int_{\mathcal{X}}xf_{X}(x\,|\,Y=y)dx$$

with $f_X(x\,|\,Y=y)=\dfrac{f_{X,Y}(x,y)}{P(Y=y)}$ (where $f_{X,Y}(x,y)$ gives the joint density of X and Y) is the conditional density of X given $Y=y$.

If both X and Y are continuous random variables, then the conditional expectation is:

$$E(X\,|\,Y=y)=\int_{\mathcal{X}}xf_{X|Y}(x\,|\,y)dx$$

where $f_{X|Y}(x\,|\,y)=\dfrac{f_{X,Y}(x,y)}{f_Y(y)}$ (where $f_Y(y)$ gives the density of Y).

Basic Properties

All the following formulas are to be understood in an almost sure sense. The sigma-algebra $\mathcal{H}$ could be replaced by a random variable Z

- Pulling out independent factors:
  - If X is independent of $\mathcal{H}$, then $E(X\,|\,\mathcal{H})=E(X)$.

Proof

Let $B\in\mathcal{H}$. Then X is independent of 1_B, so we get that

$$\int_B X \, dP = E(X1_B) = E(X)E(1_B) = E(X)P(B) = \int_B E(X) \, dP$$

Thus the definition of conditional expectation is satisfied by the constant random variable $E(X)$, as desired.

- o If X is independent of $\sigma(Y, \mathcal{H})$, then $E(XY \mid \mathcal{H}) = E(X)E(Y \mid H)$. Note that this is not necessarily the case if X is only independent of $\mathcal{H}$ and of Y.

- o If X, Y are independent, $\mathcal{G}, \mathcal{H}$ are independent, X is independent of $\mathcal{H}$ and Y is independent of $\mathcal{G}$, then $E(E(XY \mid \mathcal{G}) \mid \mathcal{H}) = E(X)E(Y) = E(E(XY \mid \mathcal{H}) \mid \mathcal{G})$.

- • Stability:

  - o If X is $\mathcal{H}$-measurable, then $E(X \mid \mathcal{H}) = X$.

  - o If Z is a random variable, then $E(f(Z) \mid Z) = f(Z)$ and in its simplest form: $E(Z \mid Z) = Z$

- • Pulling out known factors:

  - o If X is $\mathcal{H}$-measurable, then $E(XY \mid \mathcal{H}) = XE(Y \mid \mathcal{H})$.

  - o If Z is a random variable, then $E(f(Z)Y \mid Z) = f(Z)E(Y \mid Z)$

- • Law of total expectation: $E(E(X \mid \mathcal{H})) = E(X)$..

- • Tower property:

  - o For sub-σ-algebras $\mathcal{H}_1 \subset \mathcal{H}_2 \subset \mathcal{F}$ we have
    $E(E(X \mid \mathcal{H}_2) \mid \mathcal{H}_1) = E(X \mid \mathcal{H}_1) = E(E(X \mid \mathcal{H}_1) \mid \mathcal{H}_2)$.

    - ▪ A special case is when Z is a $\mathcal{H}$-measurable random variable. Then $\sigma(Z) \subset \mathcal{H}$ and thus $E(E(X \mid \mathcal{H}) \mid Z) = E(X \mid Z)$

    - ▪ Doob martingale property: the above with $Z = E(X \mid \mathcal{H})$ (which is $\mathcal{H}$-measurable), and using also $E(Z \mid Z) = Z$ gives $E(X \mid E(X \mid \mathcal{H})) = E(X \mid \mathcal{H})$.

  - o For random variables X, Y we have $E(E(X \mid Y) \mid f(Y)) = E(X \mid f(Y))$.

  - o For random variables X, Y, Z we have $E(E(X \mid Y, Z) \mid Y) = E(X \mid Y)$.

- • Linearity: we have $E(X_1 + X_2 \mid \mathcal{H}) = E(X_1 \mid \mathcal{H}) + E(X_2 \mid \mathcal{H})$ and $E(aX \mid) \quad aE(X \mid H)$ for $a \in \mathbb{R}$.

- • Positivity: If $X \geq 0$ then $E(X \mid \mathcal{H}) \geq 0$..

- • Monotonicity: If $X_1 \leq X_2$ then $E(X_1 \mid \mathcal{H}) \leq E(X_2 \mid \mathcal{H})$.

- • Monotone convergence: If $0 \leq X_n \uparrow X$ then $E(X_n \mid \mathcal{H}) \uparrow E(X \mid \mathcal{H})$.

- • Dominated convergence: If $X_n \to X$ and $\mid X_n \mid \leq Y$ with $Y \in L^1$ then $E(X_n \mid \mathcal{H}) \to E(X \mid \mathcal{H})$.

- Fatou's lemma: If $E(\inf_n X_n \mid \mathcal{H}) > -\infty$ then $E(\liminf_{n\to\infty} X_n \mid \mathcal{H}) \leq \liminf_{n\to\infty} E(X_n \mid \mathcal{H})$.

- Jensen's inequality: If $f : \mathbb{R} \to \mathbb{R}$ is a convex function, then $f(E(X \mid \mathcal{H})) \leq E(f(X) \mid \mathcal{H})$.

- Conditional variance: Using the conditional expectation we can define, by analogy with the definition of the variance as the mean square deviation from the average, the conditional variance

  - Definition: $\mathrm{Var}(X \mid \mathcal{H}) = E\big((X - E(X \mid \mathcal{H}))^2 \mid \mathcal{H}\big)$

  - Algebraic formula for the variance: $\mathrm{Var}(X \mid \mathcal{H}) = E(X^2 \mid \mathcal{H}) - \big(E(X \mid \mathcal{H})\big)^2$

  - Law of total variance: $\mathrm{Var}(X) = E(\mathrm{Var}(X \mid \mathcal{H})) + \mathrm{Var}(E(X \mid \mathcal{H}))$.

- Martingale convergence: For a random variable X, that has finite expectation, we have $E(X \mid \mathcal{H}_n) \to E(X \mid \mathcal{H})$, if either $\mathcal{H}_1 \subset \mathcal{H}_2 \subset \cdots$ is an increasing series of sub-σ-algebras and $\mathcal{H} = \sigma(\bigcup_{n=1}^{\infty} \mathcal{H}_n)$ or if $\mathcal{H}_1 \supset \mathcal{H}_2 \supset \cdots$ is a decreasing series of sub-σ-algebras and $\mathcal{H} = \bigcap_{n=1}^{\infty} \mathcal{H}_n$.

- Conditional expectation as L^2 − projection: If X, Y are in the Hilbert space of square-integrable real random variables (real random variables with finite second moment) then

  - for $\mathcal{H}$-measurable Y we have $E(Y(X - E(X \mid \mathcal{H}))) = 0$ i.e. the conditional expectation $E(X \mid \mathcal{H})$ is in the sense of the $L^2(P)$ scalar product the orthogonal projection from X to the linear subspace of $\mathcal{H}$ − measurable functions. (This allows to define and prove the existence of the conditional expectation based on the Hilbert projection theorem.)

  - the mapping $X \mapsto E(X \mid \mathcal{H})$ is self-adjoint:
  $$E(XE(Y \mid \mathcal{H})) = E\big(E(X \mid \mathcal{H})E(Y \mid \mathcal{H})\big) = E(E(X \mid \mathcal{H})Y)$$

- Conditioning is a contractive projection of L^p spaces $L^s_P(\Omega; \mathcal{F}) \to L^s_P(\Omega; \mathcal{H})$ i.e. $E \mid E(X \mid \mathcal{H}) \mid^s \leq E \mid X \mid^s$ for any $s \geq 1$.

- Doob's conditional independence property: If X, Y are conditionally independent given Z, then $P(X \in B \mid Y, Z) = P(X \in B \mid Z)$ (equivalently, $E(1_{\{X \in B\}} \mid Y, Z) = E(1_{\{X \in B\}} \mid Z)$).

Conditional Probability Distribution

In probability theory and statistics, given two jointly distributed random variables X and Y, the conditional probability distribution of Y given X is the probability distribution of Y when X is known to be a particular value; in some cases the conditional probabilities may be expressed as functions containing the unspecified value x of X as a parameter. In case that both "X" and "Y" are categorical variables, a conditional probability table is typically used to represent the conditional probability. The conditional distribution contrasts with the marginal distribution of a random variable, which is its distribution without reference to the value of the other variable.

If the conditional distribution of Y given X is a continuous distribution, then its probability density function is known as the conditional density function. The properties of a conditional distribution, such as the moments, are often referred to by corresponding names such as the conditional mean and conditional variance.

More generally, one can refer to the conditional distribution of a subset of a set of more than two variables; this conditional distribution is contingent on the values of all the remaining variables, and if more than one variable is included in the subset then this conditional distribution is the conditional joint distribution of the included variables.

Discrete Distributions

For discrete random variables, the conditional probability mass function of Y given the occurrence of the value x of X can be written according to its definition as:

$$p_Y(y \mid X = x) = P(Y = y \mid X = x) = \frac{P(X = x \cap Y = y)}{P(X = x)}.$$

Due to the occurrence of $P(X = x)$ in a denominator, this is defined only for non-zero (hence strictly positive) $P(X = x)$.

The relation with the probability distribution of X given Y is:

$$P(Y = y \mid X = x) P(X = x) = P(X = x \cap Y = y) = P(X = x \mid Y = y) P(Y = y).$$

Continuous Distributions

Similarly for continuous random variables, the conditional probability density function of Y given the occurrence of the value x of X can be written as

$$f_Y(y \mid X = x) = \frac{f_{X,Y}(x, y)}{f_X(x)},$$

where $f_{X,Y}(x, y)$ gives the joint density of X and Y, while $f_X(x)$ gives the marginal density for X. Also in this case it is necessary that $f_X(x) > 0$.

The relation with the probability distribution of X given Y is given by:

$$f_Y(y \mid X = x) f_X(x) = f_{X,Y}(x, y) = f_X(x \mid Y = y) f_Y(y).$$

The concept of the conditional distribution of a continuous random variable is not as intuitive as it might seem: Borel's paradox shows that conditional probability density functions need not be invariant under coordinate transformations.

Relation to Independence

Random variables X, Y are independent if and only if the conditional distribution of Y given X is, for all possible realizations of X, equal to the unconditional distribution of Y. For discrete random variables this means $P(Y = y \mid X = x) = P(Y = y)$ for all relevant x and y. For continuous random variables X and Y, having a joint density function, it means $f_Y(y \mid X = x) = f_Y(y)$ for all relevant x and y.

Properties

Seen as a function of y for given x, $P(Y = y \mid X = x)$ is a probability and so the sum over all y (or integral if it is a conditional probability density) is 1. Seen as a function of x for given y, it is a likelihood function, so that the sum over all x need not be 1.

Measure-Theoretic Formulation

Let $(\Omega, \mathcal{F}, P)$ be a probability space, $\mathcal{G} \subseteq \mathcal{F}$ a σ-field in $\mathcal{F}$, and $X : \Omega \to \mathbb{R}$ a real-valued random variable (measurable with respect to the Borel σ-field $\mathcal{R}^1$ on $\mathbb{R}$). It can be shown that there exists a function $\mu : \mathcal{R}^1 \times \Omega \to \mathbb{R}$ such that $\mu(\cdot, \omega)$ is a probability measure on $\mathcal{R}^1$ for each $\omega \in \Omega$ (i.e., it is regular) and $\mu(H, \cdot) = P(X \in H \mid \mathcal{G})$ (almost surely) for every $H \in \mathcal{R}^1$. For any $\omega \in \Omega$, the function $\mu(\cdot, \omega) : \mathcal{R}^1 \to \mathbb{R}$ is called a conditional probability distribution of X given $\mathcal{G}$.. In this case,

$$E[X \mid \mathcal{G}] = \int_{-\infty}^{\infty} x\,\mu(dx, \cdot)$$

almost surely.

Relation to Conditional Expectation

For any event $A \in \mathcal{A} \supseteq \mathcal{B}$,, define the indicator function:

$$\mathbf{1}_A(\omega) = \begin{cases} 1 & \text{if } \omega \in A, \\ 0 & \text{if } \omega \notin A, \end{cases}$$

which is a random variable. Note that the expectation of this random variable is equal to the probability of A itself:

$$E(\mathbf{1}_A) = P(A).$$

Then the conditional probability given $\mathcal{B}$ is a function $P(\cdot \mid \mathcal{B}) : \mathcal{A} \times \Omega \to (0,1)$ such that $P(A \mid \mathcal{B})$ is the conditional expectation of the indicator function for A:

$$P(A \mid \mathcal{B}) = E(\mathbf{1}_A \mid \mathcal{B})$$

In other words, $P(A \mid \mathcal{B})$ is a $\mathcal{B}$-measurable function satisfying

$$\int_B P(A \mid \mathcal{B})(\omega)\,\mathrm{d}P(\omega) = P(A \cap B) \qquad \text{for all} \quad A \in \mathcal{A}, B \in \mathcal{B}.$$

A conditional probability is regular if $P(\cdot \mid \mathcal{B})(\omega)$ is also a probability measure for all $\omega \in \Omega$. An expectation of a random variable with respect to a regular conditional probability is equal to its conditional expectation.

- For the trivial sigma algebra $\mathcal{B} = \{\varnothing, \Omega\}$ the conditional probability is a constant function, $P\big(A \mid \{\varnothing, \Omega\}\big) \equiv P(A)$.

- For $A \in \mathcal{B}$, as outlined above, $P(A \mid \mathcal{B}) = 1_A$...

Regular Conditional Probability

Regular conditional probability is a concept that has developed to overcome certain difficulties in formally defining conditional probabilities for continuous probability distributions. It is defined as an alternative probability measure conditioned on a particular value of a random variable.

Motivation

Normally we define the conditional probability of an event A given an event B as:

$$P(A \mid B) = \frac{P(A \cap B)}{P(B)}.$$

The difficulty with this arises when the event B is too small to have a non-zero probability. For example, suppose we have a random variable X with a uniform distribution on $[0,1]$, and B is the event that $X = 2/3$. Clearly the probability of B in this case is $P(B) = 0$, but nonetheless we would still like to assign meaning to a conditional probability such as $P(A \mid X = 2 >$ To do so rigorously requires the definition of a regular conditional probability.

Definition

Let $(\Omega, \mathcal{F}, P)$ be a probability space, and let $T : \Omega \to E$ be a random variable, defined as a Borel-measurable function from Ω to its state space $(E, \mathcal{E})$. Then a regular conditional probability is defined as a function $\nu : E \times \mathcal{F} \to [0,1]$, called a "transition probability", where $\nu(x, A)$ is a valid probability measure (in its second argument) on $\mathcal{F}$ for all $x \in E$ and a measurable function in E (in its first argument) for all $A \in \mathcal{F}$, such that for all $A \in \mathcal{F}$ and all $B \in \mathcal{E}$

$$P\left(A \cap T^{-1}(B)\right) = \int_B \nu(x, A) P\left(T^{-1}(dx)\right).$$

To express this in our more familiar notation:

$$P(A \mid T = x) = \nu(x, A),$$

where $x \in \mathrm{supp} T$, i.e. the topological support of the pushforward measure $T_* P = P\left(T^{-1}(\cdot)\right)$. As can be seen from the integral above, the value of ν for points x outside the support of the random variable is meaningless; its significance as a conditional probability is strictly limited to the support of T.

The measurable space $(\Omega, \mathcal{F})$ is said to have the regular conditional probability property if for all probability measures P on $(\Omega, \mathcal{F})$, all random variables on $(\Omega, \mathcal{F}, P)$ admit a regular conditional probability. A Radon space, in particular, has this property.

Alternate Definition

Consider a Radon space Ω (that is a probability measure defined on a Radon space endowed with the Borel sigma-algebra) and a real-valued random variable T. As discussed above, in this case there exists a regular conditional probability with respect to T. Moreover, we can alternatively define the regular conditional probability for an event A given a particular value t of the random variable T in the following manner:

$$P(A \mid T = t) = \lim_{U \supset \{T=t\}} \frac{P(A \cap U)}{P(U)},$$

where the limit is taken over the net of open neighborhoods U of t as they become smaller with respect to set inclusion. This limit is defined if and only if the probability space is Radon, and only in the support of T, as described in the article. This is the restriction of the transition probability to the support of T. To describe this limiting process rigorously:

For every $\epsilon > 0$, there exists an open neighborhood U of t, such that for every open V with $t \in V \subset U$,

$$\left| \frac{P(A \cap V)}{P(V)} - L \right| < \epsilon,$$

where $L = P(A \mid T = t)$ is the limit.

Example

To continue with our motivating example above, we consider a real-valued random variable X and write

$$P(A \mid X = x_0) = \nu(x_0, A) = \lim_{\epsilon \to 0+} \frac{P(A \cap \{x_0 - \epsilon < X < x_0 + \epsilon\})}{P(\{x_0 - \epsilon < X < x_0 + \epsilon\})},$$

(where $x_0 = 2/3$ for the example given.) This limit, if it exists, is a regular conditional probability for X, restricted to $\mathrm{supp}\, X$.

In any case, it is easy to see that this limit fails to exist for x_0 outside the support of X: since the support of a random variable is defined as the set of all points in its state space whose every neighborhood has positive probability, for every point x_0 outside the support of X (by definition) there will be an $\epsilon > 0$ such that $P(\{x_0 - \epsilon < X < x_0 + \epsilon\}) = 0$.

Thus if X is distributed uniformly on $[0,1]$, it is truly meaningless to condition a probability on "$X = 3/2$".

Disintegration Theorem

In mathematics, the disintegration theorem is a result in measure theory and probability theory. It rigorously defines the idea of a non-trivial "restriction" of a measure to a measure zero subset of the measure space in question. It is related to the existence of conditional probability measures. In a sense, "disintegration" is the opposite process to the construction of a product measure.

Motivation

Consider the unit square in the Euclidean plane $\mathbb{R}^2$, $S = [0, 1] \times [0, 1]$. Consider the probability measure μ defined on S by the restriction of two-dimensional Lebesgue measure λ^2 to S. That is, the probability of an event $E \subseteq S$ is simply the area of E. We assume E is a measurable subset of S.

Consider a one-dimensional subset of S such as the line segment $L_x = \{x\} \times [0, 1]$. L_x has μ-measure zero; every subset of L_x is a μ-null set; since the Lebesgue measure space is a complete measure space,

$$E \subseteq L_x \Rightarrow \mu(E) = 0.$$

While true, this is somewhat unsatisfying. It would be nice to say that μ "restricted to" L_x is the one-dimensional Lebesgue measure λ^1, rather than the zero measure. The probability of a "two-dimensional" event E could then be obtained as an integral of the one-dimensional probabilities of the vertical "slices" $E \cap L_x$: more formally, if μ_x denotes one-dimensional Lebesgue measure on L_x, then

$$\mu(E) = \int_{[0,1]} \mu_x(E \cap L_x) \, dx$$

for any "nice" $E \subseteq S$. The disintegration theorem makes this argument rigorous in the context of measures on metric spaces.

Statement of the Theorem

(Hereafter, $P(X)$ will denote the collection of Borel probability measures on a metric space (X, d).)

Let Y and X be two Radon spaces (i.e. separable metric spaces on which every probability measure is a Radon measure). Let $\mu \in P(Y)$, let $\pi : Y \to X$ be a Borel-measurable function, and let $\nu \in P(X)$ be the pushforward measure $\nu = \pi_*(\mu) = \mu \circ \pi^{-1}$. Then there exists a ν-almost everywhere uniquely determined family of probability measures $\{\mu_x\}_{x \in X} \subseteq P(Y)$ such that

- the function $x \mapsto \mu_x$ is Borel measurable, in the sense that $x \mapsto \mu_x(B)$ is a Borel-measurable function for each Borel-measurable set $B \subseteq Y$;

- μ_x "lives on" the fiber $\pi^{-1}(x)$: for ν-almost all $x \in X$,

  $$\mu_x\left(Y \setminus \pi^{-1}(x)\right) = 0,$$

  and so $\mu_x(E) = \mu_x(E \cap \pi^{-1}(x))$;

- for every Borel-measurable function $f : Y \to [0, \infty]$,

  $$\int_Y f(y) \, d\mu(y) = \int_X \int_{\pi^{-1}(x)} f(y) \, d\mu_x(y) \, d\nu(x).$$

  In particular, for any event $E \subseteq Y$, taking f to be the indicator function of E,

  $$\mu(E) = \int_X \mu_x(E) \, d\nu(x).$$

Applications

Product Spaces

The original example was a special case of the problem of product spaces, to which the disintegration theorem applies.

When Y is written as a Cartesian product $Y = X_1 \times X_2$ and $\pi_i : Y \to X_i$ is the natural projection, then each fibre $\pi_1^{-1}(x_1)$ can be canonically identified with X_2 and there exists a Borel family of probability measures $\{\mu_{x_1}\}_{x_1 \in X_1}$ in $P(X_2)$ (which is $(\pi_1)_*(\mu)$-almost everywhere uniquely determined) such that

$$\mu = \int_{X_1} \mu_{x_1} \, \mu\left(\pi_1^{-1}(\mathrm{d}x_1)\right) = \int_{X_1} \mu_{x_1} \, \mathrm{d}(\pi_1)_*(\mu)(x_1),$$

which is in particular

$$\int_{X_1 \times X_2} f(x_1, x_2) \mu(\mathrm{d}x_1, \mathrm{d}x_2) = \int_{X_1}\left(\int_{X_2} f(x_1, x_2) \mu(\mathrm{d}x_2 \mid x_1) \right) \mu\left(\pi_1^{-1}(\mathrm{d}x_1)\right)$$

and

$$\mu(A \times B) = \int_A \mu\left(B \mid x_1\right) \mu\left(\pi_1^{-1}(\mathrm{d}x_1)\right).$$

The relation to conditional expectation is given by the identities

$$E(f \mid \pi_1)(x_1) = \int_{X_2} f(x_1, x_2) \mu(\mathrm{d}x_2 \mid x_1),$$

$$\mu(A \times B \mid \pi_1)(x_1) = 1_A(x_1) \cdot \mu(B \mid x_1).$$

Vector Calculus

The disintegration theorem can also be seen as justifying the use of a "restricted" measure in vector calculus. For instance, in Stokes' theorem as applied to a vector field flowing through a compact surface $\Sigma \subset \mathbb{R}^3$, it is implicit that the "correct" measure on Σ is the disintegration of three-dimensional Lebesgue measure λ^3 on Σ, and that the disintegration of this measure on $\partial\Sigma$ is the same as the disintegration of λ^3 on $\partial\Sigma$.

Conditional Distributions

The disintegration theorem can be applied to give a rigorous treatment of conditional probability distributions in statistics, while avoiding purely abstract formulations of conditional probability.

Bayes' Theorem

In probability theory and statistics, Bayes' theorem (alternatively Bayes' law or Bayes' rule) describes the probability of an event, based on prior knowledge of conditions that might be related to the event. For example, if cancer is related to age, then, using Bayes' theorem, a person's age (prior knowledge) can be used to *more accurately* assess the probability that they have cancer, compared to the assessment of the probability of cancer made without prior knowledge of the person's age.

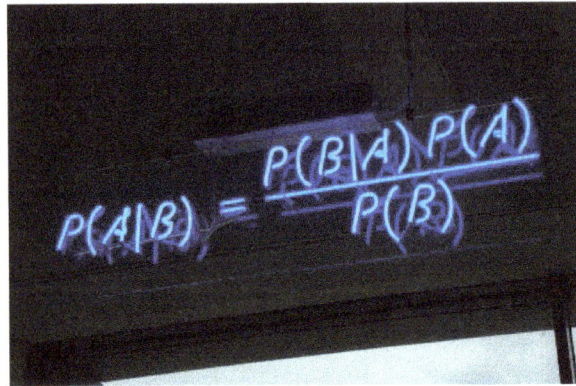

A blue neon sign, showing the simple statement of Bayes' theorem

One of the many applications of Bayes' theorem is Bayesian inference, a particular approach to statistical inference. When applied, the probabilities involved in Bayes' theorem may have different probability interpretations. With the Bayesian probability interpretation the theorem expresses how a subjective degree of belief should rationally change to account for availability of prior related evidence. Bayesian inference is fundamental to Bayesian statistics.

Bayes' theorem is named after Rev. Thomas Bayes , who first provided an equation that allows new evidence to update beliefs. It was further developed by Pierre-Simon Laplace, who first published the modern formulation in his 1812 "Théorie analytique des probabilités." Sir Harold Jeffreys put Bayes' algorithm and Laplace's formulation on an axiomatic basis. Jeffreys wrote that Bayes' theorem "is to the theory of probability what the Pythagorean theorem is to geometry."

Statement of Theorem

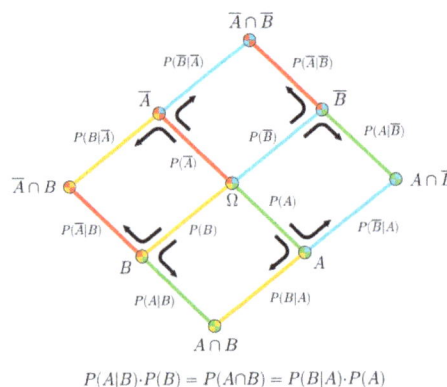

$$P(A|B) \cdot P(B) = P(A \cap B) = P(B|A) \cdot P(A)$$

Visualization of Bayes' theorem by the superposition of two decision trees

Bayes' theorem is stated mathematically as the following equation:

$$P(A \mid B) = \frac{P(B \mid A)}{P(B)},$$

where A and B are events and $P(B) \neq 0$.

- $P(A)$ and $P(B)$ are the probabilities of observing A and B without regard to each other.

- $P(A \mid B)$, a conditional probability, is the probability of observing event A given that B is true.

- $P(B \mid A)$ is the probability of observing event B given that A is true.

Examples

Cancer at Age 65

Suppose that an individual's probability of having cancer, assigned according to the general prevalence of cancer, is 1%. This is known as the "base rate" or prior (i.e. before being informed about the particular case at hand) probability of having cancer.

Next, suppose that the person is 65 years old. Let us assume that cancer and age are related. The probability that a person has cancer when they are 65 years old is known as the "current probability", where "current" refers to the theorized situation upon finding out information about the particular case at hand.

Then, suppose that the probability of being 65 years old is 0.2%, and that the probability that a person diagnosed with cancer at 65 years old is 0.5%.

Knowing this, along with the base rate, we can calculate the probability of having cancer as a 65-year-old, which is equal to the probability of being 65 years of age while having cancer, times the probability of having cancer, divided by the probability of being 65 years of age:

$$(0.5\% \times 1\%) \div 0.2\% = 2.5\%$$

Possibly more intuitively, in a community of 100,000 people, 1,000 people will have cancer and 200 people will be 65 years old. Of the 1000 people with cancer, only 5 people will be 65 years old. Thus, of the 200 people who are 65 years old, only 5 can be expected to have cancer.

It may come as a surprise that even though being 65 years old increases the risk of having cancer, that person's probability of having cancer is still fairly low. This is because the base rate of cancer (regardless of age) is low. This illustrates both the importance of base rate, as well as that it is commonly neglected. Base rate neglect leads to serious misinterpretation of statistics; therefore, special care should be taken to avoid such mistakes. Becoming familiar with Bayes' theorem is one way to combat the natural tendency to neglect base rates.

Drug Testing

The importance of specificity in this example can be seen by calculating that even if sensitivity is raised to 100% and specificity remains at 99% then the probability of the person being a drug user only rises from 33.2% to 33.4%, but if the sensitivity is held at 99% and the specificity is increased to 99.5% then probability of the person being a drug user rises to about 49.9%.

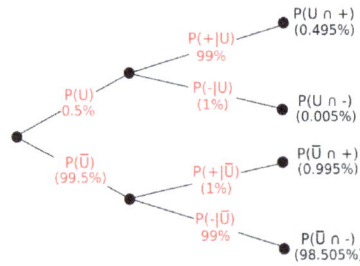

Tree diagram illustrating drug testing example. U, Ū, "+" and "−" are the events representing user, non-user, positive result and negative result. Percentages in parentheses are calculated.

Suppose a drug test is 99% sensitive and 99% specific. That is, the test will produce 99% true positive results for drug users and 99% true negative results for non-drug users. Suppose that 0.5% of people are users of the drug. If a randomly selected individual tests positive, what is the probability that he is a user?

$$P(User \mid +) = \frac{P(+\mid User)P(User)}{P(+\mid User)P(User) + P(+\mid Non\text{-}user)P(Non\text{-}user)}$$

$$= \frac{0.99 \times 0.005}{0.99 \times 0.005 + 0.01 \times 0.995}$$

$$\approx 33.2\%$$

Despite the apparent accuracy of the test, if an individual tests positive, it is more likely that they do *not* use the drug than that they do. This again illustrates the importance of base rates, and how the formation of policy can be egregiously misguided if base rates are neglected.

This surprising result arises because the number of non-users is very large compared to the number of users; thus the number of false positives (0.995%) outweighs the number of true positives (0.495%). To use concrete numbers, if 1000 individuals are tested, there are expected to be 995 non-users and 5 users. From the 995 non-users, $0.01 \times 995 \simeq 10$ false positives are expected. From the 5 users, $0.99 \times 5 \simeq 5$ true positives are expected. Out of 15 positive results, only 5, about 33%, are genuine.

A More Complicated Example

The entire output of a factory is produced on three machines. The three machines account for 20%, 30%, and 50% of the output, respectively. The fraction of defective items produced is this: for the first machine, 5%; for the second machine, 3%; for the third machine, 1%. If an item is chosen at random from the total output and is found to be defective, what is the probability that it was produced by the third machine?

A solution is as follows. Let A_i denote the event that a randomly chosen item was made by the ith machine (for $i = 1, 2, 3$). Let B denote the event that a randomly chosen item is defective. Then, we are given the following information:

$$P(A_1) = 0.2, \quad P(A_2) = 0.3, \quad P(A_3) = 0.5.$$

If the item was made by the first machine, then the probability that it is defective is 0.05; that is, $P(B \mid A_1) = 0.05$. Overall, we have

$$P(B \mid A_1) = 0.05, \quad P(B \mid A_2) = 0.03, \quad P(B \mid A_3) = 0.01.$$

To answer the original question, we first find $P(B)$. That can be done in the following way:

$$P(B) = \Sigma_i P(B \mid A_i) P(A_i) = (0.05)(0.2) + (0.03)(0.3) + (0.01)(0.5) = 0.024.$$

Hence 2.4% of the total output of the factory is defective.

We are given that B has occurred, and we want to calculate the conditional probability of A_3. By Bayes' theorem,

$$P(A_3 \mid B) = P(B \mid A_3) P(A_3)/P(B) = (0.01)(0.50)/(0.024) = 5/24.$$

Given that the item is defective, the probability that it was made by the third machine is only 5/24. Although machine 3 produces half of the total output, it produces a much smaller fraction of the defective items. Hence the knowledge that the item selected was defective enables us to replace the prior probability $P(A_3) = 1/2$ by the smaller posterior probability $P(A_3 \mid B) = 5/24$.

Once again, the answer can be reached without recourse to the formula by applying the conditions to any hypothetical number of cases. For example, in 100,000 items produced by the factory, 20,000 will be produced by Machine A, 30,000 by Machine B, and 50,000 by Machine C. Machine A will produce 1000 defective items, Machine B 900, and Machine C 500. Of the total 2400 defective items, only 500, or 5/24 were produced by Machine C.

Interpretations

The interpretation of Bayes' theorem depends on the interpretation of probability ascribed to the terms. The two main interpretations are described below.

Bayesian Interpretation

In the Bayesian (or epistemological) interpretation, probability measures a "degree of belief." Bayes' theorem then links the degree of belief in a proposition before and after accounting for evidence. For example, suppose it is believed with 50% certainty that a coin is twice as likely to land heads than tails. If the coin is flipped a number of times and the outcomes observed, that degree of belief may rise, fall or remain the same depending on the results.

For proposition A and evidence B,

- $P(A)$, the *prior*, is the initial degree of belief in A.
- $P(A \mid B)$, the "posterior," is the degree of belief having accounted for B.
- the quotient $P(B \mid A)/P(B)$ represents the support B provides for A.

Frequentist Interpretation

In the frequentist interpretation, probability measures a "proportion of outcomes." For example, suppose an experiment is performed many times. $P(A)$ is the proportion of outcomes with property A, and $P(B)$ that with property B. $P(B \mid A)$ is the proportion of outcomes with property *B out of* outcomes with property *A*, and $P(A \mid B)$ the proportion of those with *A out of* those with *B*.

The role of Bayes' theorem is best visualized with tree diagrams. The two diagrams partition the same outcomes by *A* and *B* in opposite orders, to obtain the inverse probabilities. Bayes' theorem serves as the link between these different partitionings.

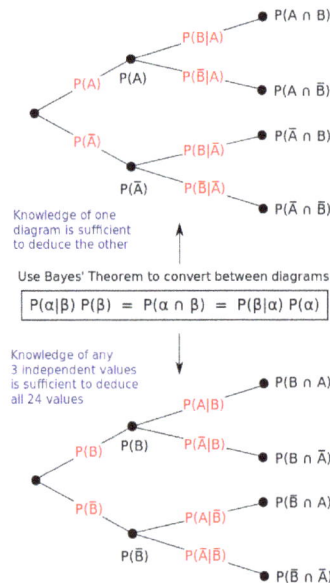

Illustration of frequentist interpretation with tree diagrams. Bayest' theoremm connects conditional probabilities to their inverses.

Example

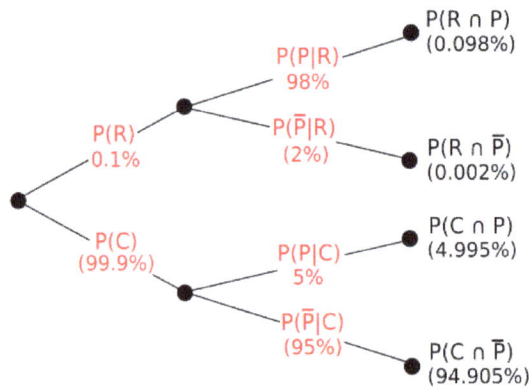

Tree diagram illustrating frequentist example. R, C, P and P bar are the events representing rare, common, pattern and no pattern. Percentages in parentheses are calculated. Note that three independent values are given, so it is possible to calculate the inverse tree.

An entomologist spots what might be a rare subspecies of beetle, due to the pattern on its back. In the rare subspecies, 98% have the pattern, or $P(\text{Pattern} \mid \text{Rare}) = 98\%$. In the common subspecies, 5% have the pattern. The rare subspecies accounts for only 0.1% of the population. How likely is the beetle having the pattern to be rare, or what is $P(\text{Rare} \mid \text{Pattern})$?

From the extended form of Bayes' theorem (since any beetle can be only rare or common),

$$P(Rare \mid Pattern) = \frac{P(Pattern \mid Rare)P(Rare)}{P(Pattern \mid Rare)P(Rare) + P(Pattern \mid Common)P(Common)}$$

$$= \frac{0.98 \times 0.001}{0.98 \times 0.001 + 0.05 \times 0.999}$$

$$\approx 1.9\%$$

Forms

Simple Form

For events A and B, provided that $P(B) \neq 0$,

$$P(A \mid B) = \frac{P(B \mid A)P(A)}{P(B)}.$$

In many applications, for instance in Bayesian inference, the event B is fixed in the discussion, and we wish to consider the impact of its having been observed on our belief in various possible events A. In such a situation the denominator of the last expression, the probability of the given evidence B, is fixed; what we want to vary is A. Bayes' theorem then shows that the posterior probabilities are proportional to the numerator:

$$P(A \mid B) \propto P(A) \cdot P(B \mid A) \text{ (proportionality over } A \text{ for given } B).$$

In words: posterior is proportional to prior times likelihood.

If events A_1, A_2, ..., are mutually exclusive and exhaustive, i.e., one of them is certain to occur but no two can occur together, and we know their probabilities up to proportionality, then we can determine the proportionality constant by using the fact that their probabilities must add up to one. For instance, for a given event A, the event A itself and its complement $\neg A$ are exclusive and exhaustive. Denoting the constant of proportionality by c we have

$$P(A \mid B) = c \cdot P(A) \cdot P(B \mid A) \ and \ P(\neg A \mid B) = c \cdot P(\neg A) \cdot P(B \mid \neg A).$$

Adding these two formulas we deduce that

$$1 = c \cdot (P(B \mid A) \cdot P(A) + P(B \mid \neg A) \cdot P(\neg A)),$$

or

$$c = \frac{1}{P(B \mid A) \cdot P(A) + P(B \mid \neg A) \cdot P(\neg A)} = \frac{1}{P(B)}.$$

Alternative Form

Another form of Bayes' Theorem that is generally encountered when looking at two competing statements or hypotheses is:

$$P(A \mid B) = \frac{P(B \mid A)P(A)}{P(B \mid A)P(A) + P(B \mid \neg A)P(\neg A)}.$$

For an epistemological interpretation:

For proposition A and evidence or background B,

- $P(A)$, the prior probability, is the initial degree of belief in A.

- $P(-A)$, is the corresponding probability of the initial degree of belief against A: $1 - P(A) = P(-A)$

- $P(B \mid A)$, the conditional probability or likelihood, is the degree of belief in B, given that the proposition A is true.

- $P(B \mid -A)$, the conditional probability or likelihood, is the degree of belief in B, given that the proposition A is false.

- $P(A \mid B)$, the posterior probability, is the probability for A after taking into account B for and against A.

Extended Form

Often, for some partition $\{A_j\}$ of the sample space, the event space is given or conceptualized in terms of $P(A_j)$ and $P(B \mid A_j)$. It is then useful to compute $P(B)$ using the law of total probability:

$$P(B) = \sum_j P(B \mid A_j)P(A_j),$$

$$\Rightarrow P(A_i \mid B) = \frac{P(B \mid A_i)P(A_i)}{\sum_j P(B \mid A_j)P(A_j)}.$$

In the special case where A is a binary variable:

$$(A \mid B) = \frac{P(B \mid A)}{P(B \mid A)P(A) + P(B \mid \neg A)P(\neg A)}.$$

Random Variables

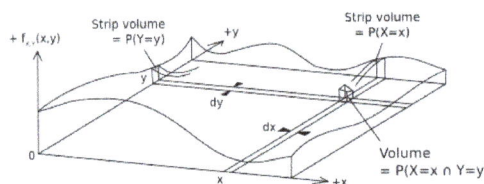

$$P(Y=y \mid X=x) = \frac{P(X=x \cap Y=y)}{P(X=x)} \qquad P(X=x \mid Y=y) = \frac{P(X=x \cap Y=y)}{P(Y=y)}$$

Diagram illustrating the meaning of Bayes' theorem as applied to an event space generated by continuous random variables X and Y. Note that there exists an instance of Bayes' theorem for each point in the domain. In practice, these instances might be parametrized by writing the specified probability densities as a function of x and y.

Consider a sample space Ω generated by two random variables X and Y. In principle, Bayes' theorem applies to the events $A = \{X = x\}$ and $B = \{Y = y\}$. However, terms become 0 at points where either variable has finite probability density. To remain useful, Bayes' theorem may be formulated in terms of the relevant densities.

Simple Form

If X is continuous and Y is discrete,

$$f_X(x \mid Y = y) = \frac{P(Y = y \mid X = x) f_X(x)}{P(Y = y)}.$$

If X is discrete and Y is continuous,

$$P(X = x \mid Y = y) = \frac{f_Y(y \mid X = x) P(X = x)}{f_Y(y)}.$$

If both X and Y are continuous,

$$f_X(x \mid Y = y) = \frac{f_Y(y \mid X = x) f_X(x)}{f_Y(y)}.$$

Extended Form

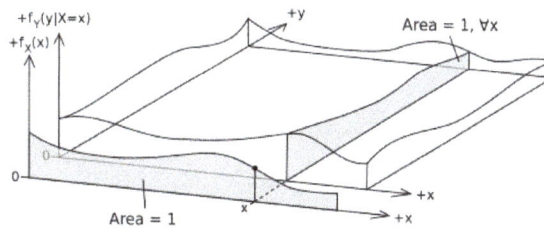

Diagram illustrating how an event space generated by continuous random variables X and Y is often conceptualized.

A continuous event space is often conceptualized in terms of the numerator terms. It is then useful to eliminate the denominator using the law of total probability. For $f_Y(y)$, this becomes an integral:

$$f_Y(y) = \int_{-\infty}^{\infty} f_Y(y \mid X = \xi) f_X(\xi) d\xi.$$

Bayes' Rule

Bayes' rule is Bayes' theorem in odds form.

$$O(A_1 : A_2 \mid B) = O(A_1 : A_2) \cdot \Lambda(A_1 : A_2 \mid B)$$

where

$$\Lambda(A_1 : A_2 \mid B) = \frac{P(B \mid A_1)}{P(B \mid A_2)}$$

is called the Bayes factor or likelihood ratio and the odds between two events is simply the ratio of the probabilities of the two events. Thus

$$O(A_1 : A_2) = \frac{P(A_1)}{P(A_2)},$$

$$O(A_1 : A_2 \mid B) = \frac{P(A_1 \mid B)}{P(A_2 \mid B)},$$

So the rule says that the posterior odds are the prior odds times the Bayes factor, or in other words, posterior is proportional to prior times likelihood.

Derivation

For Events

Bayes' theorem may be derived from the definition of conditional probability:

$$P(A \mid B) = \frac{P(A \cap B)}{P(B)}, \ if \ P(B) \neq 0,$$

$$P(B \mid A) = \frac{P(A \cap B)}{P(A)}, \ if \ P(A) \neq 0,$$

$$\Rightarrow P(A \cap B) = P(A \mid B)P(B) = P(B \mid A)P(A),$$

$$\Rightarrow P(A \mid B) = \frac{P(B \mid A)P(A)}{P(B)}, \ if \ P(B) \neq 0.$$

For Random Variables

For two continuous random variables X and Y, Bayes' theorem may be analogously derived from the definition of conditional density:

$$f_X(x \mid Y = y) = \frac{f_{X,Y}(x,y)}{f_Y(y)}$$

$$f_Y(y \mid X = x) = \frac{f_{X,Y}(x,y)}{f_X(x)}$$

$$\Rightarrow f_X(x \mid Y = y) = \frac{f_Y(y \mid X = x)f_X(x)}{f_Y(y)}.$$

History

Bayes' theorem was named after the Reverend Thomas Bayes (1701–1761), who studied how to compute a distribution for the probability parameter of a binomial distribution (in modern terminology). Bayes' unpublished manuscript was significantly edited by Richard Price before it was posthumously read at the Royal Society. Price edited Bayes' major work "An Essay towards solving a Problem in the Doctrine of Chances" (1763), which appeared in "Philosophical Transactions," and contains Bayes' Theorem. Price wrote an introduction to the paper which provides some of the philosophical basis of Bayesian statistics. In 1765 he was elected a Fellow of the Royal Society in recognition of his work on the legacy of Bayes.

The French mathematician Pierre-Simon Laplace reproduced and extended Bayes' results in 1774, apparently quite unaware of Bayes' work. Stephen Stigler suggested in 1983 that Bayes' theorem was discovered by Nicholas Saunderson, a blind English mathematician, some time before Bayes; that interpretation, however, has been disputed.

Martyn Hooper and Sharon McGrayne have argued that Richard Price's contribution was substantial:

By modern standards, we should refer to the Bayes–Price rule. Price discovered Bayes' work, recognized its importance, corrected it, contributed to the article, and found a use for it. The modern convention of employing Bayes' name alone is unfair but so entrenched that anything else makes little sense.

Rule of Succession

In probability theory, the rule of succession is a formula introduced in the 18th century by Pierre-Simon Laplace in the course of treating the sunrise problem.

The formula is still used, particularly to estimate underlying probabilities when there are few observations, or for events that have not been observed to occur at all in (finite) sample data. Assigning events a zero probability contravenes Cromwell's rule, which can never be strictly justified in physical situations, albeit sometimes must be assumed in practice.

Statement of the Rule of Succession

If we repeat an experiment that we know can result in a success or failure, n times independently, and get s successes, then what is the probability that the next repetition will succeed?

More abstractly: If X_1, ..., X_{n+1} are conditionally independent random variables that each can assume the value 0 or 1, then, if we know nothing more about them,

$$P(X_{n+1} = 1 \mid X_1 + \cdots + X_n = s) = \frac{s+1}{n+2}.$$

Interpretation

Since we have the prior knowledge that we are looking at an experiment for which both success and failure are possible, our estimate is as if we had observed one success and one failure for sure before we even started the experiments. In a sense we made $n + 2$ observations (known as pseudocounts) with $s+1$ successes. Beware: although this may seem the simplest and most reasonable assumption, which also happens to be true, it still requires a proof! Indeed, assuming a pseudocount of one per possibility is one way to generalise the binary result, but has unexpected consequences.

Nevertheless, if we had not known from the start that both success and failure are possible, then we would have had to assign

$$P'(X_{n+1} = 1 \mid X_1 + \cdots + X_n = s) = \frac{s}{n}.$$

But see Mathematical details below, for an analysis of its validity. In particular it is not valid when $s = 0$, or $s = n$.

If the number of observations increases, P and P' get more and more similar, which is intuitively clear: the more data we have, the less importance should be assigned to our prior information.

Historical Application to the Sunrise Problem

Laplace used the rule of succession to calculate the probability that the sun will rise tomorrow, given that it has risen every day for the past 5000 years. One obtains a very large factor of approximately 5000 × 365.25, which gives odds of 1826251:1 in favour of the sun rising tomorrow.

However, as the mathematical details below show, the basic assumption for using the rule of succession would be that we have no prior knowledge about the question whether the sun will or will not rise tomorrow, except that it can do either. This is not the case for sunrises.

Laplace knew this well, and he wrote to conclude the sunrise example: "But this number is far greater for him who, seeing in the totality of phenomena the principle regulating the days and seasons, realizes that nothing at the present moment can arrest the course of it." Yet Laplace was ridiculed for this calculation; his opponents gave no heed to that sentence, or failed to understand its importance.

In the 1940s, Rudolf Carnap investigated a probability-based theory of inductive reasoning, and developed measures of degree of confirmation, which he considered as alternatives to Laplace's rule of succession.

Mathematical Details

The proportion p is assigned a uniform distribution to describe the uncertainty about its true value. (This proportion is not random, but uncertain. We assign a probability distribution to p to express our uncertainty, not to attribute randomness to p. But this amounts, mathematically, to the same thing as treating p *as if* it were random).

Let X_i be 1 if we observe a "success" on the ith trial, otherwise 0, with probability p of success on each trial. Thus each X is 0 or 1; each X has a Bernoulli distribution. Suppose these Xs are conditionally independent given p.

Bayes' theorem says that to find the conditional probability distribution of p given the data X_i, $i = 1$, ..., n, one multiplies the "prior" (i.e., marginal) probability measure assigned to p by the likelihood function

$$L(p) = P(X_1 = x_1, \ldots, X_n = x_n \mid p) = \prod_{i=1}^{n} p^{x_i} (1-p)^{1-x_i} = p^s (1-p)^{n-s}$$

where $s = x_1 + \ldots + x_n$ is the number of "successes" and n is of course the number of trials, and then normalizes, to get the "posterior" (i.e., conditional on the data) probability distribution of p. (We are using capital X to denote a random variable and lower-case x either as the dummy in the definition of a function or as the data actually observed.)

The prior probability density function that expresses total ignorance of p except for the certain knowledge that it is neither 1 nor 0 (i.e., that we know that the experiment can in fact succeed or fail) is equal to 1 for $0 < p < 1$ and equal to 0 otherwise. To get the normalizing constant, we find

$$\int_0^1 p^s (1-p)^{n-s}\, dp = \frac{s!(n-s)!}{(n+1)!}$$

The posterior probability density function is therefore

$$f(p) = \frac{(n+1)!}{s!(n-s)!} p^s (1-p)^{n-s}.$$

This is a beta distribution with expected value

$$\int_0^1 p f(p)\, dp = \frac{s+1}{n+2}.$$

Since the conditional probability for success in the next experiment, given the value of p, is just p, the law of total probability tell us that the probability of success in the next experiment is just the expected value of p. Since all of this is conditional on the observed data X_i for $i = 1, \ldots, n$, we have

$$P(X_{n+1} = 1 \mid X_i = x_i \textit{ for } i = 1, \ldots, n) = \frac{s+1}{n+2}.$$

The same calculation can be performed with the prior that expresses total ignorance of p, including ignorance with regards to the question whether the experiment can succeed, or can fail. This prior, except for a normalizing constant, is $1/(p(1-p))$ for $0 \le p \le 1$ and 0 otherwise. If the calculation above is repeated with this prior, we get

$$P'(X_{n+1} = 1 \mid X_i = x_i \textit{ for } i = 1, \ldots, n) = \frac{s}{n}.$$

Thus, with the prior specifying total ignorance, the probability of success is governed by the observed frequency of success. However, the posterior distribution that led to this result is the Beta(s,$n - s$) distribution, which is not proper when $s = n$ or $s = 0$ (i.e. the normalisation constant is infinite when $s = 0$ or $s = n$). This means that we cannot use this form of the posterior distribution to calculate the probability of the next observation succeeding when $s = 0$ or $s = n$. This puts the information contained in the rule of succession in greater light: it can be thought of as expressing the prior assumption that if sampling was continued indefinitely, we would eventually observe at least one success, and at least one failure in the sample. The prior expressing total ignorance does not assume this knowledge.

To evaluate the "complete ignorance" case when $s = 0$ or $s = n$ can be dealt with by first going back to the hypergeometric distribution, denoted by $\mathrm{Hyp}(s \mid N, n, S)$. This is the approach taken in Jaynes(2003). The binomial $\mathrm{Bin}(r \mid n, p)$ can be derived as a limiting form, where $N, S \to \infty$ in such a way that their ratio $p = \frac{S}{N}$ remains fixed. One can think of S as the number of successes in the total population, of size N

The equivalent prior to $\frac{1}{p(1-p)}$ is $\frac{1}{S(N-S)}$, with a domain of $1 \le S \le N - 1$. Working conditional to N means that estimating p is equivalent to estimating S, and then dividing this estimate by N. The posterior for S can be given as:

$$P(S \mid N, n, s) \propto \frac{1}{S(N-S)} \binom{S}{s}\binom{N-S}{n-s} \propto \frac{S!(N-S)!}{S(N-S)(S-s)!(N-S-[n-s])!}$$

And it can be seen that, if $s = n$ or $s = 0$, then one of the factorials in the numerator cancels exactly with one in the denominator. Taking the $s = 0$ case, we have:

$$P(S \mid N, n, s = 0) \propto \frac{(N-S-1)!}{S(N-S-n)!} = \frac{\prod_{j=1}^{n-1}(N-S-j)}{S}$$

Adding in the normalising constant, which is always finite (because there is no singularities in the range of the posterior, and there are a finite number of terms) gives:

$$P(S \mid N, n, s = 0) = \frac{\prod_{j=1}^{n-1}(N-S-j)}{S \sum_{R=1}^{N-n} \frac{\prod_{j=1}^{n-1}(N-R-j)}{R}}$$

So the posterior expectation for $p = \frac{S}{N}$ is:

$$E\left(\frac{S}{N} \mid n, s = 0, N\right) = \frac{1}{N}\sum_{S=1}^{N-n} S P(S \mid N, n = 1, s = 0) = \frac{1}{N} \frac{\sum_{S=1}^{N-n}\prod_{j=1}^{n-1}(N-S-j)}{\sum_{R=1}^{N-n}\frac{\prod_{j=1}^{n-1}(N-R-j)}{R}}$$

An approximate analytical expression for large N is given by first making the approximation to the product term:

$$\prod_{j=1}^{n-1}(N-R-j) \approx (N-R)^{n-1}$$

and then replacing the summation in the numerator with an integral

$$\sum_{S=1}^{N-n}\prod_{j=1}^{n-1}(N-S-j) \approx \int_1^{N-n}(N-S)^{n-1}dS = \frac{(N-1)^n - n^n}{n} \approx \frac{N^n}{n}$$

The same procedure is followed for the denominator, but the process is a bit more tricky, as the integral is harder to evaluate

$$\sum_{R=1}^{N-n}\frac{\prod_{j=1}^{n-1}(N-R-j)}{R} \approx \int_1^{N-n}\frac{(N-R)^{n-1}}{R}dR$$

$$= N\int_1^{N-n}\frac{(N-R)^{n-2}}{R}dR - \int_1^{N-n}(N-R)^{n-2}dR$$

$$= N^{n-1}\left[\int_1^{N-n}\frac{dR}{R} - \frac{1}{n-1} + O\left(\frac{1}{N}\right)\right] \approx N^{n-1}\ln(N)$$

where ln is the natural logarithm plugging in these approximations into the expectation gives

$$E\left(\frac{S}{N} \mid n, s = 0, N\right) \approx \frac{1}{N}\frac{\frac{N^n}{n}}{N^{n-1}\ln(N)} = \frac{1}{n[\ln(N)]} = \frac{\log_{10}(e)}{n[\log_{10}(N)]} = \frac{0.434294}{n[\log_{10}(N)]}$$

where the base 10 logarithm has been used in the final answer for ease of calculation. For instance if the population is of size 10^k then probability of success on the next sample is given by:

$$E\left(\frac{S}{N} \mid n, s = 0, N = 10^k\right) \approx \frac{0.434294}{nk}$$

So for example, if the population be on the order of tens of billions, so that $k = 10$, and we observe $n = 10$ results without success, then the expected proportion in the population is approximately 0.43%. If the population is smaller, so that $n = 10, k = 5$ (tens of thousands), the expected proportion rises to approximately 0.86%, and so on. Similarly, if the number of observations is smaller, so that $n = 5, k = 10$, the proportion rise to approximately 0.86% again.

This probability has no lower bound, and can be made arbitrarily small for larger and larger choices of N, or k. This means that the probability depends on the size of the population from which one is sampling. In passing to the limit of infinite N (for the simpler analytic properties) we are "throwing away" a piece of very important information. Note that this ignorance relationship only holds as long as only no successes are observed. It is correspondingly revised back to the observed frequency rule $p = \frac{s}{n}$ as soon as one success is observed. The corresponding results are found for the $s=n$ case by switching labels, and then subtracting the probability from 1.

Generalization to any Number of Possibilities

This section gives a heuristic derivation to that given in *Probability Theory: The Logic of Science.*

The rule of succession has many different intuitive interpretations, and depending on which intuition one uses, the generalisation may be different. Thus, the way to proceed from here is very carefully, and to re-derive the results from first principles, rather than to introduce an intuitively sensible generalisation. The full derivation can be found in Jaynes' book, but it does admit an easier to understand alternative derivation, once the solution is known. Another point to emphasise is that the prior state of knowledge described by the rule of succession is given as an enumeration of the possibilities, with the additional information that it is possible to observe each category. This can be equivalently stated as observing each category once prior to gathering the data. To denote that this is the knowledge used, an I_m is put as part of the conditions in the probability assignments.

The rule of succession comes from setting a binomial likelihood, and a uniform prior distribution. Thus a straightforward generalisation is just the multivariate extensions of these two distributions: 1)Setting a uniform prior over the initial m categories, and 2) using the multinomial distribution as the likelihood function (which is the multivariate generalisation of the binomial distribution). It can be shown that the uniform distribution is a special case of the Dirichlet distribution with all of its parameters equal to 1 (just as the uniform is Beta(1,1) in the binary case). The Dirichlet distribution is the conjugate prior for the multinomial distribution, which means that the posterior distribution is also a Dirichlet distribution with different parameters. Let p_i denote the probability that category i will be observed, and let n_i denote the number of times category i ($i = 1, ..., m$) actually was observed. Then the joint posterior distribution of the probabilities $p_1, ..., p_m$ is given by;

$$f(p_1,...,p_m \mid n_1,...,n_m,I) = \begin{cases} \dfrac{\Gamma\left(\sum\limits_{i=1}^{m}(n_i+1)\right)}{\prod\limits_{i=1}^{m}\Gamma(n_i+1)} p_1^{n_1}\cdots p_m^{n_m}, & \sum\limits_{i=1}^{m}p_i=1 \\[2em] 0 & otherwise. \end{cases}$$

To get the generalised rule of succession, note that the probability of observing category i on the next observation, conditional on the p_i is just p_i, we simply require its expectation. Letting A_i denote the event that the next observation is in category i ($i = 1, ..., m$), and let $n = n_1 + ... + n_m$ be the total number of observations made. The result, using the properties of the dirichlet distribution is:

$$P(A_i \mid n_1,...,n_m,I_m) = \frac{n_i+1}{n+m}.$$

This solution reduces to the probability that would be assigned using the principle of indifference before any observations made (i.e. $n = 0$), consistent with the original rule of succession. It also contains the rule of succession as a special case, when $m = 2$, as a generalisation should.

Because the propositions or events A_i are mutually exclusive, it is possible to collapse the m categories into 2. Simply add up the A_i probabilities that correspond to "success" to get the probability of success. Supposing that this aggregates c categories as "success" and $m\text{-}c$ categories as "failure". Let s denote the sum of the relevant n_i values that have been termed "success". The probability of "success" at the next trial is then:

$$P(\text{success} \mid n_1,\ldots,n_m,I_m) = \frac{s+c}{n+m},$$

which is different from the original rule of succession. But note that the original rule of succession is based on I_2, whereas the generalisation is based on I_m. This means that the information contained in I_m is different from that contained in I_2. This indicates that mere knowledge of more than two outcomes we know are possible is relevant information when collapsing these categories down to just two. This illustrates the subtlety in describing the prior information, and why it is important to specify which prior information one is using.

Further Analysis

A good model is essential (i.e., a good compromise between accuracy and practicality). To paraphrase Laplace on the sunrise problem: Although we have a huge number of samples of the sun rising, there are far better models of the sun than assuming it has a certain probability of rising each day, e.g., simply having a half-life.

Given a good model, it is best to make as many observations as practicable, depending of the expected reliability of prior knowledge, cost of observations, time and resources available, and accuracy required.

One of the most difficult aspects of the rule of succession is not the mathematical formulas, but answering the question: When does the rule of succession apply? In the generalisation section, it was noted very explicitly by adding the prior information I_m into the calculations. Thus, when all that is known about a phenomenon is that there are m known possible outcomes prior to observing any data, only then does the rule of succession apply. If the rule of succession is applied in problems where this does not accurately describe the prior state of knowledge, then it may give counter-intuitive results. This is not because the rule of succession is defective, but that it is effectively answering a different question, based on different prior information.

In principle, no possibility should have its probability (or its pseudocount) set to zero, since nothing in the physical world should be assumed strictly impossible (though it may be)—even if contrary to all observations and current theories. Indeed, Bayes rule takes *absolutely* no account of an observation previously believed to have zero probability—it is still declared impossible. However, only considering the a fixed set of the possibilities is an acceptable route, one just needs to remember that the results are conditional on (or restricted to) the set being considered, and not some "universal" set. In fact Larry Bretthorst shows that including the possibility of "something else" into the hypothesis space makes no difference to the relative probabilities of the other hypothesis - it simply renormalises them to add up to a value less than 1. Until "something else" is specified, the likelihood function conditional on this "something else" is indeterminate, for how is one to determine $Pr(\text{data} \mid \text{something else}, I)$?. Thus no updating of the prior probability for "something else" can occur until it is more accurately defined.

However, it is sometimes debatable whether prior knowledge should affect the relative probabilities, or also the total weight of the prior knowledge compared to actual observations. This does not have a clear cut answer, for it depends on what prior knowledge one is considering. In fact, an alternative prior state of knowledge could be of the form "I have specified m potential categories, but I am sure that only one of them is possible prior to observing the data. However, I do not know which particular category this is." A mathematical way to describe this prior is the dirichlet distribution with all parameters equal to m^{-1}, which then gives a pseudocount of 1 to the denominator instead of m, and adds a pseudocount of m^{-1} to each category. This gives a slightly different probability in the binary case of $\frac{s+0.5}{n+1}$.

Prior probabilities are only worth spending significant effort estimating when likely to have significant effect. They may be important when there are few observations — especially when so few that there have been few, if any, observations of some possibilities – such as a rare animal, in a given region. Also important when there are many observations, where it is believed that the expectation should be heavily weighted towards the prior estimates, in spite of many observations to the contrary, such as for a roulette wheel in a well-respected casino. In the latter case, at least some of the pseudocounts may need to be very large. They are not always small, and thereby soon outweighed by actual observations, as is often assumed. However, although a last resort, for everyday purposes, prior knowledge is usually vital. So most decisions must be subjective to some extent (dependent upon the analyst and analysis used).

Conditional Independence

In probability theory, two events R and B are conditionally independent given a third event Y precisely if the occurrence or non-occurrence of R *and* the occurrence or non-occurrence of B are independent events in their conditional probability distribution given Y. In other words, R and B are conditionally independent given Y if and only if, given knowledge that Y occurs, knowledge of whether R occurs provides no information on the likelihood of B occurring, and knowledge of whether B occurs provides no information on the likelihood of R occurring.

Formal Definition

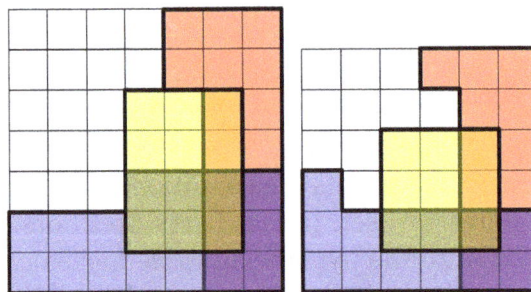

These are two examples illustrating conditional independence. Each cell represents a possible outcome. The events R, B and Y are represented by the areas shaded red, blue and yellow respectively. The overlap between the events R and B is shaded purple. The probabilities of these events are shaded areas with respect to the total area. In both examples R and B are conditionally independent given Y because: $\Pr(R \cap B \mid Y) = \Pr(R \mid Y)\Pr(B \mid Y)$ but not conditionally independent given not Y because: $\Pr(R \cap B \mid not\ Y) \neq \Pr(R \mid not\ Y)\Pr(B \mid not\ Y)$.

In the standard notation of probability theory, R and B are conditionally independent given Y if and only if

$$\Pr(R \cap B \mid Y) = \Pr(R \mid Y) \Pr(B \mid Y),$$

or equivalently,

$$\Pr(R \mid B \cap Y) = \Pr(R \mid Y).$$

Two random variables X and Y are conditionally independent given a third random variable Z if and only if they are independent in their conditional probability distribution given Z. That is, X and Y are conditionally independent given Z if and only if, given any value of Z, the probability distribution of X is the same for all values of Y and the probability distribution of Y is the same for all values of X.

Two events R and B are conditionally independent given a σ-algebra Σ if

$$\Pr(R \cap B \mid \Sigma) = \Pr(R \mid \Sigma) \Pr(B \mid \Sigma) \text{ a.s.}$$

where $\Pr(A \mid \Sigma)$ denotes the conditional expectation of the indicator function of the event A, χ_A, given the sigma algebra Σ. That is,

$$\Pr(A \mid \Sigma) := E[\chi_A \mid \Sigma].$$

Two random variables X and Y are conditionally independent given a σ-algebra Σ if the above equation holds for all R in σ(X) and B in σ(Y).

Two random variables X and Y are conditionally independent given a random variable W if they are independent given σ(W): the σ-algebra generated by W. This is commonly written:

$$X \perp\!\!\!\perp Y \mid W \text{ or}$$

$$X \perp Y \mid W$$

This is read "X is independent of Y, given W"; the conditioning applies to the whole statement: "(X is independent of Y) given W".

$$(X \perp\!\!\!\perp Y) \mid W$$

If W assumes a countable set of values, this is equivalent to the conditional independence of X and Y for the events of the form $[W = w]$. Conditional independence of more than two events, or of more than two random variables, is defined analogously.

The following two examples show that $X \perp Y$ *neither implies nor is implied by* $X \perp Y \mid W$. First, suppose W is 0 with probability 0.5 and 1 otherwise. When $W = 0$ take X and Y to be independent, each having the value 0 with probability 0.99 and the value 1 otherwise. When $W = 1$, X and Y are again independent, but this time they take the value 1 with probability 0.99. Then $X \perp Y \mid W$. But X and Y are dependent, because $\Pr(X = 0) < \Pr(X = 0 \mid Y = 0)$. This is because $\Pr(X = 0) = 0.5$, but if $Y = 0$ then it›s very likely that $W = 0$ and thus that $X = 0$ as well, so $\Pr(X = 0 \mid Y = 0) > 0.5$. For the second example, suppose $X \perp Y$, each taking the values 0 and 1 with probability 0.5. Let W be the product $X \times Y$. Then when $W = 0$, $\Pr(X = 0) = 2/3$, but $\Pr(X = 0 | Y = 0) = 1/2$, so $X \perp Y \mid W$ is false. This is also an example of Explaining Away.

An Example

The discussion on StackExchange provides a couple of useful examples.

1. Let the two events be the probabilities of persons A and B getting home in time for dinner, and the third event is the fact that a snow storm hit the city. While both A and B have a lower probability of getting home in time for dinner, the lower probabilities will still be independent of each other. That is, the knowledge that A is late does not tell you whether B will be late. (They may be living in different neighborhoods, traveling different distances, and using different modes of transportation.) However, if you have information that they live in the same neighborhood, use the same transportation, and work at the same place, then the two events are NOT conditionally independent.

2. Conditional independence depends on the nature of the third event. If you roll two dice, one may assume that the two dice behave independently of each other. Looking at the results of 1 die will not tell you about the result of the second die. (That is, the two dice are independent.) If, however, the 1st die's result is a 3, and someone tells you about a third event - that the sum of the two results is even - then this extra unit of information restricts the options for the 2nd result to an odd number. In other words, two events can be independent, but NOT conditionally independent.

3. Height and vocabulary are not independent; but they are conditionally independent if you add age.

Uses in Bayesian Inference

Let p be the proportion of voters who will vote "yes" in an upcoming referendum. In taking an opinion poll, one chooses n voters randomly from the population. For $i = 1, ..., n$, let X_i = 1 or 0 corresponding, respectively, to whether or not the ith chosen voter will or will not vote "yes".

In a frequentist approach to statistical inference one would not attribute any probability distribution to p (unless the probabilities could be somehow interpreted as relative frequencies of occurrence of some event or as proportions of some population) and one would say that $X_1, ..., X_n$ are independent random variables.

By contrast, in a Bayesian approach to statistical inference, one would assign a probability distribution to p regardless of the non-existence of any such "frequency" interpretation, and one would construe the probabilities as degrees of belief that p is in any interval to which a probability is assigned. In that model, the random variables $X_1, ..., X_n$ are *not* independent, but they are conditionally independent given the value of p. In particular, if a large number of the Xs are observed to be equal to 1, that would imply a high conditional probability, given that observation, that p is near 1, and thus a high conditional probability, given that observation, that the *next X* to be observed will be equal to 1.

Rules of Conditional Independence

A set of rules governing statements of conditional independence have been derived from the basic definition.

Note: since these implications hold for any probability space, they will still hold if one considers a sub-universe by conditioning everything on another variable, say K. For example, $X \perp\!\!\!\perp Y \Rightarrow Y \perp\!\!\!\perp X$ would also mean that $X \perp\!\!\!\perp Y \mid K \Rightarrow Y \perp\!\!\!\perp X \mid K$.

Note: below, the comma can be read as an "AND".

Symmetry

$$X \perp\!\!\!\perp Y \quad \Rightarrow \quad Y \perp\!\!\!\perp X$$

Decomposition

$$X \perp\!\!\!\perp A, B \quad \Rightarrow \quad \text{and} \begin{cases} X \perp\!\!\!\perp A \\ X \perp\!\!\!\perp B \end{cases}$$

Proof:

- $p_{X,A,B}(x,a,b) = p_X(x) p_{A,B}(a,b)$ (meaning of $X \perp\!\!\!\perp A, B$)

- $\int_B p_{X,A,B}(x,a,b) = \int_B p_X(x) p_{A,B}(a,b)$ (ignore variable B by integrating it out)

- $p_{X,A}(x,a) = p_X(x) p_A(a)$

A similar proof shows the independence of X and B.

Weak Union

$$X \perp\!\!\!\perp A, B \quad \Rightarrow \quad and \begin{cases} X \perp\!\!\!\perp A \mid B \\ X \perp\!\!\!\perp B \mid A \end{cases}$$

Proof:

- By definition, $\Pr(X) = \Pr(X \mid A, B)$.

- Due to the property of decomposition $X \perp\!\!\!\perp B$, $\Pr(X) = \Pr(X \mid B)$..

- Combining the above two equalities gives $\Pr(X \mid B) = \Pr(X \mid A, B)$ which establishes $X \perp\!\!\!\perp A \mid B$..

The second condition can be proved similarly.

Contraction

Proof:

$$\left. \begin{matrix} X \perp \\ X \perp \end{matrix} \right\} and \quad \Rightarrow X \perp\!\!\!\perp A, B$$

This property can be proved by noticing $\Pr(X \mid A, B) = \Pr(X \mid B) = \Pr(X)$, each equality of which is asserted by $X \perp\!\!\!\perp A \mid B$ and $X \perp\!\!\!\perp B$, respectively.

Contraction-weak-union-decomposition

Putting the above three together, we have:

$$
\left.\begin{array}{l} X \perp\!\!\!\perp A \mid B \\ X \perp\!\!\!\perp B \end{array}\right\} \; and \quad \Leftrightarrow \quad X \perp\!\!\!\perp A,B \quad \Rightarrow \quad and \left\{\begin{array}{l} X \perp\!\!\!\perp A \mid B \\ X \perp\!\!\!\perp B \\ X \perp\!\!\!\perp B \mid A \\ X \perp\!\!\!\perp A \end{array}\right.
$$

Intersection

For strictly positive probability distributions, the following also holds:

$$
\left.\begin{array}{l} X \perp\!\!\!\perp A \mid C,B \\ X \perp\!\!\!\perp B \mid C,A \end{array}\right\} \; and \quad \Rightarrow \quad X \perp\!\!\!\perp B \mid A,C
$$

The five rules above were termed "Graphoid Axioms" by Pearl and Paz, because they hold in graphs, if $X \perp\!\!\!\perp A \mid B$ is interpreted to mean: "All paths from X to A are intercepted by the set B".

References

- Ambrosio, L., Gigli, N. & Savaré, G. (2005). Gradient Flows in Metric Spaces and in the Space of Probability Measures. ETH Zürich, Birkhäuser Verlag, Basel. ISBN 3-7643-2428-7.

- Paulos, J.A. (1988) Innumeracy: Mathematical Illiteracy and its Consequences, Hill and Wang. ISBN 0-8090-7447-8 (p. 63 et seq.)

- Grimmett, Geoffrey; Stirzaker, David (2001). Probability and Random Processes (3rd ed.). Oxford University Press. ISBN 0-19-857222-0., pages 67–69

- Richard Allen (1999). David Hartley on Human Nature. SUNY Press. pp. 243–4. ISBN 978-0-7914-9451-6. Retrieved 16 June 2013.

- Richard Price (1991). Price: Political Writings. Cambridge University Press. p. xxiii. ISBN 978-0-521-40969-8. Retrieved 16 June 2013.

- De Vaux, Richard; Velleman, Paul; Bock, David (2016). Stats, Data and Models (4 ed.). Pearson. pp. 380–381. ISBN 978-0-321-98649-8.

- Hooper, Martyn (2013). "Richard Price, Bayes' theorem, and God". Significance. 10 (1): 36–39. doi:10.1111/j.1740-9713.2013.00638.x.

- Daniel Kahneman (25 October 2011). Thinking, Fast and Slow. Macmillan. ISBN 978-1-4299-6935-2. Retrieved 8 April 2012.

Interpretation of Probability

Probability as a word has numerous ways of being used; it has two broad categories that can be called physical probability and evidential probability. Frequentist probability, probabilistic logic, propensity probability, Bayesian probability etc. are the interpretations of probability that has been discussed within this chapter.

Probability Interpretations

The word probability has been used in a variety of ways since it was first applied to the mathematical study of games of chance. Does probability measure the real, physical tendency of something to occur or is it a measure of how strongly one believes it will occur, or does it draw on both these elements? In answering such questions, mathematicians interpret the probability values of probability theory.

There are two broad categories of probability interpretations which can be called "physical" and "evidential" probabilities. Physical probabilities, which are also called objective or frequency probabilities, are associated with random physical systems such as roulette wheels, rolling dice and radioactive atoms. In such systems, a given type of event (such as a die yielding a six) tends to occur at a persistent rate, or "relative frequency", in a long run of trials. Physical probabilities either explain, or are invoked to explain, these stable frequencies. The two main kinds of theory of physical probability are frequentist accounts (such as those of Venn, Reichenbach and von Mises) and propensity accounts (such as those of Popper, Miller, Giere and Fetzer).

Evidential probability, also called Bayesian probability, can be assigned to any statement whatsoever, even when no random process is involved, as a way to represent its subjective plausibility, or the degree to which the statement is supported by the available evidence. On most accounts, evidential probabilities are considered to be degrees of belief, defined in terms of dispositions to gamble at certain odds. The four main evidential interpretations are the classical (e.g. Laplace's) interpretation, the subjective interpretation (de Finetti and Savage), the epistemic or inductive interpretation (Ramsey, Cox) and the logical interpretation (Keynes and Carnap). There are also evidential interpretations of probability covering groups, which are often labelled as 'intersubjective' (proposed by Gillies and Rowbottom).

Some interpretations of probability are associated with approaches to statistical inference, including theories of estimation and hypothesis testing. The physical interpretation, for example, is taken by followers of "frequentist" statistical methods, such as Ronald Fisher, Jerzy Neyman and Egon Pearson. Statisticians of the opposing Bayesian school typically accept the existence and importance of physical probabilities, but also consider the calculation of evidential probabilities to be both valid and necessary in statistics. This article, however, focuses on the interpretations of probability rather than theories of statistical inference.

The terminology of this topic is rather confusing, in part because probabilities are studied within a variety of academic fields. The word "frequentist" is especially tricky. To philosophers it refers to a particular theory of physical probability, one that has more or less been abandoned. To scientists, on the other hand, "frequentist probability" is just another name for physical (or objective) probability. Those who promote Bayesian inference view "frequentist statistics" as an approach to statistical inference that recognises only physical probabilities. Also the word "objective", as applied to probability, sometimes means exactly what "physical" means here, but is also used of evidential probabilities that are fixed by rational constraints, such as logical and epistemic probabilities.

It is unanimously agreed that statistics depends somehow on probability. But, as to what probability is and how it is connected with statistics, there has seldom been such complete disagreement and breakdown of communication since the Tower of Babel. Doubtless, much of the disagreement is merely terminological and would disappear under sufficiently sharp analysis.

— (Savage, 1954, p 2)

Philosophy

The philosophy of probability presents problems chiefly in matters of epistemology and the uneasy interface between mathematical concepts and ordinary language as it is used by non-mathematicians. Probability theory is an established field of study in mathematics. It has its origins in correspondence discussing the mathematics of games of chance between Blaise Pascal and Pierre de Fermat in the seventeenth century, and was formalized and rendered axiomatic as a distinct branch of mathematics by Andrey Kolmogorov in the twentieth century. In axiomatic form, mathematical statements about probability theory carry the same sort of epistemological confidence within the philosophy of mathematics as are shared by other mathematical statements.

The mathematical analysis originated in observations of the behaviour of game equipment such as playing cards and dice, which are designed specifically to introduce random and equalized elements; in mathematical terms, they are subjects of indifference. This is not the only way probabilistic statements are used in ordinary human language: when people say that *"it will probably rain"*, they typically do not mean that the outcome of rain versus not-rain is a random factor that the odds currently favor; instead, such statements are perhaps better understood as qualifying their expectation of rain with a degree of confidence. Likewise, when it is written that "the most probable explanation" of the name of Ludlow, Massachusetts "is that it was named after Roger Ludlow", what is meant here is not that Roger Ludlow is favored by a random factor, but rather that this is the most plausible explanation of the evidence, which admits other, less likely explanations.

Thomas Bayes attempted to provide a logic that could handle varying degrees of confidence; as such, Bayesian probability is an attempt to recast the representation of probabilistic statements as an expression of the degree of confidence by which the beliefs they express are held.

Though probability initially had somewhat mundane motivations, its modern influence and use is widespread ranging from evidence-based medicine, through Six sigma, all the way to the Probabilistically checkable proof and the String theory landscape.

A summary of some interpretations of probability				
	Classical	**Frequentist**	**Subjective**	**Propensity**
Main hypothesis	Principle of indifference	Frequency of occurrence	Degree of belief	Degree of causal connection
Conceptual basis	Hypothetical symmetry	Past data and reference class	Knowledge and intuition	Present state of system
Conceptual approach	Conjectural	Empirical	Subjective	Metaphysical
Single case possible	Yes	No	Yes	Yes
Precise	Yes	No	No	Yes
Problems	Ambiguity in principle of indifference	Circular definition	Reference class problem	Disputed concept

Classical Definition

The first attempt at mathematical rigour in the field of probability, championed by Pierre-Simon Laplace, is now known as the classical definition. Developed from studies of games of chance (such as rolling dice) it states that probability is shared equally between all the possible outcomes, provided these outcomes can be deemed equally likely. (3.1)

The theory of chance consists in reducing all the events of the same kind to a certain number of cases equally possible, that is to say, to such as we may be equally undecided about in regard to their existence, and in determining the number of cases favorable to the event whose probability is sought. The ratio of this number to that of all the cases possible is the measure of this probability, which is thus simply a fraction whose numerator is the number of favorable cases and whose denominator is the number of all the cases possible.

— Pierre-Simon Laplace, A Philosophical Essay on Probabilities

The classical definition of probability works well for situations with only a finite number of equally-likely outcomes.

This can be represented mathematically as follows: If a random experiment can result in N mutually exclusive and equally likely outcomes and if N_A of these outcomes result in the occurrence of the event A, the probability of A is defined by

$$P(A) = \frac{N_A}{N}.$$

There are two clear limitations to the classical definition. Firstly, it is applicable only to situations in which there is only a 'finite' number of possible outcomes. But some important random experiments, such as tossing a coin until it rises heads, give rise to an infinite set of outcomes. And secondly, you need to determine in advance that all the possible outcomes are equally likely without relying on the notion of probability to avoid circularity—for instance, by symmetry considerations.

Frequentism

For frequentists, the probability of the ball landing in any pocket can be determined only by repeated trials in which the observed result converges to the underlying probability in the long run.

Frequentists posit that the probability of an event is its relative frequency over time, (3.4) i.e., its relative frequency of occurrence after repeating a process a large number of times under similar conditions. This is also known as aleatory probability. The events are assumed to be governed by some random physical phenomena, which are either phenomena that are predictable, in principle, with sufficient information; or phenomena which are essentially unpredictable. Examples of the first kind include tossing dice or spinning a roulette wheel; an example of the second kind is radioactive decay. In the case of tossing a fair coin, frequentists say that the probability of getting a heads is 1/2, not because there are two equally likely outcomes but because repeated series of large numbers of trials demonstrate that the empirical frequency converges to the limit 1/2 as the number of trials goes to infinity.

If we denote by n_a the number of occurrences of an event $\mathcal{A}$ in n trials, then if $\lim_{n \to +\infty} \frac{n_a}{n} = p$ we say that $P(\mathcal{A}) = p$.

The frequentist view has its own problems. It is of course impossible to actually perform an infinity of repetitions of a random experiment to determine the probability of an event. But if only a finite number of repetitions of the process are performed, different relative frequencies will appear in different series of trials. If these relative frequencies are to define the probability, the probability will be slightly different every time it is measured. But the real probability should be the same every time. If we acknowledge the fact that we only can measure a probability with some error of measurement attached, we still get into problems as the error of measurement can only be expressed as a probability, the very concept we are trying to define. This renders even the frequency definition circular; see for example "What is the Chance of an Earthquake?"

Logical, Epistemic, and Inductive Probability

It is widely recognized that the term "probability" is sometimes used in contexts where it has nothing to do with physical randomness. Consider, for example, the claim that the extinction of the dinosaurs was probably caused by a large meteorite hitting the earth. Statements such as "Hypothesis H is probably true" have been interpreted to mean that the (presently available) empirical evidence (E, say) supports H to a high degree. This degree of support of H by E has been called the logical probability of H given E, or the epistemic probability of H given E, or the inductive probability of H given E.

The differences between these interpretations are rather small, and may seem inconsequential. One of the main points of disagreement lies in the relation between probability and belief. Logical probabilities are conceived (for example in Keynes' Treatise on Probability) to be objective, logical relations between propositions (or sentences), and hence not to depend in any way upon belief. They are degrees of (partial) entailment, or degrees of logical consequence, not degrees of belief. (They do, nevertheless, dictate proper degrees of belief, as is discussed below.) Frank P. Ramsey, on the other hand, was skeptical about the existence of such objective logical relations and argued that (evidential) probability is "the logic of partial belief". (p 157) In other words, Ramsey held that epistemic probabilities simply *are* degrees of rational belief, rather than being logical relations that merely *constrain* degrees of rational belief.

Another point of disagreement concerns the *uniqueness* of evidential probability, relative to a given state of knowledge. Rudolf Carnap held, for example, that logical principles always determine a unique logical probability for any statement, relative to any body of evidence. Ramsey, by contrast, thought that while degrees of belief are subject to some rational constraints (such as, but not limited to, the axioms of probability) these constraints usually do not determine a unique value. Rational people, in other words, may differ somewhat in their degrees of belief, even if they all have the same information.

Propensity

Propensity theorists think of probability as a physical propensity, or disposition, or tendency of a given type of physical situation to yield an outcome of a certain kind or to yield a long run relative frequency of such an outcome. This kind of objective probability is sometimes called 'chance'.

Propensities, or chances, are not relative frequencies, but purported causes of the observed stable relative frequencies. Propensities are invoked to explain why repeating a certain kind of experiment will generate given outcome types at persistent rates, which are known as propensities or chances. Frequentists are unable to take this approach, since relative frequencies do not exist for single tosses of a coin, but only for large ensembles or collectives. In contrast, a propensitist is able to use the law of large numbers to explain the behaviour of long-run frequencies. This law, which is a consequence of the axioms of probability, says that if (for example) a coin is tossed repeatedly many times, in such a way that its probability of landing heads is the same on each toss, and the outcomes are probabilistically independent, then the relative frequency of heads will be close to the probability of heads on each single toss. This law allows that stable long-run frequencies are a manifestation of invariant *single-case* probabilities. In addition to explaining the emergence of stable relative frequencies, the idea of propensity is motivated by the desire to make sense of single-case probability attributions in quantum mechanics, such as the probability of decay of a particular atom at a particular time.

The main challenge facing propensity theories is to say exactly what propensity means. (And then, of course, to show that propensity thus defined has the required properties.) At present, unfortunately, none of the well-recognised accounts of propensity comes close to meeting this challenge.

A propensity theory of probability was given by Charles Sanders Peirce. A later propensity theory was proposed by philosopher Karl Popper, who had only slight acquaintance with the writings of C. S. Peirce, however. Popper noted that the outcome of a physical experiment is produced by a certain set of "generating conditions". When we repeat an experiment, as the saying goes, we really perform another experiment with a (more or less) similar set of generating conditions. To say that a set of generating conditions has propensity p of producing the outcome E means that those exact conditions, if repeated indefinitely, would produce an outcome sequence in which E occurred with limiting relative frequency p. For Popper then, a deterministic experiment would have propensity 0 or 1 for each outcome, since those generating conditions would have same outcome on each trial. In other words, non-trivial propensities (those that differ from 0 and 1) only exist for genuinely indeterministic experiments.

A number of other philosophers, including David Miller and Donald A. Gillies, have proposed propensity theories somewhat similar to Popper's.

Other propensity theorists (e.g. Ronald Giere) do not explicitly define propensities at all, but rather see propensity as defined by the theoretical role it plays in science. They argue, for example, that physical magnitudes such as electrical charge cannot be explicitly defined either, in terms of more basic things, but only in terms of what they do (such as attracting and repelling other electrical charges). In a similar way, propensity is whatever fills the various roles that physical probability plays in science.

What roles does physical probability play in science? What are its properties? One central property of chance is that, when known, it constrains rational belief to take the same numerical value. David Lewis called this the *Principal Principle*, (3.3 & 3.5) a term that philosophers have mostly adopted. For example, suppose you are certain that a particular biased coin has propensity 0.32 to land heads every time it is tossed. What is then the correct price for a gamble that pays \$1 if the coin lands heads, and nothing otherwise? According to the Principal Principle, the fair price is 32 cents.

Subjectivism

Subjectivists, also known as Bayesians or followers of epistemic probability, give the notion of probability a subjective status by regarding it as a measure of the 'degree of belief' of the individual assessing the uncertainty of a particular situation. Epistemic or subjective probability is sometimes called credence, as opposed to the term chance for a propensity probability.

Some examples of epistemic probability are to assign a probability to the proposition that a proposed law of physics is true, and to determine how probable it is that a suspect committed a crime, based on the evidence presented.

Gambling odds reflect the average bettor's 'degree of belief' in the outcome.

Gambling odds don't reflect the bookies' belief in a likely winner, so much as the other bettors' belief, because the bettors are actually betting against one another. The odds are set based on how many people have bet on a possible winner, so that even if the high odds players always win, the bookies will always make their percentages anyway.

The use of Bayesian probability raises the philosophical debate as to whether it can contribute valid justifications of belief.

Bayesians point to the work of Ramsey (p 182) and de Finetti (p 103) as proving that subjective beliefs must follow the laws of probability if they are to be coherent. Evidence casts doubt that humans will have coherent beliefs.

The use of Bayesian probability involves specifying a prior probability. This may be obtained from consideration of whether the required prior probability is greater or lesser than a reference probability associated with an urn model or a thought experiment. The issue is that for a given problem, multiple thought experiments could apply, and choosing one is a matter of judgement: different people may assign different prior probabilities, known as the reference class problem. The "sunrise problem" provides an example.

Prediction

An alternative account of probability emphasizes the role of *prediction* – predicting future observations on the basis of past observations, not on unobservable parameters. In its modern form, it is mainly in the Bayesian vein. This was the main function of probability before the 20th century, but fell out of favor compared to the parametric approach, which modeled phenomena as a physical system that was observed with error, such as in celestial mechanics. The modern predictive approach was pioneered by Bruno de Finetti, with the central idea of exchangeability – that future observations should behave like past observations. This view came to the attention of the Anglophone world with the 1974 translation of de Finetti's book, and has since been propounded by such statisticians as Seymour Geisser.

Axiomatic Probability

The mathematics of probability can be developed on an entirely axiomatic basis that is independent of any interpretation: the scetion on probability theory and probability axioms.

Classical Definition of Probability

The classical definition or interpretation of probability is identified with the works of Jacob Bernoulli and Pierre-Simon Laplace. As stated in Laplace's *Théorie analytique des probabilités*,

> The probability of an event is the ratio of the number of cases favorable to it, to the number of all cases possible when nothing leads us to expect that any one of these cases should occur more than any other, which renders them, for us, equally possible.

This definition is essentially a consequence of the principle of indifference. If elementary events are assigned equal probabilities, then the probability of a disjunction of elementary events is just the number of events in the disjunction divided by the total number of elementary events.

The classical definition of probability was called into question by several writers of the nineteenth century, including John Venn and George Boole. The frequentist definition of probability became widely accepted as a result of their criticism, and especially through the works of R.A. Fisher. The classical definition enjoyed a revival of sorts due to the general interest in Bayesian probability, because Bayesian methods require a prior probability distribution and the principle of indifference offers one source of such a distribution. Classical probability can offer prior probabilities that reflect ignorance which often seems appropriate before an experiment is conducted.

History

As a mathematical subject, the theory of probability arose very late—as compared to geometry for example—despite the fact that we have prehistoric evidence of man playing with dice from cultures from all over the world. One of the earliest writers on probability was Gerolamo Cardano. He perhaps produced the earliest known definition of classical probability.

The sustained development of probability began in the year 1654 when Blaise Pascal had some correspondence with his father's friend Pierre de Fermat about two problems concerning games of chance he had heard from the Chevalier de Méré earlier the same year, whom Pascal happened to accompany during a trip. One problem was the so-called problem of points, a classic problem already then (treated by Luca Pacioli as early as 1494, and even earlier in an anonymous manuscript in 1400), dealing with the question how to split the money at stake *in a fair way* when the game at hand is interrupted half-way through. The other problem was one about a mathematical rule of thumb that seemed not to hold when extending a game of dice from using one die to two dice. This last problem, or paradox, was the discovery of Méré himself and showed, according to him, how dangerous it was to apply mathematics to reality. They discussed other mathematical-philosophical issues and paradoxes as well during the trip that Méré thought was strengthening his general philosophical view.

Pascal, in disagreement with Méré's view of mathematics as something beautiful and flawless but poorly connected to reality, determined to prove Méré wrong by solving these two problems within pure mathematics. When he learned that Fermat, already recognized as a distinguished mathematician, had reached the same conclusions, he was convinced they had solved the problems conclusively. This correspondence circulated among other scholars at the time, in particular, to Huygens, Roberval and indirectly Caramuel, and marks the starting point for when mathematicians in general began to study problems from games of chance. The correspondence did not mention "probability"; It focused on fair prices.

Half a century later, Bernoulli showed a sophisticated grasp of probability. He showed facility with permutations and combinations, discussed the concept of probability with examples beyond the classical definition (such as personal, judicial and financial decisions) and showed that probabilities could be estimated by repeated trials with uncertainty diminished as the number of trials increased.

The 1765 volume of Diderot and d'Alembert's classic Encyclopédie contains a lengthy discussion of probability and summary of knowledge up to that time. A distinction is made between probabilities "drawn from the consideration of nature itself" (physical) and probabilities "founded only on the experience in the past which can make us confidently draw conclusions for the future" (evidential).

The source of a clear and lasting definition of probability was Laplace. As late as 1814 he stated:

The theory of chance consists in reducing all the events of the same kind to a certain number of cases equally possible, that is to say, to such as we may be equally undecided about in regard to their existence, and in determining the number of cases favorable to the event whose probability is sought. The ratio of this number to that of all the cases possible is the measure of this probability, which is thus simply a fraction whose numerator is the number of favorable cases and whose denominator is the number of all the cases possible.

— Pierre-Simon Laplace, A Philosophical Essay on Probabilities

This description is what would ultimately provide the classical definition of probability. Laplace published several editions of multiple documents (technical and a popularization) on probability over a half-century span. Many of his predecessors (Cardano, Bernoulli, Bayes) published a single document posthumously.

Criticism

The classical definition of probability assigns equal probabilities to events based on physical symmetry which is natural for coins, cards and dice.

- Some mathematicians object that the definition is circular. The probability for a "fair" coin is... A "fair" coin is defined by a probability of...

- The definition is very limited. It says nothing about cases where no physical symmetry exists. Insurance premiums, for example, can only be rationally priced by measured rates of loss.

- It is not trivial to justify the principle of indifference except in the simplest and most idealized of cases (an extension of the problem limited definition). Coins are not truly symmetric. Can we assign equal probabilities to each side? Can we assign equal probabilities to any real world experience?

However limiting, the definition is accompanied with substantial confidence. A casino which observes a marked departure from classical probability is confident that its assumptions have been violated (somebody is cheating). Much of the mathematics of probability was developed on the basis of this simplistic definition. Alternative interpretations of probability (for example frequentist and subjective) also have problems.

Mathematical probability theory deals in abstractions, avoiding the limitations and philosophical complications of any probability interpretation.

Frequentist Probability

John Venn

Frequentist probability or frequentism is a standard interpretation of probability; it defines an event's probability as the limit of its relative frequency in a large number of trials. This interpretation supports the statistical needs of experimental scientists and pollsters; probabilities can be found (in principle) by a repeatable objective process (and are thus ideally devoid of opinion). It does not support all needs; gamblers typically require estimates of the odds without experiments.

The development of the frequentist account was motivated by the problems and paradoxes of the previously dominant viewpoint, the classical interpretation. In the classical interpretation, probability was defined in terms of the principle of indifference, based on the natural symmetry of a problem, so, *e.g.* the probabilities of dice games arise from the natural symmetric 6-sidedness of the cube. This classical interpretation stumbled at any statistical problem that has no natural symmetry for reasoning.

Definition

In the frequentist interpretation, probabilities are discussed only when dealing with well-defined random experiments (or random samples). The set of all possible outcomes of a random experiment is called the sample space of the experiment. An event is defined as a particular subset of the sample space to be considered. For any given event, only one of two possibilities may hold: it occurs or it does not. The relative frequency of occurrence of an event, observed in a number of repetitions of the experiment, is a measure of the probability of that event. This is the core conception of probability in the frequentist interpretation.

Thus, if n_t is the total number of trials and n_x is the number of trials where the event $P(x)$ occurred, the probability $P(x)$ of the event occurring will be approximated by the relative frequency as follows:

$$P(x) \approx \frac{n_x}{n_t}.$$

Clearly, as the number of trials is increased, one might expect the relative frequency to become a better approximation of a "true frequency".

A claim of the frequentist approach is that in the "long run," as the number of trials approaches infinity, the relative frequency will converge *exactly* to the true probability:

$$P(x) = \lim_{n_t \to \infty} \frac{n_x}{n_t}.$$

Scope

The frequentist interpretation is a philosophical approach to the definition and use of probabilities; it is one of several such approaches. It does not claim to capture all connotations of the concept 'probable' in colloquial speech of natural languages.

As an interpretation, it is not in conflict with the mathematical axiomatization of probability theory; rather, it provides guidance for how to apply mathematical probability theory to real-world situations. It offers distinct guidance in the construction and design of practical experiments, especially when contrasted with the Bayesian interpretation. As to whether this guidance is useful, or is apt to mis-interpretation, has been a source of controversy. Particularly when the frequency interpretation of probability is mistakenly assumed to be the only possible basis for frequentist inference. So, for example, a list of mis-interpretations of the meaning of p-values accompanies the article on p-values; controversies are detailed in the article on statistical hypothesis testing. The Jeffreys–Lindley paradox shows how different interpretations, applied to the same data set, can lead to different conclusions about the 'statistical significance' of a result.

As William Feller noted:

There is no place in our system for speculations concerning the probability that the sun will rise tomorrow. Before speaking of it we should have to agree on an (idealized) model which would presumably run along the lines "out of infinitely many worlds one is selected at random..." Little imagination is required to construct such a model, but it appears both uninteresting and meaningless.

Feller's comment was criticism of Laplace, who published a solution to the sunrise problem using an alternative probability interpretation. Despite Laplace's explicit and immediate disclaimer in the source, based on expertise in astronomy as well as probability, two centuries of criticism have followed.

History

The frequentist view may have been foreshadowed by Aristotle, in *Rhetoric*, when he wrote:

the probable is that which for the most part happens

Poisson clearly distinguished between objective and subjective probabilities in 1837. Soon thereafter a flurry of nearly simultaneous publications by Mill, Ellis ("On the Foundations of the Theory of Probabilities" and "Remarks on the Fundamental Principles of the Theory of Probabilities"), Cournot (*Exposition de la théorie des chances et des probabilités*) and Fries introduced the frequentist view. Venn provided a thorough exposition (*The Logic of Chance: An Essay on the Foundations and Province of the Theory of Probability* (published editions in 1866, 1876, 1888)) two decades later. These were further supported by the publications of Boole and Bertrand. By the end of the 19th century the frequentist interpretation was well established and perhaps dominant in the sciences. The following generation established the tools of classical inferential statistics (significance testing, hypothesis testing and confidence intervals) all based on frequentist probability.

Alternatively, Jacob Bernoulli (AKA James or Jacques) understood the concept of frequentist probability and published a critical proof (the weak law of large numbers) posthumously in *1713*. He is also credited with some appreciation for subjective probability (prior to and without Bayes theorem). Gauss and Laplace used frequentist (and other) probability in derivations of the least squares method a century later, a generation before Poisson. Laplace considered the probabilities of testimonies, tables of mortality, judgments of tribunals, etc. which are unlikely candidates for classical probability. In this view, Poisson's contribution was his sharp criticism of the alternative "inverse" (subjective, Bayesian) probability interpretation. Any criticism by Gauss and Laplace was muted and implicit. (Their later derivations did not use inverse probability.)

Major contributors to "classical" statistics in the early 20th century included Fisher, Neyman and Pearson. Fisher contributed to most of statistics and made significance testing the core of experimental science; Neyman formulated confidence intervals and contributed heavily to sampling theory; Neyman and Pearson paired in the creation of hypothesis testing. All valued objectivity, so the best interpretation of probability available to them was frequentist. All were suspicious of "inverse probability" (the available alternative) with prior probabilities chosen by the using the principle of indifference. Fisher said, "...the theory of inverse probability is founded upon an error, [referring to Bayes theorem] and must be wholly rejected." (from his Statistical Methods for Research Workers). While Neyman was a pure frequentist, Fisher's views of probability were unique; Both had nuanced view of probability. von Mises offered a combination of mathematical and philosophical support for frequentism in the era.

Etymology

According to the *Oxford English Dictionary*, the term 'frequentist' was first used by M. G. Kendall in 1949, to contrast with Bayesians, whom he called "non-frequentists". He observed

> 3....we may broadly distinguish two main attitudes. One takes probability as 'a degree of rational belief', or some similar idea...the second defines probability in terms of frequencies of occurrence of events, or by relative proportions in 'populations' or 'collectives'.

...

12. It might be thought that the differences between the frequentists and the non-frequentists (if I may call them such) are largely due to the differences of the domains which they purport to cover.

...

I assert that this is not so ... The essential distinction between the frequentists and the non-frequentists is, I think, that the former, in an effort to avoid anything savouring of matters of opinion, seek to define probability in terms of the objective properties of a population, real or hypothetical, whereas the latter do not.

"The Frequency Theory of Probability" was used a generation earlier as a chapter title in Keynes (1921).

The historical sequence: probability concepts were introduced and much of probability mathematics derived (prior to the 20th century), classical statistical inference methods were developed, the mathematical foundations of probability were solidified and current terminology was introduced (all in the 20th century). The primary historical sources in probability and statistics did not use the current terminology of classical, subjective (Bayesian) and frequentist probability.

Alternative Views

Probability theory is a branch of mathematics. While its roots reach centuries into the past, it reached maturity with the axioms of Andrey Kolmogorov in 1933. The theory focuses on the valid operations on probability values rather than on the initial assignment of values; the mathematics is largely independent of any interpretation of probability.

Applications and interpretations of probability are considered by philosophy, the sciences and statistics. All are interested in the extraction of knowledge from observations—inductive reasoning. There are a variety of competing interpretations; All have problems. Major interpretations include classical probability, subjective probability and frequency interpretations.

- Classical probability assigns probabilities based on physical idealized symmetry (dice, coins, cards). The classical definition is at risk of circularity; Probabilities are defined by assuming equality of probabilities. In the absence of symmetry the utility of the definition is limited.

- Subjective probability (a family of competing interpretations) considers degrees of belief. All practical "subjective" probability interpretations are so constrained to rationality as to avoid most subjectivity. Real subjectivity is repellent to the sciences which strive for results independent of the observer and analyst. The historical roots of this concept extended to such non-numeric applications as legal evidence.

- Frequency interpretations are empirical—they are defined by a ratio from an infinite series of trials. This is a very natural interpretation for scientific experiments. Mathematicians are dubious of the convergence properties of the non-mathematical series.

The frequentist interpretation does resolve difficulties with the classical interpretation, such as any problem where the natural symmetry of outcomes is not known. It does not address other issues, such as the dutch book. Propensity probability is an alternative physicalist approach.

Probabilistic Logic

The aim of a probabilistic logic (also probability logic and probabilistic reasoning) is to combine the capacity of probability theory to handle uncertainty with the capacity of deductive logic to exploit structure of formal argument. The result is a richer and more expressive formalism with a broad range of possible application areas. Probabilistic logics attempt to find a natural extension of traditional logic truth tables: the results they define are derived through probabilistic expressions instead. A difficulty with probabilistic logics is that they tend to multiply the computational complexities of their probabilistic and logical components. Other difficulties include the possibility of counter-intuitive results, such as those of Dempster-Shafer theory. The need to deal with a broad variety of contexts and issues has led to many different proposals.

Historical Context

There are numerous proposals for probabilistic logics. Very roughly, they can be categorized into two different classes: those logics that attempt to make a probabilistic extension to logical entailment, such as Markov logic networks, and those that attempt to address the problems of uncertainty and lack of evidence (evidentiary logics).

That probability and uncertainty are not quite the same thing may be understood by noting that, despite the mathematization of probability in the Enlightenment, mathematical probability theory remains, to this very day, entirely unused in criminal courtrooms, when evaluating the "probability" of the guilt of a suspected criminal.

More precisely, in evidentiary logic, there is a need to distinguish the truth of a statement from the confidence in its truth: thus, being uncertain of a suspect's guilt is not the same as assigning a numerical probability to the commission of the crime. A single suspect may be guilty or not guilty, just as a coin may be flipped heads or tails. Given a large collection of suspects, a certain percentage may be guilty, just as the probability of flipping "heads" is one-half. However, it is incorrect to take this law of averages with regard to a single criminal (or single coin-flip): the criminal is no more "a little bit guilty" than a single coin flip is "a little bit heads and a little bit tails": we are merely uncertain as to which it is. Conflating probability and uncertainty may be acceptable when making scientific measurements of physical quantities, but it is an error, in the context of "common sense" reasoning and logic. Just as in courtroom reasoning, the goal of employing uncertain inference is to gather evidence to strengthen the confidence of a proposition, as opposed to performing some sort of probabilistic entailment.

Historically, attempts to quantify probabilistic reasoning date back to antiquity. There was a particularly strong interest starting in the 12th century, with the work of the Scholastics, with the invention of the half-proof (so that two half-proofs are sufficient to prove guilt), the elucidation of moral certainty (sufficient certainty to act upon, but short of absolute certainty), the development of Catholic probabilism (the idea that it is always safe to follow the established rules of doctrine or the opinion of experts, even when they are less probable), the case-based reasoning of casuistry, and the scandal of Laxism (whereby probabilism was used to give support to almost any statement at all, it being possible to find an expert opinion in support of almost any proposition.).

Modern Proposals

Below is a list of proposals for probabilistic and evidentiary extensions to classical and predicate logic.

- The term *"probabilistic logic"* was first used in a paper by Nils Nilsson published in 1986, where the truth values of sentences are probabilities. The proposed semantical generalization induces a probabilistic logical entailment, which reduces to ordinary logical entailment when the probabilities of all sentences are either 0 or 1. This generalization applies to any logical system for which the consistency of a finite set of sentences can be established.

- The central concept in the theory of subjective logic are *opinions* about some of the propositional variables involved in the given logical sentences. A binomial opinion applies to a single proposition and is represented as a 3-dimensional extension of a single probability value to express various degrees of ignorance about the truth of the proposition. For the computation of derived opinions based on a structure of argument opinions, the theory proposes respective operators for various logical connectives, such as e.g. multiplication (AND), comultiplication (OR), division (UN-AND) and co-division (UN-OR) of opinions as well as conditional deduction (MP) and abduction (MT).

- Approximate reasoning formalism proposed by fuzzy logic can be used to obtain a logic in which the models are the probability distributions and the theories are the lower envelopes. In such a logic the question of the consistency of the available information is strictly related with the one of the coherence of partial probabilistic assignment and therefore with Dutch book phenomenon.

- Markov logic networks implement a form of uncertain inference based on the maximum entropy principle—the idea that probabilities should be assigned in such a way as to maximize entropy, in analogy with the way that Markov chains assign probabilities to finite state machine transitions.

- Systems such as Pei Wang's Non-Axiomatic Reasoning System (NARS) or Ben Goertzel's Probabilistic Logic Networks (PLN) add an explicit confidence ranking, as well as a probability to atoms and sentences. The rules of deduction and induction incorporate this uncertainty, thus side-stepping difficulties in purely Bayesian approaches to logic (including Markov logic), while also avoiding the paradoxes of Dempster-Shafer theory. The implementation of PLN attempts to use and generalize algorithms from logic programming, subject to these extensions.

- In the theory of probabilistic argumentation, probabilities are not directly attached to logical sentences. Instead it is assumed that a particular subset W of the variables V involved in the sentences defines a probability space over the corresponding sub-σ-algebra. This induces two distinct probability measures with respect to V, which are called *degree of support* and *degree of possibility*, respectively. Degrees of support can be regarded as non-additive *probabilities of provability*, which generalizes the concepts of ordinary logical entailment (for $V = \{\}$) and classical posterior probabilities (for $V = W$). Mathematically, this view is compatible with the Dempster-Shafer theory.

- The theory of evidential reasoning also defines non-additive *probabilities of probability* (or *epistemic probabilities*) as a general notion for both logical entailment (provability) and probability. The idea is to augment standard propositional logic by considering an epistemic operator K that represents the state of knowledge that a rational agent has about the world. Probabilities are then defined over the resulting *epistemic universe* Kp of all propositional sentences p, and it is argued that this is the best information available to an analyst. From this view, Dempster-Shafer theory appears to be a generalized form of probabilistic reasoning.

Possible Application Areas

- Argumentation theory
- Artificial intelligence
- Artificial general intelligence
- Bioinformatics
- Formal epistemology
- Game theory
- Philosophy of science
- Psychology
- Statistics

Propensity Probability

The propensity theory of probability is one interpretation of the concept of probability. Theorists who adopt this interpretation think of probability as a physical propensity, or disposition, or tendency of a given type of physical situation to yield an outcome of a certain kind, or to yield a long run relative frequency of such an outcome.

Propensities are not relative frequencies, but purported *causes* of the observed stable relative frequencies. Propensities are invoked to *explain why* repeating a certain kind of experiment will generate a given outcome type at a persistent rate. A central aspect of this explanation is the law of large numbers. This law, which is a consequence of the axioms of probability, says that if (for example) a coin is tossed repeatedly many times, in such a way that its probability of landing heads is the same on each toss, and the outcomes are probabilistically independent, then the relative frequency of heads will (with high probability) be close to the probability of heads on each single toss. This law suggests that stable long-run frequencies are a manifestation of invariant *single-case* probabilities. Frequentists are unable to take this approach, since relative frequencies do not exist for single tosses of a coin, but only for large ensembles or collectives. Hence, these single-case probabilities are known as propensities or chances.

In addition to explaining the emergence of stable relative frequencies, the idea of propensity is motivated by the desire to make sense of single-case probability attributions in quantum mechanics, such as the probability of decay of a particular atom at a particular time.

The main challenge facing propensity theories is to say exactly what propensity *means*, and to show that propensity thus defined has the required properties.

History

A propensity theory of probability was given by Charles Sanders Peirce.

Karl Popper

A later propensity theory was proposed by philosopher Karl Popper, who had only slight acquaintance with the writings of Charles S. Peirce, however. Popper noted that the outcome of a physical experiment is produced by a certain set of "generating conditions". When we repeat an experiment, as the saying goes, we really perform another experiment with a (more or less) similar set of generating conditions. To say that a set of generating conditions has propensity p of producing the outcome E means that those exact conditions, if repeated indefinitely, would produce an outcome sequence in which E occurred with limiting relative frequency p. For Popper then, a deterministic experiment would have propensity 0 or 1 for each outcome, since those generating conditions would have same outcome on each trial. In other words, non-trivial propensities (those that differ from 0 and 1) only exist for genuinely indeterministic experiments.

Popper's propensities, while they are not relative frequencies, are yet defined in terms of relative frequency. As a result, they face many of the serious problems that plague frequency theories. First, propensities cannot be empirically ascertained, on this account, since the limit of a sequence is a tail event, and is thus independent of its finite initial segments. Seeing a coin land heads every time for the first million tosses, for example, tells one nothing about the limiting proportion of heads on Popper's view. Moreover, the use of relative frequency to define propensity *assumes* the existence of stable relative frequencies, so one cannot then use propensity to *explain* the existence of stable relative frequencies, via the Law of large numbers.

Recent Work

A number of other philosophers, including David Miller and Donald A. Gillies, have proposed propensity theories somewhat similar to Popper's, in that propensities are defined in terms of either long-run or infinitely long-run relative frequencies.

Other propensity theorists (*e.g.* Ronald Giere) do not explicitly define propensities at all, but rather see propensity as defined by the theoretical role it plays in science. They argue, for example, that physical magnitudes such as electrical charge cannot be explicitly defined either, in terms of more basic things, but only in terms of what they do (such as attracting and repelling other electrical charges). In a similar way, propensity is whatever fills the various roles that physical probability plays in science.

Other theories have been offered by D. H. Mellor, and Ian Hacking

Principal Principle of David Lewis

What roles does physical probability play in science? What are its properties? One central property of chance is that, when known, it constrains rational belief to take the same numerical value. David Lewis called this the *Principal Principle*, a term that philosophers have mostly adopted. For example, suppose you are certain that a particular biased coin has propensity 0.32 to land heads every time it is tossed. What is then the correct price for a gamble that pays \$1 if the coin lands heads, and nothing otherwise? According to the Principal Principle, the fair price is 32 cents. It is argued that Propensity Theories fail to meet the Principal Principle.

Bayesian Probability

 Bayesian probability is an interpretation of the concept of probability, in which, instead of frequency or propensity of some phenomenon, assigned probabilities represent states of knowledge or belief.

The Bayesian interpretation of probability can be seen as an extension of propositional logic that enables reasoning with hypotheses, i.e., the propositions whose truth or falsity is uncertain. In the Bayesian view, a probability is assigned to a hypothesis, whereas under frequentist inference, a hypothesis is typically tested without being assigned a probability.

Bayesian probability belongs to the category of evidential probabilities; to evaluate the probability of a hypothesis, the Bayesian probabilist specifies some prior probability, which is then updated to a posterior probability in the light of new, relevant data (evidence). The Bayesian interpretation provides a standard set of procedures and formulae to perform this calculation.

The term "Bayesian" derives from the 18th century mathematician and theologian Thomas Bayes, who provided the first mathematical treatment of a non-trivial problem of Bayesian inference. Mathematician Pierre-Simon Laplace pioneered and popularised what is now called Bayesian probability.

Broadly speaking, there are two views on Bayesian probability that interpret the *probability* concept in different ways. According to the *objectivist view*, the rules of Bayesian statistics can be justified by requirements of rationality and consistency and interpreted as an extension of logic. According to the *subjectivist view*, probability quantifies a "personal belief".

Bayesian Methodology

Bayesian methods are characterized by concepts and procedures as follows:

- The use of random variables or, more generally, unknown quantities, to model all sources of uncertainty in statistical models. This also includes uncertainty resulting from lack of information.

- The need to determine the *prior probability distribution* taking into account the available (prior) information.

- The *sequential use of Bayes' formula*: when more data become available, calculate the *posterior distribution* using Bayes' formula; subsequently, the posterior distribution becomes the next prior.

- While for the frequentist a hypothesis is a proposition (which must be either true or false), so that the frequentist probability of a hypothesis is either 0 or 1, in Bayesian statistics the probability that can be assigned to a hypothesis can also be in a range from 0 to 1 if the truth value is uncertain.

Objective and Subjective Bayesian Probabilities

Broadly speaking, there are two interpretations on Bayesian probability. For *objectivists*, *probability* objectively measures the plausibility of propositions, i.e., probability corresponds to a reasonable belief everyone (even a "robot") sharing the same knowledge should share in accordance with the rules of Bayesian statistics, which can be justified by requirements of rationality and consistency. For *subjectivists*, probability corresponds to a "personal belief"; rationality and consistency allow for substantial variation within the constraints they pose. The objective and subjective variants of Bayesian probability differ mainly in their interpretation and construction of the prior probability.

History

The term *Bayesian* refers to Thomas Bayes (1702–1761), who proved a special case of what is now called Bayes' theorem in a paper titled "An Essay towards solving a Problem in the Doctrine of Chances". In that special case, the prior and posterior distributions were Beta distributions and the data came from Bernoulli trials. It was Pierre-Simon Laplace (1749–1827) who introduced a general version of the theorem and used it to approach problems in celestial mechanics, medical statistics, reliability, and jurisprudence. Early Bayesian inference, which used uniform priors following Laplace's principle of insufficient reason, was called "inverse probability" (because it infers backwards from observations to parameters, or from effects to causes). After the 1920s, "inverse probability" was largely supplanted by a collection of methods that came to be called frequentist statistics.

In the 20th century, the ideas of Laplace were further developed in two different directions, giving rise to *objective* and *subjective* currents in Bayesian practice. Harold Jeffreys' *Theory of Probability* (first published in 1939) played an important role in the revival of the Bayesian view of probability, followed by works by Abraham Wald (1950) and Leonard J. Savage (1954). The adjective *Bayesian* itself dates to the 1950s; the derived *Bayesianism*, *neo-Bayesianism* is of 1960s coinage. In the objectivist stream, the statistical analysis depends on only the model assumed and the data analysed. No subjective decisions need to be involved. In contrast, "subjectivist" statisticians deny the possibility of fully objective analysis for the general case.

In the 1980s there was a dramatic growth in research and applications of Bayesian methods, most-ly attributed to the discovery of Markov chain Monte Carlo methods and the consequent removal of many of the computational problems, and to an increasing interest in nonstandard, complex applications. While frequentist statistics remains strong (as seen by the fact that most undergrad-uate teaching is still based on it), Bayesian methods are widely accepted and used, e.g., in the field of machine learning.

Justification of Bayesian Probabilities

The use of Bayesian probabilities as the basis of Bayesian inference has been supported by several arguments, such as Cox axioms, the Dutch book argument, arguments based on decision theory and de Finetti's theorem.

Axiomatic Approach

Richard T. Cox showed that Bayesian updating follows from several axioms, including two func-tional equations and a hypothesis of differentiability. The assumption of differentiability or even continuity is controversial; Halpern found a counterexample based on his observation that the Boolean algebra of statements may be finite. Other axiomatizations have been suggested by vari-ous authors with the purpose of making the theory more rigorous.

Dutch Book Approach

The Dutch book argument was proposed by de Finetti; it is based on betting. A Dutch book is made when a clever gambler places a set of bets that guarantee a profit, no matter what the outcome of the bets. If a bookmaker follows the rules of the Bayesian calculus in the construction of his odds, a Dutch book cannot be made.

However, Ian Hacking noted that traditional Dutch book arguments did not specify Bayesian up-dating: they left open the possibility that non-Bayesian updating rules could avoid Dutch books. For example, Hacking writes "And neither the Dutch book argument, nor any other in the person-alist arsenal of proofs of the probability axioms, entails the dynamic assumption. Not one entails Bayesianism. So the personalist requires the dynamic assumption to be Bayesian. It is true that in consistency a personalist could abandon the Bayesian model of learning from experience. Salt could lose its savour."

In fact, there are non-Bayesian updating rules that also avoid Dutch books (as discussed in the literature on "probability kinematics" following the publication of Richard C. Jeffreys' rule, which is itself regarded as Bayesian). The additional hypotheses sufficient to (uniquely) specify Bayesian updating are substantial, complicated, and unsatisfactory.

Decision Theory Approach

A decision-theoretic justification of the use of Bayesian inference (and hence of Bayesian probabil-ities) was given by Abraham Wald, who proved that every admissible statistical procedure is either a Bayesian procedure or a limit of Bayesian procedures. Conversely, every Bayesian procedure is admissible.

Personal Probabilities and Objective Methods for Constructing Priors

Following the work on expected utility theory of Ramsey and von Neumann, decision-theorists have accounted for rational behavior using a probability distribution for the agent. Johann Pfanzagl completed the *Theory of Games and Economic Behavior* by providing an axiomatization of subjective probability and utility, a task left uncompleted by von Neumann and Oskar Morgenstern: their original theory supposed that all the agents had the same probability distribution, as a convenience. Pfanzagl's axiomatization was endorsed by Oskar Morgenstern: "Von Neumann and I have anticipated" the question whether probabilities "might, perhaps more typically, be subjective and have stated specifically that in the latter case axioms could be found from which could derive the desired numerical utility together with a number for the probabilities (cf. p. 19 of The Theory of Games and Economic Behavior). We did not carry this out; it was demonstrated by Pfanzagl ... with all the necessary rigor".

Ramsey and Savage noted that the individual agent's probability distribution could be objectively studied in experiments. The role of judgment and disagreement in science has been recognized since Aristotle and even more clearly with Francis Bacon. The objectivity of science lies not in the psychology of individual scientists, but in the process of science and especially in statistical methods, as noted by C. S. Peirce. Recall that the objective methods for falsifying propositions about personal probabilities have been used for a half century, as noted previously. Procedures for testing hypotheses about probabilities (using finite samples) are due to Ramsey (1931) and de Finetti (1931, 1937, 1964, 1970). Both Bruno de Finetti and Frank P. Ramsey acknowledge their debts to pragmatic philosophy, particularly (for Ramsey) to Charles S. Peirce.

The "Ramsey test" for evaluating probability distributions is implementable in theory, and has kept experimental psychologists occupied for a half century. This work demonstrates that Bayesian-probability propositions can be falsified, and so meet an empirical criterion of Charles S. Peirce, whose work inspired Ramsey. (This falsifiability-criterion was popularized by Karl Popper.)

Modern work on the experimental evaluation of personal probabilities uses the randomization, blinding, and Boolean-decision procedures of the Peirce-Jastrow experiment. Since individuals act according to different probability judgments, these agents' probabilities are "personal" (but amenable to objective study).

Personal probabilities are problematic for science and for some applications where decision-makers lack the knowledge or time to specify an informed probability-distribution (on which they are prepared to act). To meet the needs of science and of human limitations, Bayesian statisticians have developed "objective" methods for specifying prior probabilities.

Indeed, some Bayesians have argued the prior state of knowledge defines *the* (unique) prior probability-distribution for "regular" statistical problems; cf. well-posed problems. Finding the right method for constructing such "objective" priors (for appropriate classes of regular problems) has been the quest of statistical theorists from Laplace to John Maynard Keynes, Harold Jeffreys, and Edwin Thompson Jaynes: These theorists and their successors have suggested several methods for constructing "objective" priors:

- Maximum entropy

- Transformation group analysis

- Reference analysis

Each of these methods contributes useful priors for "regular" one-parameter problems, and each prior can handle some challenging statistical models (with "irregularity" or several parameters). Each of these methods has been useful in Bayesian practice. Indeed, methods for constructing "objective" (alternatively, "default" or "ignorance") priors have been developed by avowed subjective (or "personal") Bayesians like James Berger (Duke University) and José-Miguel Bernardo (Universitat de València), simply because such priors are needed for Bayesian practice, particularly in science. The quest for "the universal method for constructing priors" continues to attract statistical theorists.

Thus, the Bayesian statistician needs either to use informed priors (using relevant expertise or previous data) or to choose among the competing methods for constructing "objective" priors.

References

- Spanos, Aris (1986). Statistical foundations of econometric modelling. Cambridge New York: Cambridge University Press. ISBN 978-0521269124.

- Peterson, Martin (2009). An introduction to decision theory. Cambridge, UK New York: Cambridge University Press. p. 140. ISBN 978-0521716543.

- Burks, Arthur W. (1978). Chance, Cause and Reason: An Inquiry into the Nature of Scientific Evidence. University of Chicago Press. pp. 694 pages. ISBN 0-226-08087-0.

- Ronald N. Giere (1973). "Objective Single Case Probabilities and the Foundations of Statistics". Studies in Logic and the Foundations of Mathematics. 73. Elsevier. ISBN 978-0-444-10491-5.

- Jaynes, E. T. (2003). Probability theory the logic of science. Cambridge, UK New York, NY: Cambridge University Press. ISBN 978-0521592710.

- James Franklin, The Science of Conjecture: Evidence and Probability before Pascal (2001) The Johns Hopkins University Press ISBN 0-8018-7109-3

- James Franklin, The Science of Conjecture: Evidence and Probability before Pascal, 2001 The Johns Hopkins Press, ISBN 0-8018-7109-3

- Burks, Arthur W. (1978). Chance, Cause and Reason: An Inquiry into the Nature of Scientific Evidence. University of Chicago Press. pp. 694 pages. ISBN 0-226-08087-0.

- Ronald N. Giere (1973). "Objective Single Case Probabilities and the Foundations of Statistics". Studies in Logic and the Foundations of Mathematics. 73. Elsevier. ISBN 978-0-444-10491-5.

- Popper, Karl. (2002) The Logic of Scientific Discovery 2nd Edition, Routledge ISBN 0-415-27843-0 (Reprint of 1959 translation of 1935 original) Page 57.

Stochastic Process: An Overview

Stochastic process is an important model in probability. It is used to describe a time sequence of any system and helps in representing the evolution of that system by a variable. This section is an overview of the subject matter incorporating all the major aspects of stochastic process.

Stochastic Process

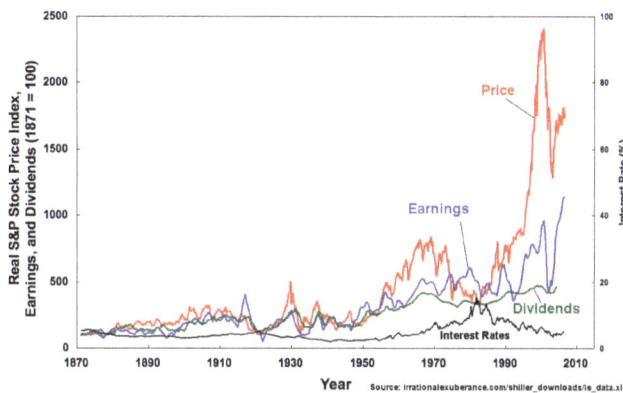

Stock market fluctuations have been modeled by stochastic processes.

A stochastic process (or random process) is a probability model used to describe phenomena that evolve over time or space. More specifically, in probability theory, a stochastic process is a time sequence representing the evolution of some system represented by a variable whose change is subject to a random variation. This is the probabilistic counterpart to a deterministic process (or deterministic system). Instead of describing a process which can only evolve in one way (as in the case, for example, of solutions of an ordinary differential equation), in a stochastic, or random process, there is some indeterminacy: even if the initial condition (or starting point) is known, there are several (often infinitely many) directions in which the process may evolve. In many stochastic processes, the movement to the next state or position depends on only the current state, and is independent from prior states or values the process has taken.

In the simple case of discrete time, as opposed to continuous time, a stochastic process is a sequence of random variables. The random variables corresponding to various times may be completely different, the only requirement being that these different random quantities all take values in the same space (the codomain of the function). One approach may be to model these random variables as random functions of one or several deterministic arguments (in most cases, the time parameter). Although the random values of a stochastic process at different times may be independent random variables, in most commonly considered situations they exhibit complicated statistical dependence.

Familiar examples of stochastic processes include stock market and exchange rate fluctuations; signals such as speech; audio and video; medical data such as a patient's EKG, EEG, blood pressure or temperature; and random movement such as Brownian motion or random walks.

A generalization, the random field, is defined by letting the variables be parametrized by members of a topological space instead of time. Examples of random fields include static images, random terrain (landscapes), wind waves and composition variations of a heterogeneous material.

Formal Definition and Basic Properties

Definition

Given a probability space $(\Omega, \mathcal{F}, P)$ and a measurable space (S, Σ), an S-valued stochastic process is a collection of S-valued random variables on Ω, indexed by a totally ordered set T ("time"). That is, a stochastic process X is a collection

$$\{X_t : t \in T\}$$

where each X_t is an S-valued random variable on Ω. The space S is then called the state space of the process.

Finite-dimensional Distributions

Let X be an S-valued stochastic process. For every finite sequence $T' = (t_1, \ldots, t_k) \in T^k$, the k-tuple $X_{T'} = (X_{t_1}, X_{t_2}, \ldots, X_{t_k})$ is a random variable taking values in S^k. The distribution $\mathbb{P}_{T'}(\cdot) = \mathbb{P}(X_{T'}^{-1}(\cdot))$ of this random variable is a probability measure on S^k. This is called a finite-dimensional distribution of X.

Under suitable topological restrictions, a suitably "consistent" collection of finite-dimensional distributions can be used to define a stochastic process.

History of Stochastic Processes

Stochastic processes were first studied rigorously in the late 19th century to aid in understanding financial markets and Brownian motion. The first person to describe the mathematics behind Brownian motion was Thorvald N. Thiele in a paper on the method of least squares published in 1880. This was followed independently by Louis Bachelier in 1900 in his PhD thesis "The theory of speculation", in which he presented a stochastic analysis of the stock and option markets. Albert Einstein (in one of his 1905 papers) and Marian Smoluchowski (1906) brought the solution of the problem to the attention of physicists, and presented it as a way to indirectly confirm the existence of atoms and molecules. Their equations describing Brownian motion were subsequently verified by the experimental work of Jean Baptiste Perrin in 1908.

An excerpt from Einstein's paper describes the fundamentals of a stochastic model:

"It must clearly be assumed that each individual particle executes a motion which is independent of the motions of all other particles; it will also be considered that the movements of one and the same particle in different time intervals are independent processes, as long as these time intervals are not chosen too small.

We introduce a time interval τ into consideration, which is very small compared to the observable time intervals, but nevertheless so large that in two successive time intervals τ, the motions executed by the particle can be thought of as events which are independent of each other".

Construction

In the ordinary axiomatization of probability theory by means of measure theory, the problem is to construct a sigma-algebra of measurable subsets of the space of all functions, and then put a finite measure on it. For this purpose one traditionally uses a method called Kolmogorov extension.

Kolmogorov Extension

The Kolmogorov extension proceeds along the following lines: assuming that a probability measure on the space of all functions $f : X \to Y$ exists, then it can be used to specify the joint probability distribution of finite-dimensional random variables $f(x_1),\ldots,f(x_n)$. Now, from this n-dimensional probability distribution we can deduce an $(n-1)$-dimensional marginal probability distribution for $f(x_1),\ldots,f(x_{n-1})$. Note that the obvious compatibility condition, namely, that this marginal probability distribution be in the same class as the one derived from the full-blown stochastic process, is not a requirement. Such a condition only holds, for example, if the stochastic process is a Wiener process (in which case the marginals are all gaussian distributions of the exponential class) but not in general for all stochastic processes. When this condition is expressed in terms of probability densities, the result is called the Chapman–Kolmogorov equation.

The Kolmogorov extension theorem guarantees the existence of a stochastic process with a given family of finite-dimensional probability distributions satisfying the Chapman–Kolmogorov compatibility condition.

Separability, or What the Kolmogorov Extension Does Not Provide

Recall that in the Kolmogorov axiomatization, measurable sets are the sets which have a probability or, in other words, the sets corresponding to yes/no questions that have a probabilistic answer.

The Kolmogorov extension starts by declaring to be measurable all sets of functions where finitely many coordinates $[f(x_1),\ldots,f(x_n)]$ are restricted to lie in measurable subsets of Y_n. In other words, if a yes/no question about f can be answered by looking at the values of at most finitely many coordinates, then it has a probabilistic answer.

In measure theory, if we have a countably infinite collection of measurable sets, then the union and intersection of all of them is a measurable set. For our purposes, this means that yes/no questions that depend on countably many coordinates have a probabilistic answer.

The good news is that the Kolmogorov extension makes it possible to construct stochastic processes with fairly arbitrary finite-dimensional distributions. Also, every question that one could ask about a sequence has a probabilistic answer when asked of a random sequence. The bad news is that certain questions about functions on a continuous domain don't have a probabilistic answer. One might hope that the questions that depend on uncountably many values of a function be of little interest, but the really bad news is that virtually all concepts of calculus are of this sort. For example:

1. boundedness

2. continuity

3. differentiability

all require knowledge of uncountably many values of the function.

One solution to this problem is to require that the stochastic process be separable. In other words, that there be some countable set of coordinates $\{f(x_i)\}$ whose values determine the whole random function f.

The Kolmogorov continuity theorem guarantees that processes that satisfy certain constraints on the moments of their increments have continuous modifications and are therefore separable.

Filtrations

Given a probability space $(\Omega, \mathcal{F}, P)$, a filtration is a weakly increasing collection of sigma-algebras on Ω, $\{\mathcal{F}_t, t \in T\}$, indexed by some totally ordered set T, and bounded above by $\mathcal{F}$, i.e. for $s, t \in T$ with $s < t$,

$$\mathcal{F}_s \subseteq \mathcal{F}_t \subseteq \mathcal{F}.$$

A stochastic process X on the same time set T is said to be adapted to the filtration if, for every $t \in T$, X_t is $\mathcal{F}_t$-measurable.

Natural Filtration

Given a stochastic process $X = \{X_t : t \in T\}$,, the natural filtration for (or induced by) this process is the filtration where $\mathcal{F}_t$ is generated by all values of X_s up to time $s = t$, i.e.
$$\mathcal{F}_t = \sigma(\{X_s^{-1}(A) : s \leq t, A \in \Sigma\}).\cdot$$

A stochastic process is always adapted to its natural filtration.

Classification

Stochastic processes can be classified according to the cardinality of its index set (usually interpreted as time) and state space.

Discrete Time and Discrete State Space

If both t and X_t belong to $\mathbb{N}$, the set of natural numbers, then we have models which lead to Markov chains. For example:

(a) If X_t means the bit (0 or 1) in position t of a sequence of transmitted bits, then X_t can be modelled as a Markov chain with two states. This leads to the error-correcting Viterbi algorithm in data transmission.

(b) If X_t represents the combined genotype of a breeding couple in the tth generation in an inbreeding model, it can be shown that the proportion of heterozygous individuals in the population approaches zero as t goes to ∞.

Continuous Time and Continuous State Space

The paradigm of continuous stochastic process is that of the Wiener process. In its original form the problem was concerned with a particle floating on a liquid surface, receiving "kicks" from the molecules of the liquid. The particle is then viewed as being subject to a random force which, since the molecules are very small and very close together, is treated as being continuous and since the particle is constrained to the surface of the liquid by surface tension, is at each point in time a vector parallel to the surface. Thus, the random force is described by a two-component stochastic process; two real-valued random variables are associated to each point in the index set, time, (note that since the liquid is viewed as being homogeneous the force is independent of the spatial coordinates) with the domain of the two random variables being R, giving the x and y components of the force. A treatment of Brownian motion generally also includes the effect of viscosity, resulting in an equation of motion known as the Langevin equation.

Discrete Time and Continuous State Space

If the index set of the process is N (the natural numbers), and the range is R (the real numbers), there are some natural questions to ask about the sample sequences of a process $\{X_i\}_{i \in N}$, where a sample sequence is $\{X_i(\omega)\}_{i \in N}$.

1. What is the probability that each sample sequence is bounded?

2. What is the probability that each sample sequence is monotonic?

3. What is the probability that each sample sequence has a limit as the index approaches ∞?

4. What is the probability that the series obtained from a sample sequence from converges?

5. What is the probability distribution of the sum?

Main applications of discrete time continuous state stochastic models include Markov chain Monte Carlo (MCMC) and the analysis of Time Series.

Continuous Time and Discrete State Space

Similarly, if the index space I is a finite or infinite interval, we can ask about the sample paths $\{X_t(\omega)\}_{t \in I}$

1. What is the probability that it is bounded/integrable...?

2. What is the probability that it has a limit at ∞

3. What is the probability distribution of the integral?

Wiener Process

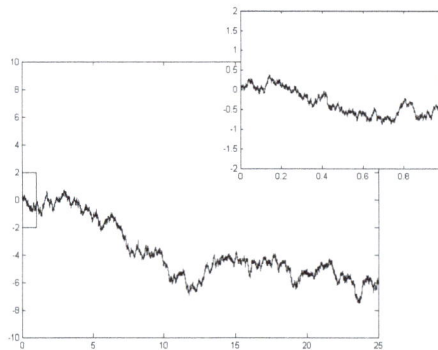

A single realization of a one-dimensional Wiener process

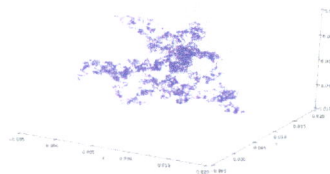

A single realization of a three-dimensional Wiener process

In mathematics, the Wiener process is a continuous-time stochastic process named in honor of Norbert Wiener. It is often called standard Brownian motion, after Robert Brown. It is one of the best known Lévy processes (càdlàg stochastic processes with stationary independent increments) and occurs frequently in pure and applied mathematics, economics, quantitative finance, and physics.

The Wiener process plays an important role both in pure and applied mathematics. In pure mathematics, the Wiener process gave rise to the study of continuous time martingales. It is a key process in terms of which more complicated stochastic processes can be described. As such, it plays a vital role in stochastic calculus, diffusion processes and even potential theory. It is the driving process of Schramm–Loewner evolution. In applied mathematics, the Wiener process is used to represent the integral of a white noise Gaussian process, and so is useful as a model of noise in electronics engineering, instrument errors in filtering theory and unknown forces in control theory.

The Wiener process has applications throughout the mathematical sciences. In physics it is used to study Brownian motion, the diffusion of minute particles suspended in fluid, and other types of diffusion via the Fokker–Planck and Langevin equations. It also forms the basis for the rigorous path integral formulation of quantum mechanics (by the Feynman–Kac formula, a solution to the Schrödinger equation can be represented in terms of the Wiener process) and the study of eternal inflation in physical cosmology. It is also prominent in the mathematical theory of finance, in particular the Black–Scholes option pricing model.

Characterisations of the Wiener process

The Wiener process W_t is characterised by the following properties:

1. $W_0 = 0$ a.s.

2. W has independent increments: $W_{t+u} - W_t$ is independent of $\sigma(W_s : s \le t)$ for $u \ge 0$

3. W has Gaussian increments: $W_{t+u} - W_t$ is normally distributed with mean 0 and variance u, $W_{t+u} - W_t \sim N(0, u)$

4. W has continuous paths: With probability 1, W_t is continuous in t.

The independent increments means that if $0 \le s_1 < t_1 \le s_2 < t_2$ then $W_{t1} - W_{s1}$ and $W_{t2} - W_{s2}$ are independent random variables, and the similar condition holds for n increments.

An alternative characterisation of the Wiener process is the so-called *Lévy characterisation* that says that the Wiener process is an almost surely continuous martingale with $W_0 = 0$ and quadratic variation $[W_t, W_t] = t$ (which means that $W_t^2 - t$ is also a martingale).

A third characterisation is that the Wiener process has a spectral representation as a sine series whose coefficients are independent $N(0, 1)$ random variables. This representation can be obtained using the Karhunen–Loève theorem.

Another characterisation of a Wiener process is the Definite integral (from zero to time t) of a zero mean, unit variance, delta correlated ("white") Gaussian process.

The Wiener process can be constructed as the scaling limit of a random walk, or other discrete-time stochastic processes with stationary independent increments. This is known as Donsker's theorem. Like the random walk, the Wiener process is recurrent in one or two dimensions (meaning that it returns almost surely to any fixed neighborhood of the origin infinitely often) whereas it is not recurrent in dimensions three and higher. Unlike the random walk, it is scale invariant, meaning that

$$\alpha^{-1} W_{\alpha^2 t}$$

is a Wiener process for any nonzero constant α. The Wiener measure is the probability law on the space of continuous functions g, with $g(0) = 0$, induced by the Wiener process. An integral based on Wiener measure may be called a Wiener integral.

Wiener Process as a Limit of Random Walk

Let $\xi_1, \xi_2, \ldots$ be i.i.d. random variables with mean 0 and variance 1. For each n, define a continuous time stochastic process

$$W_n(t) = \frac{1}{\sqrt{n}} \sum_{1 \le k \le \lfloor nt \rfloor} \xi_k$$

This is a random step function. Increments of W_n are independent because the   are independent. For large n, $W_n(t) - W_n(s)$ is close to $N(0, t - s)$ by the central limit theorem. It's tempting to believe that as $n \to \infty$, W_n will approach a Wiener process, and this temptation bears true fruit. The proof is provided by Donsker's theorem. This formulation explained why Brownian motion is ubiquitous.

Properties of a One-dimensional Wiener Process

Basic Properties

The unconditional probability density function, which follows normal distribution with mean = 0 and variance = t, at a fixed time t:

$$f_{W_t}(x) = \frac{1}{\sqrt{2\pi t}} e^{-x^2/(2t)}.$$

The expectation is zero:

$$E[W_t] = 0.$$

The variance, using the computational formula, is t:

$$\mathrm{Var}(W_t) = E\left[W_t^2\right] - E^2[W_t] = E\left[W_t^2\right] - 0 = E\left[W_t^2\right] = t.$$

Covariance and Correlation

The covariance and correlation:

$$\mathrm{cov}(W_s, W_t) = \min(s, t),$$

$$\mathrm{corr}(W_s, W_t) = \frac{\mathrm{cov}(W_s, W_t)}{\sigma_{W_s} \sigma_{W_t}} = \frac{\min(s, t)}{\sqrt{st}} = \sqrt{\frac{\min(s, t)}{\max(s, t)}}.$$

The results for the expectation and variance follow immediately from the definition that increments have a normal distribution, centered at zero. Thus

$$W_t = W_t - W_0 \sim N(0, t).$$

The results for the covariance and correlation follow from the definition that non-overlapping increments are independent, of which only the property that they are uncorrelated is used. Suppose that $t_1 < t_2$.

$$\mathrm{cov}(W_{t_1}, W_{t_2}) = \mathrm{E}\left[(W_{t_1} - \mathrm{E}[W_{t_1}]) \cdot (W_{t_2} - \mathrm{E}[W_{t_2}])\right] = \mathrm{E}\left[W_{t_1} \cdot W_{t_2}\right].$$

Substituting

$$W_{t_2} = (W_{t_2} - W_{t_1}) + W_{t_1}$$

we arrive at:

$$\mathrm{E}[W_{t_1} \cdot W_{t_2}] = \mathrm{E}\left[W_{t_1} \cdot ((W_{t_2} - W_{t_1}) + W_{t_1})\right]$$
$$= \mathrm{E}\left[W_{t_1} \cdot (W_{t_2} - W_{t_1})\right] + \mathrm{E}\left[W_{t_1}^2\right].$$

Since $W(t_1) = W(t_1) - W(t_0)$ and $W(t_2) - W(t_1)$, are independent,

$$\mathrm{E}\left[W_{t_1} \cdot (W_{t_2} - W_{t_1})\right] = \mathrm{E}[W_{t_1}] \cdot \mathrm{E}[W_{t_2} - W_{t_1}] = 0.$$

Thus

$$\mathrm{cov}(W_{t_1}, W_{t_2}) = \mathrm{E}\left[W_{t_1}^2\right] = t_1.$$

Wiener Representation

Wiener (1923) also gave a representation of a Brownian path in terms of a random Fourier series. If ξ_n are independent Gaussian variables with mean zero and variance one, then

$$W_t = \xi_0 t + \sqrt{2}\sum_{n=1}^{\infty} \xi_n \frac{\sin \pi n t}{\pi n}$$

and

$$W_t = \sqrt{2}\sum_{n=1}^{\infty} \xi_n \frac{\sin\left(\left(n - \frac{1}{2}\right)\pi t\right)}{\left(n - \frac{1}{2}\right)\pi}$$

represent a Brownian motion on $[0,1]$. The scaled process

$$\sqrt{c}\, W\left(\frac{t}{c}\right)$$

is a Brownian motion on $[0,c]$ (cf. Karhunen–Loève theorem).

Running Maximum

The joint distribution of the running maximum

$$M_t = \max_{0 \le s \le t} W_s$$

and W_t is

$$f_{M_t, W_t}(m, w) = \frac{2(2m - w)}{t\sqrt{2\pi t}} e^{-\frac{(2m-w)^2}{2t}}, \qquad m \ge 0, w \le m.$$

To get the unconditional distribution of f_{M_t}, integrate over $-\infty < w \le m$:

$$f_{M_t}(m) = \int_{-\infty}^{m} f_{M_t, W_t}(m, w)\, dw = \int_{-\infty}^{m} \frac{2(2m - w)}{t\sqrt{2\pi t}} e^{-\frac{(2m-w)^2}{2t}}\, dw$$

$$= \sqrt{\frac{2}{\pi t}}\, e^{-\frac{m^2}{2t}}, \qquad m \ge 0.$$

And the expectation

$$E[M_t] = \int_0^\infty m f_{M_t}(m)\,dm = \int_0^\infty m \sqrt{\frac{2}{\pi t}} e^{-\frac{m^2}{2t}}\,dm = \sqrt{\frac{2t}{\pi}}$$

Self-similarity

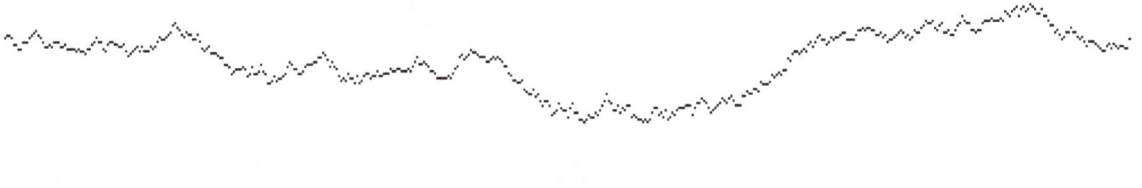

A demonstration of Brownian scaling, showing $V_t = (1/\sqrt{c})W_{ct}$ for decreasing c. Note that the average features of the function do not change while zooming in, and note that it zooms in quadratically faster horizontally than vertically.

Brownian Scaling

For every $c > 0$ the process $V_t = (1/\sqrt{c})W_{ct}$ is another Wiener process.

Time Reversal

The process $V_t = W_1 - W_{1-t}$ for $0 \le t \le 1$ is distributed like W_t for $0 \le t \le 1$.

Time Inversion

The process $V_t = tW_{1/t}$ is another Wiener process.

A class of Brownian Martingales

If a polynomial $p(x, t)$ satisfies the PDE

$$\left(\frac{\partial}{\partial t} + \frac{1}{2}\frac{\partial^2}{\partial x^2} \right) p(x,t) = 0$$

then the stochastic process

$$M_t = p(W_t, t)$$

is a martingale.

Example: $W_t^2 - t$ is a martingale, which shows that the quadratic variation of W on $[0, t]$ is equal to t. It follows that the expected time of first exit of W from $(-c, c)$ is equal to c^2.

More generally, for every polynomial $p(x, t)$ the following stochastic process is a martingale:

$$M_t = p(W_t,t) - \int_0^t a(W_s,s)ds,$$

where a is the polynomial

$$a(x,t) = \left(\frac{\partial}{\partial t} + \frac{1}{2}\frac{\partial^2}{\partial x^2} \right) p(x,t).$$

Example: $p(x,t) = (x^2 - t)^2$, $a(x,t) = 4x^2$; the process

$$(W_t^2 - t)^2 - 4\int_0^t W_s^2 \, ds$$

is a martingale, which shows that the quadratic variation of the martingale $W_t^2 - t$ on [0, t] is equal to

$$4\int_0^t W_s^2 \, ds.$$

About functions $p(xa, t)$ more general than polynomials.

Some Properties of Sample Paths

The set of all functions w with these properties is of full Wiener measure. That is, a path (sample function) of the Wiener process has all these properties almost surely.

Qualitative Properties

- For every $\varepsilon > 0$, the function w takes both (strictly) positive and (strictly) negative values on (0, ε).

- The function w is continuous everywhere but differentiable nowhere (like the Weierstrass function).

- Points of local maximum of the function w are a dense countable set; the maximum values are pairwise different; each local maximum is sharp in the following sense: if w has a local maximum at t then

$$\lim_{s \to t} \frac{|w(s) - w(t)|}{|s - t|} \to \infty.$$

  The same holds for local minima.

- The function w has no points of local increase, that is, no $t > 0$ satisfies the following for some ε in (0, t): first, $w(s) \le w(t)$ for all s in ($t - \varepsilon$, t), and second, $w(s) \ge w(t)$ for all s in (t, $t + \varepsilon$). (Local increase is a weaker condition than that w is increasing on ($t - \varepsilon$, $t + \varepsilon$).) The same holds for local decrease.

- The function w is of unbounded variation on every interval.

- The quadratic variation of w over [0,t] is t.

- Zeros of the function w are a nowhere dense perfect set of Lebesgue measure 0 and Hausdorff dimension 1/2 (therefore, uncountable).

Quantitative Properties

Law of the Iterated Logarithm

$$\limsup_{t \to +\infty} \frac{|w(t)|}{\sqrt{2t \log \log t}} = 1, \quad \text{almost surely.}$$

Modulus of Continuity

Local modulus of continuity:

$$\limsup_{\varepsilon \to 0+} \frac{|w(\varepsilon)|}{\sqrt{2\varepsilon \log \log(1/\varepsilon)}} = 1, \qquad \text{almost surely.}$$

Global modulus of continuity (Lévy):

$$\limsup_{\varepsilon \to 0+} \sup_{0 \le s < t \le 1, t-s \le \varepsilon} \frac{|w(s) - w(t)|}{\sqrt{2\varepsilon \log(1/\varepsilon)}} = 1, \textit{almost surely.}$$

Local Time

The image of the Lebesgue measure on [0, t] under the map w (the pushforward measure) has a density $L_t(\cdot)$. Thus,

$$\int_0^t f(w(s))ds = \int_{-\infty}^{+\infty} f(x)L_t(x)dx$$

for a wide class of functions f (namely: all continuous functions; all locally integrable functions; all non-negative measurable functions). The density L_t is (more exactly, can and will be chosen to be) continuous. The number $L_t(x)$ is called the local time at x of w on [0, t]. It is strictly positive for all x of the interval (a, b) where a and b are the least and the greatest value of w on [0, t], respectively. (For x outside this interval the local time evidently vanishes.) Treated as a function of two variables x and t, the local time is still continuous. Treated as a function of t (while x is fixed), the local time is a singular function corresponding to a nonatomic measure on the set of zeros of w.

These continuity properties are fairly non-trivial. Consider that the local time can also be defined (as the density of the pushforward measure) for a smooth function. Then, however, the density is discontinuous, unless the given function is monotone. In other words, there is a conflict between good behavior of a function and good behavior of its local time. In this sense, the continuity of the local time of the Wiener process is another manifestation of non-smoothness of the trajectory.

Related Processes

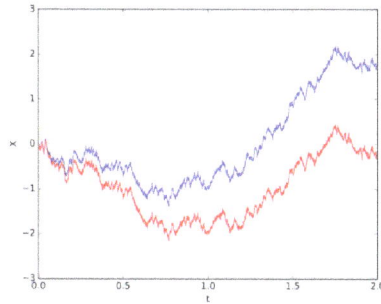

Wiener processes with drift (blue) and without drift (ref).

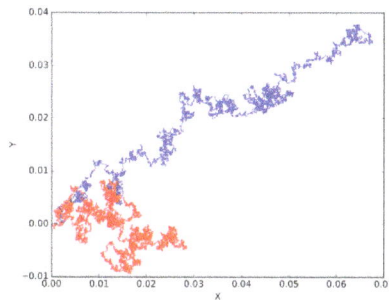

2D Wiener processes with drift (blue) and without drift (ref).

The generator of a Brownian motion is ½ times the Laplace–Beltrami operator. The image above is of the
Brownian motion on a special manifold: the surface of a sphere.

The stochastic process defined by

$$X_t = \mu t + \sigma W_t$$

is called a Wiener process with drift μ and infinitesimal variance σ^2. These processes exhaust continuous Lévy processes.

Two random processes on the time interval [0, 1] appear, roughly speaking, when conditioning the Wiener process to vanish on both ends of [0,1]. With no further conditioning, the process takes both positive and negative values on [0, 1] and is called Brownian bridge. Conditioned also to stay positive on (0, 1), the process is called Brownian excursion. In both cases a rigorous treatment involves a limiting procedure, since the formula $P(A|B) = P(A \cap B)/P(B)$ does not apply when $P(B) = 0$.

A geometric Brownian motion can be written

$$\mu t - \!\!\!\!\! - \!\!\!\!\! - \!\!\!\!\! + \sigma W$$

It is a stochastic process which is used to model processes that can never take on negative values, such as the value of stocks.

The stochastic process

$$X_t = e^{-t} W_{e^{2t}}$$

is distributed like the Ornstein–Uhlenbeck process.

The time of hitting a single point $x > 0$ by the Wiener process is a random variable with the Lévy distribution. The family of these random variables (indexed by all positive numbers x) is a left-continuous modification of a Lévy process. The right-continuous modification of this process is given by times of first exit from closed intervals $[0, x]$.

The local time $L = (L^x_t)_{x \in R, t \geq 0}$ of a Brownian motion describes the time that the process spends at the point x. Formally

$$L^x(t) = \int_0^t \delta(x - B_t)ds$$

where δ is the Dirac delta function. The behaviour of the local time is characterised by Ray–Knight theorems.

Brownian Martingales

Let A be an event related to the Wiener process (more formally: a set, measurable with respect to the Wiener measure, in the space of functions), and X_t the conditional probability of A given the Wiener process on the time interval $[0, t]$ (more formally: the Wiener measure of the set of trajectories whose concatenation with the given partial trajectory on $[0, t]$ belongs to A). Then the process X_t is a continuous martingale. Its martingale property follows immediately from the definitions, but its continuity is a very special fact – a special case of a general theorem stating that all Brownian martingales are continuous. A Brownian martingale is, by definition, a martingale adapted to the Brownian filtration; and the Brownian filtration is, by definition, the filtration generated by the Wiener process.

Integrated Brownian Motion

The time-integral of the Wiener process

$$W^{(-1)}(t) := \int_0^t W(s)ds$$

is called integrated Brownian motion or integrated Wiener process. It arises in many applications and can be shown to have the distribution $N(0, t^3/3)$, calculus lead using the fact that the covariance of the Wiener process is $t \wedge s = \min(t, s)$.

Time Change

Every continuous martingale (starting at the origin) is a time changed Wiener process.

Example: $2W_t = V(4t)$ where V is another Wiener process (different from W but distributed like W).

Example. $W_t^2 - t = V_{A(t)}$ where $A(t) = 4 \int_0^t W_s^2 \, ds$ and V is another Wiener process.

In general, if M is a continuous martingale then $M_t - M_0 = V_{A(t)}$ where $A(t)$ is the quadratic variation of M on $[0, t]$, and V is a Wiener process.

Corollary. Let M_t be a continuous martingale, and

$$M_\infty^- = \liminf_{t \to \infty} M_t,$$

$$M_\infty^+ = \limsup_{t \to \infty} M_t.$$

Then only the following two cases are possible:

$$-\infty < M_\infty^- = M_\infty^+ < +\infty,$$

$$-\infty = M_\infty^- < M_\infty^+ = +\infty;$$

other cases (such as $M_\infty^- = M_\infty^+ = +\infty$, $M_\infty^- < M_\infty^+ < +\infty$ etc.) are of probability 0.

Especially, a nonnegative continuous martingale has a finite limit (as $t \to \infty$) almost surely.

All stated (in this subsection) for martingales holds also for local martingales.

Change of Measure

A wide class of continuous semimartingales (especially, of diffusion processes) is related to the Wiener process via a combination of time change and change of measure.

Using this fact, the qualitative properties stated above for the Wiener process can be generalized to a wide class of continuous semimartingales.

Complex-valued Wiener Process

The complex-valued Wiener process may be defined as a complex-valued random process of the form $Z_t = X_t + iY_t$ where X_t, Y_t are independent Wiener processes (real-valued).

Self-similarity

Brownian scaling, time reversal, time inversion: the same as in the real-valued case.

Rotation invariance: for every complex number c such that $|c| = 1$ the process cZ_t is another complex-valued Wiener process.

Time Change

If f is an entire function then the process $f(Z_t) - f(0)$ is a time-changed complex-valued Wiener process.

Example: $Z_t^2 = (X_t^2 - Y_t^2) + 2X_t Y_t i = U_{A(t)}$ where

$$A(t) = 4 \int_0^t |Z_s|^2 \, ds$$

and U is another complex-valued Wiener process.

In contrast to the real-valued case, a complex-valued martingale is generally not a time-changed complex-valued Wiener process. For example, the martingale $2X_t + iY_t$ is not (here X_t, Y_t are independent Wiener processes, as before).

Ornstein–Uhlenbeck Process

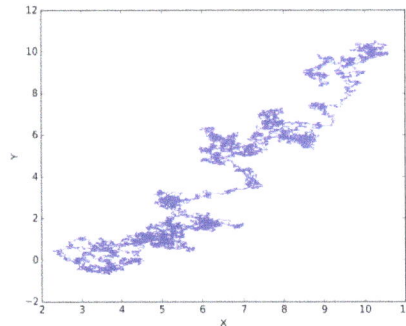

A simulation with $\theta = 1.0$, $\sigma = 300$ and $\mu = (0, 0)$. Initially at the position $(10, 10)$, the particle tends to move to the central point μ.

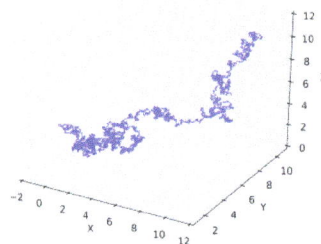

A 3D simulation with $\theta = 1.0$, $\sigma = 300$, $\mu = (0, 0, 0)$ and the initial position $(10, 10, 10)$.

In mathematics, the Ornstein–Uhlenbeck process (named after Leonard Ornstein and George Eugene Uhlenbeck), is a stochastic process that, roughly speaking, describes the velocity of a massive

Brownian particle under the influence of friction. The process is stationary Gauss–Markov process (which means that it is both a Gaussian and Markovian process), and is the only nontrivial process that satisfies these three conditions, up to allowing linear transformations of the space and time variables. Over time, the process tends to drift towards its long-term mean: such a process is called mean-reverting.

The process can be considered to be a modification of the random walk in continuous time, or Wiener process, in which the properties of the process have been changed so that there is a tendency of the walk to move back towards a central location, with a greater attraction when the process is further away from the centre. The Ornstein–Uhlenbeck process can also be considered as the continuous-time analogue of the discrete-time AR(1) process.

Representation Via a Stochastic Differential Equation

An Ornstein–Uhlenbeck process, x_t, satisfies the following stochastic differential equation:

$$dx_t = \theta(\mu - x_t)dt + \sigma dW_t$$

where $\theta > 0$, μ and $\sigma > 0$ are parameters and W_t denotes the Wiener process.

The above representation can be taken as the primary definition of an Ornstein–Uhlenbeck process or sometimes also mentioned as the Vasicek model.

Fokker–Planck Equation Representation

The probability density function $f(x, t)$ of the Ornstein–Uhlenbeck process satisfies the Fokker–Planck equation

$$\frac{\partial f}{\partial t} = \theta \frac{\partial}{\partial x}[(x - \mu)f] + \frac{\sigma^2}{2}\frac{\partial^2 f}{\partial x^2}$$

The Green function of this linear parabolic partial differential equation, taking $\mu = 0$ and $D = \sigma^2 / 2$ for simplicity, and the initial condition consisting of a unit point mass at location is

$$f(x,t) = \sqrt{\frac{\theta}{2\pi D(1 - e^{-2\theta t})}} \exp\left\{\frac{-\theta}{2D}\left[\frac{(x - ye^{-\theta t})^2}{1 - e^{-2\theta t}}\right]\right\}$$

The stationary solution of this equation is the limit for time tending to infinity which is a Gaussian distribution with mean μ and variance $\sigma^2 / (2\theta)$

$$f_s(x) = \sqrt{\frac{\theta}{\pi\sigma^2}}e^{-\theta(x-\mu)^2/\sigma^2}.$$

Application in Physical Sciences

The Ornstein–Uhlenbeck process is a prototype of a noisy relaxation process. Consider for example a Hookean spring with spring constant k whose dynamics is highly overdamped with friction coefficient γ. In the presence of thermal fluctuations with temperature T, the length $x(t)$ of the

---

---

spring will fluctuate stochastically around the spring rest length x_0; its stochastic dynamic is described by an Ornstein–Uhlenbeck process with:

$$\theta = k/\gamma,$$
$$\mu = x_0,$$
$$\sigma = \sqrt{2k_B T/\gamma},$$

where σ is derived from the Stokes–Einstein equation $D = \sigma^2/2 = k_B T/\gamma$ for the effective diffusion constant.

In physical sciences, the stochastic differential equation of an Ornstein–Uhlenbeck process is rewritten as a Langevin equation

$$\dot{x}(t) = -\frac{k}{\gamma}(x(t) - x_0) + \xi(t)$$

where $\xi(t)$ is white Gaussian noise with $\langle \xi(t_1)\xi(t_2)\rangle = 2k_B T/\gamma\, \delta(t_1 - t_2)$.

At equilibrium, the spring stores an average energy $\langle E \rangle = k\langle (x - x_0)^2\rangle/2 = k_B T/2$ in accordance with the equipartition theorem.

Application in Financial Mathematics

The Ornstein–Uhlenbeck process is one of several approaches used to model (with modifications) interest rates, currency exchange rates, and commodity prices stochastically. The parameter μ represents the equilibrium or mean value supported by fundamentals; σ the degree of volatility around it caused by shocks, and θ the rate by which these shocks dissipate and the variable reverts towards the mean. One application of the process is a trading strategy known as pairs trade.

Mathematical Properties

The Ornstein–Uhlenbeck process is an example of a Gaussian process that has a bounded variance and admits a stationary probability distribution, in contrast to the Wiener process; the difference between the two is in their "drift" term. For the Wiener process the drift term is constant, whereas for the Ornstein–Uhlenbeck process it is dependent on the current value of the process: if te current value of the process is less than the (long-term) mean, the drift will be positive; if the current value of the process is greater than the (long-term) mean, the drift will be negative. In other words, the mean acts as an equilibrium level for the process. This gives the process its informative name, "mean-reverting." The stationary (long-term) variance is given by

$$\text{var}(x_t) = \frac{\sigma^2}{2\theta}.$$

The Ornstein–Uhlenbeck process is the continuous-time analogue of the discrete-time AR(1) process.

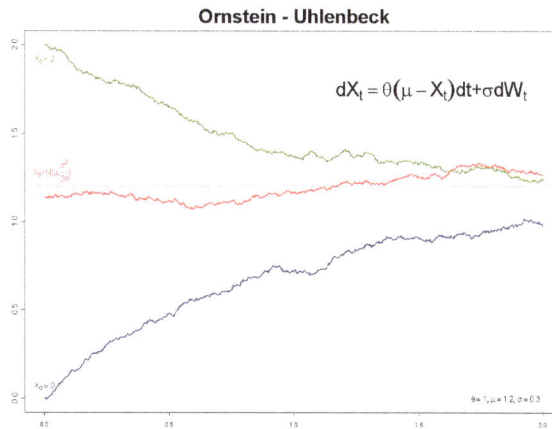

Ornstein - Uhlenbeck

$$dX_t = \theta(\mu - X_t)dt + \sigma dW_t$$

three sample paths of different OU-processes with $\theta = 1$, $\mu = 1.2$, $\sigma = 0.3$:
blue: initial value $a = 0$ (a.s.)
green: initial value $a = 2$ (a.s.)
red: initial value normally distributed so that the process has invariant measure

Solution

This stochastic differential equation is solved by variation of parameters. Changing variable

$$f(x_t, t) = x_t e^{\theta t}$$

we get

$$
\begin{aligned}
df(x_t, t) &= \theta x_t e^{\theta t} dt + e^{\theta t} dx_t \\
&= e^{\theta t} \theta \mu dt + \sigma e^{\theta t} dW_t.
\end{aligned}
$$

Integrating from 0 to t we get

$$x_t e^{\theta t} = x_0 + \int_0^t e^{\theta s} \theta \mu ds + \int_0^t \sigma e^{\theta s} dW_s$$

whereupon we see

$$x_t = x_0 e^{-\theta t} + \mu(1 - e^{-\theta t}) + \sigma \int_0^t e^{-\theta(t-s)} dW_s.$$

Formulas for Moments of Nonstationary Processes

From this representation, the first moment is given by (assuming that x_0 is a constant)

$$E(x_t) = x_0 e^{-\theta t} + \mu(1 - e^{-\theta t})$$

The Itō isometry can be used to calculate the covariance function by

$$\mathrm{cov}(x_s, x_t) = E[(x_s - E[x_s])(x_t - E[x_t])]$$

$$= E\left[\int_0^s \sigma e^{\theta(u-s)} \, dW_u \int_0^t \sigma e^{\theta(v-t)} \, dW_v\right]$$

$$= \sigma^2 e^{-\theta(s+t)} E\left[\int_0^s e^{\theta u} \, dW_u \int_0^t e^{\theta v} \, dW_v\right]$$

$$= \frac{\sigma^2}{2\theta} e^{-\theta(s+t)} (e^{2\theta \min(s,t)} - 1).$$

Thus if $s < t$ (so that $\min(s, t) = s$), then we have

$$\mathrm{cov}(x_s, x_t) = \frac{\sigma^2}{2\theta}\left(e^{-\theta(t-s)} - e^{-\theta(t+s)}\right).$$

Alternative Representation for Nonstationary Processes

It is also possible (and often convenient) to represent x_t (unconditionally, i.e. as $t \to \infty$) as a scaled time-transformed Wiener process:

$$x_t = \mu + \frac{\sigma}{\sqrt{2\theta}} e^{-\theta t} W_{e^{2\theta t}}$$

or conditionally (given x_0) as

$$x_t = x_0 e^{-\theta t} + \mu(1 - e^{-\theta t}) + \frac{\sigma}{\sqrt{2\theta}} e^{-\theta t} W_{e^{2\theta t} - 1}.$$

The time integral of this process can be used to generate noise with a $1/f$ power spectrum.

Scaling Limit Interpretation

The Ornstein–Uhlenbeck process can be interpreted as a scaling limit of a discrete process, in the same way that Brownian motion is a scaling limit of random walks. Consider an urn containing n blue and yellow balls. At each step a ball is chosen at random and replaced by a ball of the opposite colour (equivalently, a ball chosen uniformly at random changes color). Let X_n be the number of blue balls in the urn after n steps. Then $\dfrac{X_{[nt]} - n/2}{\sqrt{n}}$ converges in law to an Ornstein–Uhlenbeck process as n tends to infinity.

Generalizations

It is possible to extend Ornstein–Uhlenbeck processes to processes where the background driving process is a Lévy process (instead of a simple Brownian motion).These processes are widely studied by Ole Barndorff-Nielsen and Neil Shephard, and others.

In addition, in finance, stochastic processes are used the volatility increases for larger values of X.. In particular, the CKLS (Chan–Karolyi–Longstaff–Sanders) process with the volatility term replaced by $x \, dW$ can be solved in closed form for $\gamma = 1/2$ or 1, as well as for $\gamma = 0$, which corresponds to the conventional OU process.

Random Walk

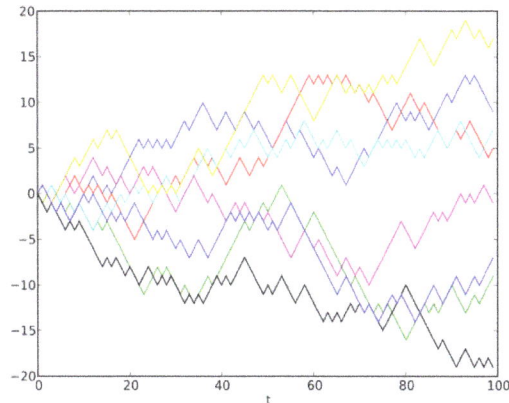

Example of eight random walks in one dimension starting at 0. The plot shows the current position on the line (vertical axis) versus the time steps (horizontal axis).

A random walk is a mathematical object which describes a path that consists of a succession of random steps. For example, the path traced by a molecule as it travels in a liquid or a gas, the search path of a foraging animal, superstring behavior, the price of a fluctuating stock and the financial status of a gambler can all be approximated by random walk models, even though they may not be truly random in reality. As illustrated by those examples, random walks have applications to many scientific fields including ecology, psychology, computer science, physics, chemistry, and biology, and also to economics. Random walks explain the observed behaviors of many processes in these fields, and thus serve as a fundamental model for the recorded stochastic activity. As a more mathematical application, the value of pi can be approximated by the usage of random walk in agent-based modelling environment. The term *random walk* was first introduced by Karl Pearson in 1905.

Various different types of random walks are of interest, which can differ in several ways. The term itself most often refers to a special category of Markov chains or Markov processes, but many time-dependent processes are referred to as random walks, with a modifier indicating their specific properties. Random walks (Markov or not) can also take place on a variety of spaces: commonly studied ones include graphs, others on the integers or the real line, in the plane or in higher-dimensional vector spaces, on curved surfaces or higher-dimensional Riemannian manifolds, and also on groups finite, finitely generated or Lie. The time parameter can also be tinkered with. In the simplest context the walk is in discrete time, that is a sequence of random variables $(Xt) = (X_1, X_2,...)$ indexed by the natural numbers. However, it is also possible to define random walks which take their steps at random times, and in that case the position X t has to be defined for all times $t \in [0,+\infty]$. Specific cases or limits of random walks include the Lévy flight and diffusion models such as Brownian motion.

Random walks are a fundamental topic in discussions of Markov processes. Their mathematical study has been extensive. Several properties, including dispersal distributions, first-passage or hitting times, encounter rates, recurrence or transience, have been introduced to quantify their behaviour.

Lattice Random Walk

A popular random walk model is that of a random walk on a regular lattice, where at each step the location jumps to another site according to some probability distribution. In a simple random walk, the location can only jump to neighboring sites of the lattice, forming a lattice path. In simple symmetric random walk on a locally finite lattice, the probabilities of the location jumping to each one of its immediate neighbours are the same. The best studied example is of random walk on the d-dimensional integer lattice (sometimes called the hypercubic lattice) . $\mathbb{Z}^d$.

If the state space is limited to finite dimensions, the random walk model is called bordered symmetric random walk and the transition probabilities depend on the location of the state, because on margin and corner states the movement is limited.

One-dimensional Random Walk

An elementary example of a random walk is the random walk on the integer number line, $\mathbb{Z}$, which starts at 0 and at each step moves +1 or −1 with equal probability.

This walk can be illustrated as follows. A marker is placed at zero on the number line and a fair coin is flipped. If it lands on heads, the marker is moved one unit to the right. If it lands on tails, the marker is moved one unit to the left. After five flips, the marker could now be on 1, −1, 3, −3, 5, or −5. With five flips, three heads and two tails, in any order, will land on 1. There are 10 ways of landing on 1 (by flipping three heads and two tails), 10 ways of landing on −1 (by flipping three tails and two heads), 5 ways of landing on 3 (by flipping four heads and one tail), 5 ways of landing on −3 (by flipping four tails and one head), 1 way of landing on 5 (by flipping five heads), and 1 way of landing on −5 (by flipping five tails). See the figure below for an illustration of the possible outcomes of 5 flips.

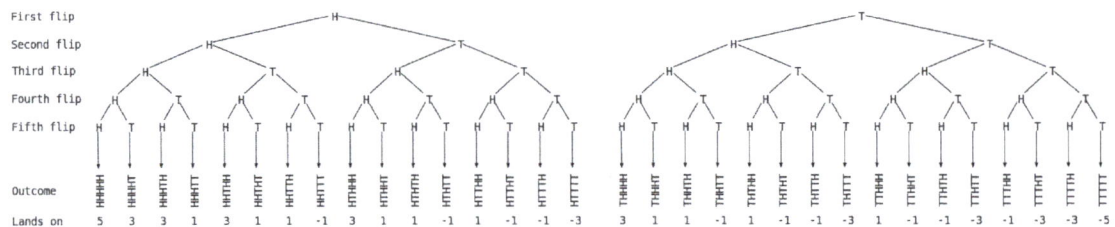

All possible random walk outcomes after 5 flips of a fair coin

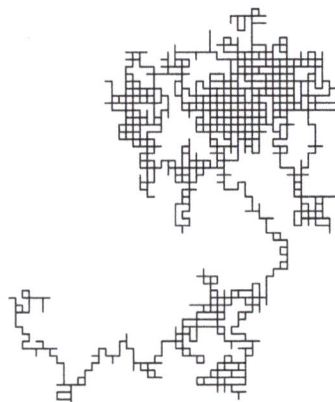

Random walk in two dimensions (animated version)

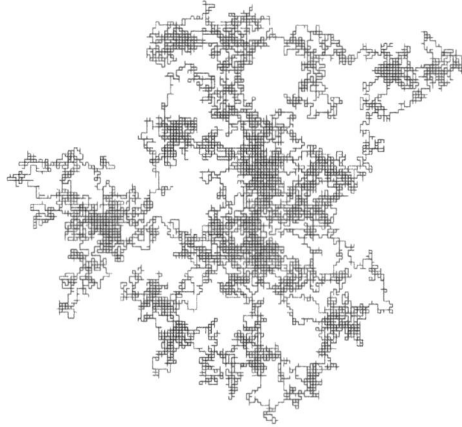

Random walk in two dimensions with 25 thousand steps (animated version)

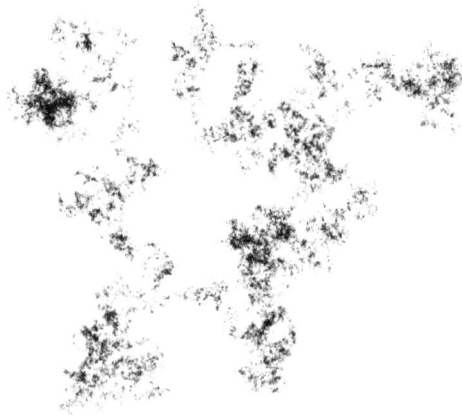

Random walk in two dimensions with two million even smaller steps. This image was generated in such a way that points that are more frequently traversed are darker. In the limit, for very small steps, one obtains Brownian motion.

To define this walk formally, take independent random variables $Z_1, Z_2, \ldots$, where each variable is either 1 or −1, with a 50% probability for either value, and set $S_0 = 0$ and $S_n = \sum Z_j$. The series $\{S_n\}$ is called the simple random walk on $\mathbb{Z}$. This series (the sum of the sequence of −1s and 1s) gives the distance walked, if each part of the walk is of length one. The expectation $E(S_n)$ of S_n is zero. That is, the mean of all coin flips approaches zero as the number of flips increases. This follows by the finite additivity property of expectation:

$$E(S_n) = \sum_{j=1}^{n} E(Z_j) = 0.$$

A similar calculation, using the independence of the random variables and the fact that $E(Z_n^2) = 1$, , shows that:

$$E(S_n^2) = \sum_{i=1}^{n}\sum_{j=1}^{n} E(Z_j Z_i) = n.$$

This hints that $E(|S_n|)$, the expected translation distance after n steps, should be of the order of

$\sqrt{n}$. In fact,

$$\lim_{n\to\infty}\frac{E(|S_n|)}{\sqrt{n}}=\sqrt{\frac{2}{\pi}}.$$

This result shows that diffusion is ineffective for mixing because of the way the square root behaves for large N.

How many times will a random walk cross a boundary line if permitted to continue walking forever? A simple random walk on $\mathbb{Z}$ will cross every point an infinite number of times. This result has many names: the *level-crossing phenomenon*, *recurrence* or the *gambler's ruin*. The reason for the last name is as follows: a gambler with a finite amount of money will eventually lose when playing *a fair game* against a bank with an infinite amount of money. The gambler's money will perform a random walk, and it will reach zero at some point, and the game will be over.

If a and b are positive integers, then the expected number of steps until a one-dimensional simple random walk starting at 0 first hits b or $-a$ is ab. The probability that this walk will hit b before $-a$ is $a/(a+b)$, which can be derived from the fact that simple random walk is a martingale.

Some of the results mentioned above can be derived from properties of Pascal's triangle. The number of different walks of n steps where each step is $+1$ or -1 is 2^n. For the simple random walk, each of these walks are equally likely. In order for S_n to be equal to a number k it is necessary and sufficient that the number of $+1$ in the walk exceeds those of -1 by k. The number of walks which satisfy $S_n = k$ is equally the number of ways of choosing $(n - k)/2$ with n is the number of allowed moves, denoted $\binom{n}{(n-k)/2}$. For this to have meaning, it is necessary that n and k be even numbers. Therefore, the probability that $S_n = k$ is equal to $2^{-n}\binom{n}{(n-k)/2}$. By representing entries of Pascal's triangle in terms of factorials and using Stirling's formula, one can obtain good estimates for these probabilities for large values of n. .

If the space is confined to $\mathbb{Z}+$ for brevity, the number of ways in which a random walk will land on any given number having five flips can be shown as $\{0,5,0,4,0,1\}$.

This relation with Pascal's triangle is demonstrated for small values of n. At zero turns, the only possibility will be to remain at zero. However, at one turn, there is one chance of landing on -1 or one chance of landing on 1. At two turns, a marker at 1 could move to 2 or back to zero. A marker at -1, could move to -2 or back to zero. Therefore, there is one chance of landing on -2, two chances of landing on zero, and one chance of landing on 2.

k	-5	-4	-3	-2	-1	0	1	2	3	4	5
$P[S_0 = k]$						1					
$2P[S_1 = k]$					1		1				
$2^2 P[S_2 = k]$				1		2		1			
$2^3 P[S_3 = k]$			1		3		3		1		

$2^4 P[S_4 = k]$		1		4		6		4		1	
$2^5 P[S_5 = k]$	1		5		10		10		5		1

The central limit theorem and the law of the iterated logarithm describe important aspects of the behavior of simple random walks on $\mathbb{Z}$. In particular, the former entails that as n increases, the probabilities (proportional to the numbers in each row) approach a normal distribution.

As a direct generalization, one can consider random walks on crystal lattices (infinite-fold abelian covering graphs over finite graphs). Actually it is possible to establish the central limit theorem and large deviation theorem in this setting.

As a Markov Chain

A one-dimensional random walk can also be looked at as a Markov chain whose state space is given by the integers $i = 0, \pm 1, \pm 2, \ldots$. For some number p satisfying $0 < p < 1$, , the transition probabilities (the probability $P_{i,j}$ of moving from state i to state j) are given by

$$P_{i,i+1} = p = 1 - P_{i,i-1}.$$

Higher Dimensions

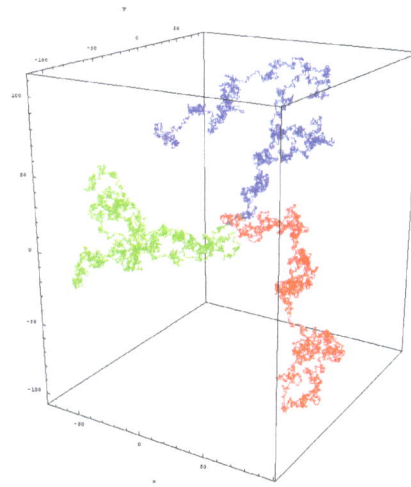

Three random walks in three dimensions

In higher dimensions, the set of randomly walked points has interesting geometric properties. In fact, one gets a discrete fractal, that is, a set which exhibits stochastic self-similarity on large scales. On small scales, one can observe "jaggedness" resulting from the grid on which the walk is performed. Two books of Lawler referenced below are a good source on this topic. The trajectory of a random walk is the collection of points visited, considered as a set with disregard to *when* the walk arrived at the point. In one dimension, the trajectory is simply all points between the minimum height and the maximum height the walk achieved (both are, on average, on the order of $\sqrt{n}$).

To visualize the two dimensional case, one can imagine a person walking randomly around a city. The city is effectively infinite and arranged in a square grid of sidewalks. At every intersection, the person randomly chooses one of the four possible routes (including the one originally traveled from). Formally, this is a random walk on the set of all points in the plane with integer coordinates.

Will the person ever get back to the original starting point of the walk? This is the 2-dimensional equivalent of the level crossing problem discussed above. It turns out that the person almost surely will in a 2-dimensional random walk, but for 3 dimensions or higher, the probability of returning to the origin decreases as the number of dimensions increases. In 3 dimensions, the probability decreases to roughly 34%.

The asymptotic function for a two dimensional random walk as the number of steps increases is given by a Rayleigh distribution. The probability distribution is a function of the radius from the origin and the step length is constant for each step.

$$P(r) = \frac{2r}{N} e^{-r^2/N}$$

Relation to Wiener Process

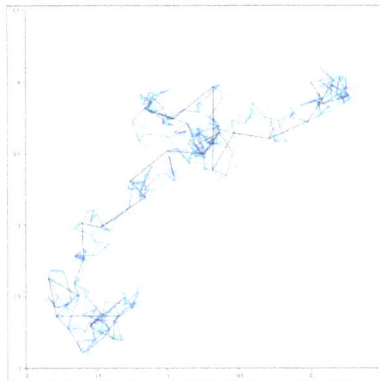

Simulated steps approximating a Wiener process in two dimensions

A Wiener process is a stochastic process with similar behaviour to Brownian motion, the physical phenomenon of a minute particle diffusing in a fluid. (Sometimes the Wiener process is called "Brownian motion", although this is strictly speaking a confusion of a model with the phenomenon being modeled.)

A Wiener process is the scaling limit of random walk in dimension 1. This means that if you take a random walk with very small steps you get an approximation to a Wiener process (and, less accurately, to Brownian motion). To be more precise, if the step size is ε, one needs to take a walk of length L/ε^2 to approximate a Wiener length of L. As the step size tends to 0 (and the number of steps increases proportionally) random walk converges to a Wiener process in an appropriate sense. Formally, if B is the space of all paths of length L with the maximum topology, and if M is the space of measure over B with the norm topology, then the convergence is in the space M. Similarly, a Wiener process in several dimensions is the scaling limit of random walk in the same number of dimensions.

A random walk is a discrete fractal (a function with integer dimensions; 1, 2, ...), but a Wiener process trajectory is a true fractal, and there is a connection between the two. For example, take a random walk until it hits a circle of radius r times the step length. The average number of steps it performs is r^2. This fact is the *discrete version* of the fact that a Wiener process walk is a fractal of Hausdorff dimension 2.

In two dimensions, the average number of points the same random walk has on the *boundary* of its trajectory is $r^{4/3}$. This corresponds to the fact that the boundary of the trajectory of a Wiener process is a fractal of dimension 4/3, a fact predicted by Mandelbrot using simulations but proved only in 2000 by Lawler, Schramm and Werner.

A Wiener process enjoys many symmetries random walk does not. For example, a Wiener process walk is invariant to rotations, but random walk is not, since the underlying grid is not (random walk is invariant to rotations by 90 degrees, but Wiener processes are invariant to rotations by, for example, 17 degrees too). This means that in many cases, problems on random walk are easier to solve by translating them to a Wiener process, solving the problem there, and then translating back. On the other hand, some problems are easier to solve with random walks due to its discrete nature.

Random walk and Wiener process can be *coupled*, namely manifested on the same probability space in a dependent way that forces them to be quite close. The simplest such coupling is the Skorokhod embedding, but there exist more precise couplings, such as Komlós–Major–Tusnády approximation theorem.

The convergence of a random walk toward the Wiener process is controlled by the central limit theorem, and by Donsker's theorem. For a particle in a known fixed position at $t = 0$, the central limit theorem tells us that after a large number of independent steps in the random walk, the walker's position is distributed according to a normal distribution of total variance:

$$\sigma^2 = \frac{t}{\delta t}\varepsilon^2,$$

where t is the time elapsed since the start of the random walk, ε is the size of a step of the random walk, and δt is the time elapsed between two successive steps.

This corresponds to the Green function of the diffusion equation that controls the Wiener process, which suggests that, after a large number of steps, the random walk converges toward a Wiener process.

In 3D, the variance corresponding to the Green's function of the diffusion equation is:

$$\sigma^2 = 6Dt$$

By equalizing this quantity with the variance associated to the position of the random walker, one obtains the equivalent diffusion coefficient to be considered for the asymptotic Wiener process toward which the random walk converges after a large number of steps:

$$D = \frac{\varepsilon^2}{6\delta t} \text{ (valid only in 3D)}$$

Remark: the two expressions of the variance above correspond to the distribution associated to the vector $\vec{R}$ that links the two ends of the random walk, in 3D. The variance associated to each component R_x, R_y or R_z is only one third of this value (still in 3D).

For 2D:

$$D = \frac{\varepsilon^2}{4\delta t}$$

For 1D:

$$D = \frac{\varepsilon^2}{2\delta t}$$

Gaussian Random Walk

A random walk having a step size that varies according to a normal distribution is used as a model for real-world time series data such as financial markets. The Black–Scholes formula for modeling option prices, for example, uses a Gaussian random walk as an underlying assumption.

Here, the step size is the inverse cumulative normal distribution $\Phi^{-1}(z, \mu, \sigma)$ where $0 \leq z \leq 1$ is a uniformly distributed random number, and μ and σ are the mean and standard deviations of the normal distribution, respectively.

If μ is nonzero, the random walk will vary about a linear trend. If v_s is the starting value of the random walk, the expected value after n steps will be $v_s + n\mu$.

For the special case where μ is equal to zero, after n steps, the translation distance's probability distribution is given by $N(0, n\sigma^2)$, where $N()$ is the notation for the normal distribution, n is the number of steps, and σ is from the inverse cumulative normal distribution as given above.

Proof: The Gaussian random walk can be thought of as the sum of a series of independent and identically distributed random variables, X_i from the inverse cumulative normal distribution with mean equal zero and σ of the original inverse cumulative normal distribution:

$$Z = \sum_{i=0}^{n} X_i,$$

but we have the distribution for the sum of two independent normally distributed random variables, $Z = X + Y$, is given by

$$N(\mu_X + \mu_Y, \sigma^2_X + \sigma^2_Y)$$

In our case, $\mu_X = \mu_Y = 0$ and $\sigma^2_X = \sigma^2_Y = \sigma^2$ yield

$$N(0, 2\sigma^2)$$

By induction, for n steps we have

$$Z \sim N(0, n\sigma^2).$$

For steps distributed according to any distribution with zero mean and a finite variance (not necessarily just a normal distribution), the root mean square translation distance after n steps is

$$\sqrt{E \mid S_n^2 \mid} = \sigma \sqrt{n}.$$

But for the Gaussian random walk, this is just the standard deviation of the translation distance's distribution after n steps. Hence, if μ is equal to zero, and since the root mean square(rms) translation distance is one standard deviation, there is 68.27% probability that the rms translation distance after n steps will fall between $\pm\, \sigma \sqrt{n}$. Likewise, there is 50% probability that the translation distance after n steps will fall between $\pm\, 0.6745 \sigma \sqrt{n}$.

Anomalous Diffusion

In disordered systems such as porous media and fractals σ^2 may not be proportional to t but to . The exponent d_w is called the anomalous diffusion exponent and can be larger or smaller than 2. Anomalous diffusion may also be expressed as $\sigma_r^2 \sim Dt^\alpha$ where α is the anomaly parameter. Some diffusions in random environment are even proportional to a power of the logarithm of the time.

Number of Distinct Sites

The number of distinct sites visited by a single random walker $S(t)$ has been studied extensively for square and cubic lattices and for fractals. This quantity is useful for the analysis of problems of trapping and kinetic reactions. It is also related to the vibrational density of states, diffusion reactions processes and spread of populations in ecology. The generalization of this problem to the number of distinct sites visited by N random walkers, $S_N(t)$, has recently been studied for d-dimensional Euclidean lattices. The number of distinct sites visited by N walkers is not simply related to the number of distinct sites visited by each walker.

Applications

Antony Gormley's *Quantum Cloud* sculpture in London was designed by a computer using a random walk algorithm.

As mentioned the range of natural phenomena which have been subject to attempts at description by some flavour random walks is considerable, in particular in physics and chemistry, materials science, biology and various other fields. The following are some specific applications of random walk:

- In financial economics, the "random walk hypothesis" is used to model shares prices and other factors. Empirical studies found some deviations from this theoretical model, especially in short term and long term correlations.

- In population genetics, random walk describes the statistical properties of genetic drift

- In physics, random walks are used as simplified models of physical Brownian motion and diffusion such as the random movement of molecules in liquids and gases. For example diffusion-limited aggregation. Also in physics, random walks and some of the self interacting walks play a role in quantum field theory.

- In mathematical ecology, random walks are used to describe individual animal movements, to empirically support processes of biodiffusion, and occasionally to model population dynamics.

- In polymer physics, random walk describes an ideal chain. It is the simplest model to study polymers.

- In other fields of mathematics, random walk is used to calculate solutions to Laplace's equation, to estimate the harmonic measure, and for various constructions in analysis and combinatorics.

- In computer science, random walks are used to estimate the size of the Web. In the World Wide Web conference-2006, Bar-Yossef et al. published their findings and algorithms for the same.

- In image segmentation, random walks are used to determine the labels (i.e., "object" or "background") to associate with each pixel. This algorithm is typically referred to as the random walker segmentation algorithm.

In all these cases, random walk is often substituted for Brownian motion.

- In brain research, random walks and reinforced random walks are used to model cascades of neuron firing in the brain.

- In vision science, ocular drift tends to behave like a random walk. According to some authors, fixational eye movements in general are also well described by a random walk.

- In psychology, random walks explain accurately the relation between the time needed to make a decision and the probability that a certain decision will be made.

- Random walks can be used to sample from a state space which is unknown or very large, for example to pick a random page off the internet or, for research of working conditions, a random worker in a given country.

  - When this last approach is used in computer science it is known as Markov Chain

Monte Carlo or MCMC for short. Often, sampling from some complicated state space also allows one to get a probabilistic estimate of the space's size. The estimate of the permanent of a large matrix of zeros and ones was the first major problem tackled using this approach.

- Random walks have also been used to sample massive online graphs such as online social networks.

- In wireless networking, a random walk is used to model node movement.

- Motile bacteria engage in a biased random walk.

- Random walks are used to model gambling.

- In physics, random walks underlie the method of Fermi estimation.

- On the web, the Twitter website uses random walks to make suggestions of who to follow

- Dave Bayer and Persi Diaconis have proven that 7 riffle shuffles are enough to mix a pack of cards. This result translates to a statement about random walk on the symmetric group which is what they prove, with a crucial use of the group structure via Fourier analysis.

Variants of Random Walks

A number of types of stochastic processes have been considered that are similar to the pure random walks but where the simple structure is allowed to be more generalized. The *pure* structure can be characterized by the steps being defined by independent and identically distributed random variables.

Random Walk on Graphs

A random walk of length k on a possibly infinite graph G with a root o is a stochastic process with random variables $X_1, X_2, \ldots, X_k$ such that $X_1 = 0$ and X_{i+1} is a vertex chosen uniformly at random from the neighbors of X_i. Then the number $p_{v,w,k}(G)$ is the probability that a random walk of length k starting at v ends at w. In particular, if G is a graph with root o, $p_{0,0,2k}$ is the probability that a $2k$ − step random walk returns to o.

Building on the analogy from the earlier section on higher dimensions, assume now that our city is no longer a perfect square grid. When our person reaches a certain junction he picks between the various available roads with equal probability. Thus, if the junction has seven exits the person will go to each one with probability one seventh. This is a random walk on a graph. Will our person reach his home? It turns out that under rather mild conditions, the answer is still yes. For example, if the lengths of all the blocks are between a and b (where a and b are any two finite positive numbers), then the person will, almost surely, reach his home. Notice that we do not assume that the graph is planar, i.e. the city may contain tunnels and bridges. One way to prove this result is using the connection to electrical networks. Take a map of the city and place a one ohm resistor on every block. Now measure the "resistance between a point and infinity". In other words, choose some number R and take all the points in the electrical network with distance bigger than R from our point and wire them together. This is now a finite electrical network and we may measure the resistance from our point to the wired points. Take R to infinity. The limit is called the *resistance between a point and infinity*. It turns out that the following is true (an elementary proof can be found in the book by Doyle and Snell):

Theorem: *a graph is transient if and only if the resistance between a point and infinity is finite. It is not important which point is chosen if the graph is connected.*

In other words, in a transient system, one only needs to overcome a finite resistance to get to infinity from any point. In a recurrent system, the resistance from any point to infinity is infinite.

This characterization of recurrence and transience is very useful, and specifically it allows us to analyze the case of a city drawn in the plane with the distances bounded.

A random walk on a graph is a very special case of a Markov chain. Unlike a general Markov chain, random walk on a graph enjoys a property called *time symmetry* or *reversibility*. Roughly speaking, this property, also called the principle of detailed balance, means that the probabilities to traverse a given path in one direction or in the other have a very simple connection between them (if the graph is regular, they are just equal). This property has important consequences.

Starting in the 1980s, much research has gone into connecting properties of the graph to random walks. In addition to the electrical network connection described above, there are important connections to isoperimetric inequalities, functional inequalities such as Sobolev and Poincaré inequalities and properties of solutions of Laplace's equation. A significant portion of this research was focused on Cayley graphs of finitely generated groups. In many cases these discrete results carry over to, or are derived from manifolds and Lie groups.

In the context of Random graphs, particularly that of the Erdős–Rényi model, analytical results to some properties of random walkers have been obtained. These include the distribution of first and last hitting times of the walker, where the first hitting time is given by the first time the walker steps into a previously visited site of the graph, and the last hitting time corresponds the first time the walker cannot perform an additional move without revisiting a previously visited site.

A good reference for random walk on graphs is the online book by Aldous and Fill. If the transition kernel $p(x, y)$ is itself random (based on an environment ω) then the random walk is called a "random walk in random environment". When the law of the random walk includes the randomness of   the law is called the annealed law; on the other hand, if ω is seen as fixed, the law is called a quenched law.

We can think about choosing every possible edge with the same probability as maximizing uncertainty (entropy) locally. We could also do it globally – in maximal entropy random walk (MERW) we want all paths to be equally probable, or in other words: for each two vertexes, each path of given length is equally probable. This random walk has much stronger localization properties.

Self-interacting Random Walks

There are a number of interesting models of random paths in which each step depends on the past in a complicated manner. All are more complex for solving analytically than the usual random walk; still, the behavior of any model of a random walker is obtainable using computers. Examples include:

- The self-avoiding walk.

The self-avoiding walk of length n on $Z^{\wedge}d$ is the random n-step path which starts at the origin, makes transitions only between adjacent sites in $Z^{\wedge}d$, never revisits a site, and is chosen uniformly among all such paths. In two dimensions, due to self-trapping, a typical self-avoiding walk is very short, while in higher dimension it grows beyond all bounds. This model has often been used in polymer physics (since the 1960s).

- The loop-erased random walk.
- The reinforced random walk.
- The exploration process.
- The multiagent random walk.

Long-range Correlated Walks

Long-range correlated time series are found in many biological, climatological and economic systems.

- Heartbeat records
- Non-coding DNA sequences
- Volatility time series of stocks
- Temperature records around the globe

Poisson Point Process

In probability, statistics and related fields, a Poisson point process or Poisson process (also called a Poisson random measure, Poisson random point field or Poisson point field) is a type of random mathematical object that consists of points randomly located on a mathematical space. The process has convenient mathematical properties, which has led to it being frequently defined in Euclidean space and used as a mathematical model for seemingly random processes in numerous disciplines such as astronomy, biology, ecology, geology, physics, image processing, and telecommunications.

The Poisson point process is often defined on the real line. For example, in queueing theory it is used to model random events, such as the arrival of customers at a store or phone calls at an exchange, distributed in time. In the plane, the point process, also known as a spatial Poisson process, may represent scattered objects such as transmitters in a wireless network, particles colliding into a detector, or trees in a forest. In this setting, the process is often used in mathematical models and in the related fields of spatial point processes, stochastic geometry, spatial statistics and continuum percolation theory. In more abstract spaces, the Poisson point process serves as an object of mathematical study in its own right.

In all settings, the Poisson point process has the property that each point is stochastically independent to all the other points in the process, which is why it is sometimes called a *purely* or *completely* random process. Despite its wide use as a stochastic model of phenomena representable as points, the inherent nature of the process implies that it does not adequately describe phenomena in which there is sufficiently strong interaction between the points. This has sometimes led to the overuse of the point process in mathematical models, and has inspired other point processes, some of which are constructed via the Poisson point process, that seek to capture this interaction.

The process is named after French mathematician Siméon Denis Poisson despite Poisson never having studied the process. Its name derives from the fact that if a collection of random points in some space forms a Poisson process, then the number of points in a region of finite size is a random variable with a Poisson distribution. The process was discovered independently in several different settings.

The process depends on a single parameter, which, depending on the context, may be a constant, a locally integrable function or, in more general settings, a Radon measure. This parameter represents the expected value of the density of the points in the Poisson process, also known as the intensity; a constant parameter represents constant intensity, a locally integrable function represents variable (location-dependent) intensity, and a Radon measure represents singular intensity. In the first case, the resulting process is called a homogeneous or stationary Poisson point process, and in the other two cases the process is called an inhomogeneous or nonhomogeneous Poisson point process. The word *point* is often omitted, but there are other *Poisson processes* of objects, which, instead of points, consist of more complicated mathematical objects such as lines and polygons, and such processes can be based on the Poisson point process.

History

Poisson Distribution

Despite its name, the Poisson point process was neither discovered nor studied by the French mathematician Siméon Denis Poisson; the name is cited as an example of Stigler's law. The name stems from its inherent relation to the Poisson distribution, derived by Poisson as a limiting case of the binomial distribution. This describes the probability of the sum of n Bernoulli trials with probability p, often likened to the number of heads (or tails) after n biased flips of a coin with the probability of a head (or tail) occurring being p. For some positive constant $\Lambda > 0$, as n increases towards infinity and p decreases towards zero such that the product $np = \Lambda$ is fixed, the Poisson distribution more closely approximates that of the binomial.

Poisson derived the Poisson distribution, published in 1841, by examining the binomial distribution in the limit of p (to zero) and n (to infinity). It only appears once in all of Poisson's work, and the result was not well-known during his time, even though over the following years a number of people would use the distribution without citing Poisson including Philipp Ludwig von Seidel and

Ernst Abbe. The distribution would be studied years after Poisson at the end of the 19th century in a different setting by Ladislaus Bortkiewicz who did cite Poisson and used the distribution with real data to study the number of deaths from horse kicks in the Prussian army.

Discovery

There are a number of claims for early uses or discoveries of the Poisson point process. It has been proposed that the earliest use of the Poisson point process was by John Michell in 1767, a decade before Poisson was born. Michell was interested in the probability a star being within a certain region of another star under the assumption that the stars were "scattered by mere chance", and studied an example consisting of the six brightest stars in the Pleiades, without deriving the Poisson distribution. This work inspired Simon Newcomb to study the problem and to calculate the Poisson distribution as an approximation for the binomial distribution.

At the beginning of the 20th century the Poisson point process would arise independently during the same period in three different situations. In 1909 the Danish mathematician and engineer A.K. Erlang derived the Poisson distribution when developing a mathematical model for the number of incoming phone calls in a finite time interval. Erlang, not at the time aware of Poisson's earlier work, assumed that the number phone calls arriving in each interval of time were independent to each other, and then found the limiting case, which is effectively recasting the Poisson distribution as a limit of the binomial distribution. In 1910 physicists Ernest Rutherford and Hans Geiger, after conducting an experiment in counting the number of alpha particles, published their results in which English mathematician Harry Bateman derived the Poisson probabilities as a solution to a family of differential equations, although Bateman acknowledged that the solutions had been previously solved by others. This experimental work by Rutherford and Geiger partly inspired physicist Norman Campbell who in 1909 and 1910 published two key papers on thermionic noise, also known as shot noise, in vacuum tubes, where it is believed he independently discovered and used the Poisson process. In Campbell's work, he also outlined a form of Campbell's theorem, a key result in the theory of point processes, but Campbell credited the proof to the mathematician G. H. Hardy. The three above discoveries and applications of the Poisson point process has motivated some to say that 1909 should be considered the discovery year of the Poisson point process.

Early Applications

The years after 1909 led to a number of studies and applications of the Poisson point process, however, its early history is complex, which has been explained by the various applications of the process in numerous fields by biologists, ecologists, engineers and others working in the physical sciences. The early results were published in different languages and in different settings, with no standard terminology and notation used. For example, in 1922 Swedish chemist and Nobel Laureate Theodor Svedberg proposed a model in which a spatial Poisson point process is the underlying process in order to study how plants are distributed in plant communities. A number of mathematicians started studying the process in the early 1930s, and important contributions were made by Andrey Kolmogorov, William Feller and Aleksandr Khinchin, among others. As an application, Kolmogorov used a spatial Poisson point process to model the formation of crystals in metals. In the field of teletraffic engineering, where a lot of the early researchers were Danes, such as Erlang, and Swedes, mathematicians and statisticians studied and used Poisson and other point processes.

History of Terms

The Swede Conny Palm in his 1943 dissertation studied the Poisson and other point processes in the one-dimensional setting by examining them in terms of the statistical or stochastic dependence between the points in time. In his work exists the first known recorded use of the term *point process* as *Punktprozesse* in German.

It is believed that William Feller, who also made the term random variable popular over a competing term *chance variable* through a coin flip with Joseph Doob, was the first in print to refer to it as the *Poisson process* in a 1940 paper. Although the Swedish statistician Ove Lundberg used the term *Poisson process* in his 1940 PhD dissertation, in which Feller was acknowledged as an influence, it has been claimed that Feller coined the term before 1940. It has been remarked that both Feller and Lundberg used the term as though it were well-known, implying it was already in spoken use. Feller worked from 1936 to 1939 alongside Swedish mathematician and statistician Harald Cramér at Stockholm University, where Lundberg was a PhD student under Cramér who did not use the term *Poisson process* in a book by him, finished in 1936, but did in subsequent editions, which his has led to the speculation that the term *Poisson process* was coined sometime between 1936 and 1939 at the Stockholm University.

Overview of Definitions

The Poisson point process is one of the most studied point processes, in both the field of probability and in more applied disciplines concerning random phenomena, due to its convenient properties as a mathematical model as well as being mathematically interesting. Depending on the setting, the process has several equivalent definitions as well definitions of varying generality owing to its many applications and characterizations. It may be defined, studied and used in one dimension (on the real line) where it can be interpreted as a counting process or part of a queueing model; in higher dimensions such as the plane where it plays a role in stochastic geometry and spatial statistics; or on more abstract mathematical spaces. Consequently, the notation, terminology and level of mathematical rigour used to define and study the Poisson point process and points processes in general vary according to the context. Despite its different forms and varying generality, the Poisson point process has two key properties.

First Key Property: Poisson Distributed Number of Points

The Poisson point process is related to the Poisson distribution, which implies that the probability of a Poisson random variable N is equal to n is given by:

$$P\{N = n\} = \frac{\Lambda^n}{n!}e^{-\Lambda}$$

where $n!$ denotes n factorial and Λ is the single Poisson parameter that is used to define the Poisson distribution. If a Poisson point process is defined on some underlying mathematical space, called a *state space* or *carrier space*, then the number of points in a bounded region of the space will be a Poisson random variable with some parameter whose form will depend on the setting.

Second key property: Complete Independence

The other key property is that for a collection of disjoint and bounded subregions of the underlying space, the number of points in each bounded subregion will be completely independent of all the others. This property is known under several names such as *complete randomness*, *complete independence*, or *independent scattering* and is common to all Poisson point processes. In other words, there is a lack of interaction between different regions and the points in general, which motivates the Poisson process being sometimes called a *purely* or *completely* random process.

Different Definitions

The Poisson point process is often defined on the real line in the homogeneous setting, and then extended to a more general settings with more mathematical rigour. For all the instances of the Poisson point process, the two key properties of the Poisson distribution and complete independence play an important role.

Homogeneous Poisson Point Process

If a Poisson point process has a constant parameter, say, λ, then it is called a homogeneous or stationary Poisson point process. The parameter, called *rate* or *intensity*, is related to the expected (or average) number of Poisson points existing in some bounded region. In fact, the parameter λ can be interpreted as the average number of points per some unit of extent such as length, area, volume, or time, depending on the underlying mathematical space, hence it is sometimes called the *mean density*.

Defined on the Real Line

Consider two real numbers a and b, where $a \leq b$, and which may represent points in time. Denote by $N(a,b]$ the random number of points of a homogeneous Poisson point process existing with values greater than a but less than or equal to b, or in other words, the number of points of the process in the interval $(a,b]$. If the points form or belong to a homogeneous Poisson process with parameter $\lambda > 0$, then the probability of n points existing in the above interval $(a,b]$ is given by:

$$P\{N(a,b] = n\} = \frac{[\lambda(b-a)]^n}{n!} e^{-\lambda(b-a)},$$

In other words, $N(a,b]$ is a Poisson random variable with mean $\lambda(b-a)$. Furthermore, the number of points in any two disjoint intervals, say, $(a_1,b_1]$ and $(a_2,b_2]$ are independent of each other, and this extends to any finite number of disjoint intervals. In the queueing theory context, one can consider a point existing (in an interval) as an *event*, but this is different to the word event in the probability theory sense. It follows that λ is the expected number of *arrivals* that occur per unit of time, and it is sometimes called the *rate parameter*.

For a more formal definition of a stochastic process, such as a point process, one can use the Kolmogorov theorem, which in this context gives the joint probability of some number of points existing in each disjoint finite interval. More specifically, let $N(a_i,b_i]$ denote the number of points of (a point process) happening in the half-open interval $(a_i, b_i]$, where the real numbers $a_i < b_i \leq a_{i+1}$. Then for some positive integer k, the homogeneous Poisson point process on the real line with parameter $\lambda > 0$ is defined with the finite-dimensional distribution:

$$P\{N(a_i, b_i] = n_i, i = 1, \ldots, k\} = \prod_{i=1}^{k} \frac{[\lambda(b_i - a_i)]^{n_i}}{n_i!} e^{-\lambda(b_i - a_i)},$$

Key Properties

The above definition has two important features pertaining to the Poisson point processes in general:

- the number of points in each finite interval has a Poisson distribution;

- the number of points in disjoint intervals are independent random variables.

Furthermore, it has a third feature related to just the homogeneous Poisson process:

- the distribution of each interval $(a + t, b + t]$ only depends on the interval's length $b - a$..

In other words, for any finite $t > 0$, the random variable $N(a + t, b + t]$ is independent of t, and, hence, the process is Stationary process, which is why it is sometimes called the *stationary Poisson process*.

Law of Large Numbers

The quantity $\lambda(b_i - a_i)$ can be interpreted as the expected or average number of points occurring in the interval $(a_i, b_i]$, namely:

$$E\{N(a_i, b_i]\} = \lambda(b_i - a_i),$$

where E denotes the expectation operator. In other words, the parameter λ of the Poisson process coincides with the *density* of points. Furthermore, the homogeneous Poisson point process adheres to its own form of the (strong) law of large numbers. More specifically, with probability one:

$$\lim_{t \to \infty} \frac{N(t)}{t} = \lambda,$$

where lim denotes the limit of a function.

Memoryless Property

The distance between two consecutive points of a point process on the real line will be an exponential random variable with parameter λ (or equivalently, mean $1/\lambda$). This implies that the points have the memoryless property: the existence of one point existing in a finite interval does not affect the probability (distribution) of other points existing. This property is directly related to the complete independence of the Poisson process, however, it has no natural equivalence when the Poisson process is defined in higher dimensions.

Orderliness and Simplicity

A stochastic process with stationary increments is sometimes said to be *orderly*, *ordinary* or *regular* if

$$P\{N(t,t+\delta]>1\}=o(\delta),$$

where little-o notation is used. A point process is called a simple point process when the probability of any of its two points coinciding in the same position (on the underlying state space) is zero. For point processes in general on the real line, the (probability distribution) property of orderliness implies that the process is simple or has the (sample path) property of *simplicity*, which is the case for the homogeneous Poisson point process.

Relationship to Other Processes

On the real line, the Poisson point process is a type of continuous-time Markov process known as a birth-death process (with just births and zero deaths) and is called a *pure* or *simple* birth process. More complicated processes with the Markov property, such as Markov arrival processes, have been defined where the Poisson process is a special case.

Counting Process Interpretation

The homogeneous Poisson point process, when considered on the positive half-line, is sometimes defined as a counting process, which can be denoted as $\{N(t),t\geq 0\}$. A counting process represents the total number of occurrences or events that have happened up to and including time t. A counting process is a Poisson counting process with rate $\lambda>0$ if it has the following three properties:

- $N(0)=0$;

- has independent increments; and

- the number of events (or points) in any interval of length t is a Poisson random variable with parameter (or mean) λt.

The last property implies

$$E[N(t)]=\lambda t.$$

The Poisson counting process can also be defined by stating that the time differences between events of the counting process are exponential variables with mean $1/\lambda$. The time differences between the events or arrivals are known as interarrival or interoccurence times. These two definitions of the Poisson counting process agree with the previous definition of the Poisson point process.

Martingale Characterization

On the real line, the homogeneous Poisson point process has a connection to the theory of martingales via the following characterization: a point process is the homogeneous Poisson point process if and only if

$$N(-\infty, t] - t,$$

is a martingale.

Restricted to the Half-line

If the homogeneous Poisson point process is considered just on the half-line $[0, \infty)$, which is often the case when t represents time, as it does for the previous counting process, then the resulting process is not truly invariant under translation. In that case the process is no longer stationary, according to some definitions of stationarity.

Applications

There have been many applications of the homogeneous Poisson point process on the real line in an attempt to model seemingly random and independent events occurring. It has a fundamental role in queueing theory, which is the probability field of developing suitable stochastic models to represent the random arrival and departure of certain phenomena. For example, customers arriving and being served or phone calls arriving at a phone exchange can be both studied with techniques from queueing theory. In the original paper proposing the online payment system known as Bitcoin featured a mathematical model based on a homogeneous Poisson point process.

Generalizations

The Poisson counting process or, more generally, the homogeneous Poisson point process on the real line is considered one of the simplest stochastic processes for counting random numbers of points. The process can be generalized in a number of ways. One possible generalization is to extend the distribution of interarrival times from the exponential distribution to other distributions, which introduces the stochastic process known as a renewal process. Another generalization is to define it on higher dimensional spaces such as the plane.

Spatial Poisson Point Process

A spatial Poisson process is a Poisson point process defined in the plane $\mathbf{R}^2$. For its definition, consider a bounded, open or closed (or more precisely, Borel measurable) region B of the plane. Denote by $N(B)$ the (random) number of points of N existing in this region $B \subset \mathbf{R}^2$. If the points belong to a homogeneous Poisson process with parameter $\lambda > 0$, then the probability of n points existing in B is given by:

$$P\{N(B) = n\} = \frac{(\lambda \mid B \mid)^n}{n!} e^{-\lambda |B|}$$

where $\mid B \mid$ denotes the area of B.

More formally, for some finite integer $k \geq 1$, consider a collection of disjoint, bounded Borel (measurable) sets $B_1, \ldots, B_k$. Let $N(B_i)$ denote the number of points of existing in $N(B_i)$. Then the homogeneous Poisson point process with parameter B_i has the finite-dimensional distribution

$$P\{N(B_i)=n_i, i=1,\ldots,k\}=\prod_{i=1}^{k}\frac{(\lambda\,|\,B_i\,|)^{n_i}}{n_i!}e^{-\lambda|B_i|}.$$

Applications

According to one statistical study, the positions of cellular or mobile phone base stations in the Australian city Sydney, pictured above, resemble a Poisson point process, while in many other cities around the world they do not and other point processes are required.

The spatial Poisson point process features prominently in spatial statistics, stochastic geometry, and continuum percolation theory. This process is applied in various physical sciences such as a model developed for alpha particles being detected. In recent years, it has been frequently used to model seemingly disordered spatial configurations of certain wireless communication networks. For example, models for cellular or mobile phone networks have been developed where it is assumed the phone network transmitters, known as base stations, are positioned according to a homogeneous Poisson point process.

Defined in Higher Dimensions

The previous homogeneous Poisson point process immediately extends to higher dimensions by replacing the notion of area with (high dimensional) volume. For some bounded region B of Euclidean space $\mathbf{R}^d$, if the points form a homogeneous Poisson process with parameter $\lambda > 0$, then the probability of n points existing in $B \subset \mathbf{R}^d$ is given by:

$$P\{N(B)=n\}=\frac{(\lambda\,|\,B\,|)^{n}}{n!}e^{-\lambda|B|}$$

where $|B|$ now denotes the n-dimensional volume of B.. Furthermore, for a collection of disjoint, bounded Borel sets $B_1,\ldots,B_k \subset \mathbf{R}^d$, let $N(B_i)$ denote the number of points of N existing in B_i. Then the corresponding homogeneous Poisson point process with parameter $\lambda > 0$ has the finite-dimensional distribution

$$P\{N(B_i)=n_i, i=1,\ldots,k\}=\prod_{i=1}^{k}\frac{(\lambda\,|\,B_i\,|)^{n_i}}{n_i!}e^{-\lambda|B_i|}.$$

Homogeneous Poisson point processes do not depend on the position of the underlying state space through its parameter λ, which implies it is both a stationary process (invariant to translation) and an isotropic (invariant to rotation) stochastic process. Similarly to the one-dimensional case, the homogeneous point process is restricted to some bounded subset of $\mathbf{R}^d$, then depending on some definitions of stationarity, the process is no longer stationary.

Points are Uniformly Distributed

If the homogeneous point process is defined on the real line as a mathematical model for occurrences of some phenomenon, then it has the characteristic that the positions of these occurrences or events on the real line (often interpreted as time) will be uniformly distributed. More specifically, if an event occurs (according to this process) in an interval $(a - b]$ where $a \leq b$, then its location will be a uniform random variable defined on that interval. Furthermore, the homogeneous point process is sometimes called the *uniform* Poisson point process. This uniformity property extends to higher dimensions in the Cartesian coordinate, but it does not hold in other coordinate systems (for example, polar or spherical).

Inhomogeneous Poisson Point Process

The inhomogeneous or nonhomogeneous Poisson point process is a Poisson point process with a Poisson parameter set as some location-dependent function in the underlying space on which the Poisson process is defined. For Euclidean space R^d, this is achieved by introducing a locally integrable positive function $\lambda(x)$, where x is a d-dimensional point located in R^d, such that for any bounded region B the (d – dimensional) volume integral of $\lambda(x)$ over region B is finite. In other words, if this integral, denoted by $\Lambda(B)$, is:

$$\Lambda(B) = \int_B \lambda(x)dx < \infty,$$

where dx is a (d-dimensional) volume element,[c] then for any collection of disjoint bounded Borel measurable sets $B_1, \ldots, B_k$, an inhomogeneous Poisson process with (intensity) function $\lambda(x)$ has the finite-dimensional distribution:

$$P\{N(B_i) = n_i, i = 1, \ldots, k\} = \prod_{i=1}^{k} \frac{(\Lambda(B_i))^{n_i}}{n_i!} e^{-\Lambda(B_i)}.$$

Furthermore, $\Lambda(B)$ has the interpretation of being the expected number of points of the Poisson process located in the bounded region B, namely

$$\Lambda(B) = E[N(B)].$$

Defined on the Real Line

On the real line, the inhomogeneous or non-homogeneous Poisson point process has mean measure given by a one-dimensional integral. For two real numbers a and b, where $a \leq b$, denote by $N(a,b]$ the number points of an inhomogeneous Poisson process with intensity function $\lambda(t)$ with values greater than a but less than or equal to b. The probability of n points existing in the above interval $(a,b]$ is given by:

$$P\{N(a,b]=n\} = \frac{[\Lambda(a,b)]^n}{n!} e^{-\Lambda(a,b)}.$$

where the mean or intensity measure is:

$$\Lambda(a,b) = \int_a^b \lambda(t)dt,$$

which means that the random variable $N(a,b]$ is a Poisson random variable with mean $E\{N(a,b]\} = \Lambda(a,b)$.

A feature of the one-dimension setting considered useful is that an inhomogeneous Poisson point process can be transformed into a homogeneous by a monotone transformation or mapping, which is achieved with the inverse of Λ.

Counting Process Interpretation

The inhomogeneous Poisson point process, when considered on the positive half-line, is also sometimes defined as a counting process. With this interpretation, the process, which is sometimes written as $\{N(t),t \geq 0\}$, represents the total number of occurrences or events that have happened up to and including time t. A counting process is said to be an inhomogeneous Poisson counting process if it has the four properties:

- $N(0) = 0$;

- has independent increments;

- $P\{N(t+h) - N(t) = 1\} = \lambda(t)h + o(h)$; and

- $P\{N(t+h) - N(t) \geq 2\} = o(h)$,

where $o(h)$ is asymptotic or little-o notation for $o(h)/h \to 0$ as $h \to 0$. In the case of point processes with refractoriness (e.g., neural spike trains) a stronger version of property 4 applies: $P(N(t+h) - N(t) \geq 2) = o(h^2)$.

The above properties imply that $N(t+h) - N(t)$ is a Poisson random variable with the parameter (or mean)

$$E[N(t+h) - N(t)] = \int_t^{t+h} \lambda(s)ds,$$

which implies

$$E[N(h)] = \int_0^h \lambda(s)ds.$$

Spatial Poisson Point Process

An inhomogeneous Poisson process, just like a homogeneous Poisson process, defined in the plane $\mathbf{R}^2$ is called a spatial Poisson point process. Calculating its intensity measure requires performing an area integral of its intensity function over some region. For example, its intensity function (as a function of Cartesian coordinates x and y) may be

$$\lambda(x, y) = e^{-(x^2+y^2)},$$

hence it has an intensity measure given by the area integral

$$\Lambda(B) = \int_B e^{-(x^2+y^2)} dx dy,$$

where B is some bounded region in the plane R^2. The previous intensity function can be re-written, via a change of coordinates, in polar coordinates as

$$\lambda(r, \theta) = e^{-r^2},$$

which reveals that the intensity function in this example is independent of the angular coordinate θ, or, in other words, it is isotropic or rotationally invariant. The intensity measure is then given by the area integral

$$\Lambda(B') = \int_{B'} e^{-r^2} r dr dd\theta,$$

where B' is some bounded region in the plane R^2.

In Higher Dimensions

In the plane, $\Lambda(B)$ corresponds to an area integral while in $\mathbf{R}^d$ the integral becomes a (d-dimensional) volume integral.

Applications

The real line, as mentioned earlier, is often interpreted as time and in this setting the inhomogeneous process is used in the fields of counting processes and in queueing theory. Examples of phenomena which have been represented by or appear as an inhomogeneous Poisson point process include:

- Goals being scored in a soccer game.

- Defects in a circuit board

In the plane, the Poisson point process is of fundamental importance in the related disciplines of stochastic geometry and spatial statistics. This point process is not stationary owing to the fact that its distribution is dependent on the location of underlying space or state space. Hence, it can be used to model phenomena with a density that varies over some region. In other words, the phenomena can be represented as points that have a location-dependent density. Uses for this process as a mathematical model are diverse and have appeared across various disciplines including the study of salmon and sea lice in the oceans, forestry, and search problems.

Interpretation of the Intensity Function

The Poisson intensity function $\lambda(x)$ has an interpretation, considered intuitive, with the volume element dx in the infinitesimal sense: $\lambda(x)dx$ is the infinitesimal probability of a point of a Poisson point process existing in a region of space with volume dx located at x.

For example, given a homogeneous Poisson point process on the real line, the probability of finding a single point of the process in a small interval of width δ is approximately $\lambda\delta x$. In fact, such intuition is how the Poisson point process is sometimes introduced and its distribution derived.

Simple Point Process

If a Poisson point process has an intensity measure that is a locally finite and diffuse (or non-atomic), then it is a simple point process. For a simple point process, the probability of a point existing at a single point or location in the underlying (state) space is either zero or one. This implies that, with probability one, no two (or more) points of a Poisson point process coincide in location in the underlying space.

Simulation

Simulating a Poisson point process on a computer is usually done in a bounded region of space, known as a simulation *window*, and requires two steps: appropriately creating a random number of points and then suitably placing the points in a random manner. Both these two steps depend on the specific Poisson point process that is being simulated.

Step 1: Number of Points

The number of points N in the window, denoted here by W, needs to be simulated, which is done by using a (pseudo)-random number generating function capable of simulating Poisson random variables.

Homogeneous Case

For the homogeneous case with the constant λ, the mean of the Poisson random variable N is set to $\lambda|W|$ where $|W|$ is the length, area or (d-dimensional) volume of W.

Inhomogeneous Case

For the inhomogeneous case, $\lambda|W|$ is replaced with the (d-dimensional) volume integral

$$\Lambda(W) = \int_W \lambda(x)dx$$

Step 2: Positioning of Points

The second stage requires randomly placing the N points in the window W.

Homogeneous Case

For the homogeneous case in one dimension, all points are uniformly and independently placed in the window or interval W. For higher dimensions in a Cartesian coordinate system, each coordinate is uniformly and independently placed in the window W. If the window is not a subspace of Cartesian space (for example, inside a unit sphere or on the surface of a unit sphere), then the points will not be uniformly placed in W, and suitable change of coordinates (from Cartesian) are needed.

Inhomogeneous Case

For the inhomogeneous, a couple of different methods can be used depending on the nature of the intensity function $\lambda(x)$.. If the intensity function is sufficiently simple, then independent and random non-uniform (Cartesian or other) coordinates of the points can be generated. For example, simulating a Poisson point process on a circular window can be done for an isotropic intensity function (in polar coordinates r and θ), implying it is rotationally variant or independent of θ but dependent on , by a change of variable in r if the intensity function is sufficiently simple.

For more complicated intensity functions, one can use an acceptance-rejection method, which consists of using (or 'accepting') only certain random points and not using (or 'rejecting') the other points, based on the ratio

$$\frac{\lambda(x_i)}{\Lambda(W)} = \frac{\lambda(x_i)}{\int_W \lambda(x)dx}.$$

where x_i is the point under consideration for acceptance or rejection.

General Poisson Point Process

The Poisson point process can be further generalized to what is sometimes known as the general Poisson point process by using a Radon measure Λ, hence this measure is locally finite. The Radon measure can be atomic, that is it can have atoms at points in the underlying state space, while some researchers assume the converse where the Radon measure Λ is diffuse or non-atomic. If the measure is atomic Λ, then the number of points at x is a Poisson random variable with mean $\Lambda(x)$.

Assuming that the underlying space of the Poisson point process is $\mathbf{R}^d$ (the space can be more general), then $\Lambda(\{x\}) = 0$ for any single point x in $\mathbf{R}^d$ and $\Lambda(B)$ is finite for any bounded subset B of $\mathbf{R}^d$. Then a point process N is a general Poisson point process with intensity $\ddot{E}$ if it has the two following properties:

- the number of points in a bounded Borel set B is a Poisson random variable with mean $\Lambda(B)$. In other words, denote the total number of points located in B by $N(B)$, then the probability that the random variable $N(B)$ is equal to n is given by:

$$P\{N(B) = n\} = \frac{(\Lambda(B))^n}{n!} e^{-\Lambda(B)}$$

- the number of points in n disjoint Borel sets forms n independent random variables.

The Radon measure $\ddot{E}$ maintains its previous interpretation of being the expected number of points of N located in the bounded region B, namely

$$\Lambda(B) = E[N(B)].$$

Furthermore, if Λ is absolutely continuous such that it has a density (or more precisely, a Radon–Nikodym density or derivative) with respect to the Lebesgue measure, then for all Borel sets B it can be written as:

$$\Lambda(B) = \int_B \lambda(x)dx,$$

where the density $\lambda(x)$ is known, among other terms, as the intensity function.

Terminology

In addition to the word *point* often being omitted, the terminology of the Poisson point process and point process theory varies, which has been criticized. The homogeneous Poisson (point) process is also called a *stationary* Poisson (point) process, sometimes the *uniform* Poisson (point) process, and in the past it was, by William Feller and others, referred to as a *Poisson ensemble* of points. The term *point process* has been criticized and some authors prefer the term *random point field*, hence the terms *Poisson random point field* or *Poisson point field* are also used. A point process is considered, and sometimes called, a random counting measure, hence the Poisson point process is also referred to as a *Poisson random measure*, a term used in the study of Lévy processes, but some choose to use the two terms for slightly different random objects.

The inhomogenous Poisson point process, as well as be being called *nonhomogeneous* or *non-homogeneous*, is sometimes referred to as the *non-stationary*, *heterogeneous* or *spatially dependent* Poisson (point) process.

The measure Λ is sometimes called the *parameter measure* or *intensity measure* or *mean measure*. If Λ has a derivative or density, denoted by $\lambda(x)$, it may be called the *intensity function* of the general Poisson point process or simply the *rate* or *intensity*, since there are no standard terms. For the homogeneous Poisson point process, the intensity is simply a constant $\lambda > 0$, which can be referred to as the *mean rate* or *mean density* or *rate parameter*. For $\lambda = 1$, the corresponding process is sometimes referred to as the *standard Poisson* (point) process.

The underlying mathematical space on which the point process, Poisson or other, is defined is known as a *state space* or *carrier space*.

Notation

The notation of the Poisson point process depends on its setting and the field it is being applied in. For example, on the real line, the Poisson process, both homogeneous or inhomogeneous, is sometimes interpreted as a counting process, and the notation $\{N(t), t \geq 0\}$ is used to represent the Poisson process.

Another reason for varying notation is due to the theory of point processes, which has a couple of mathematical interpretations. For example, a simple Poisson point process may be considered as a random set, which suggests the notation $x \in N$, implying that N is a random point belonging to or being an element of the Poisson point process N. Another, more general, interpretation is to consider a Poisson or any other point process as a random counting measure, so one can write the number of points of a Poisson point process N being found or located in some (Borel measurable) region B as $N(B)$, which is a random variable. These different interpretations results in notation being used from mathematical fields such as measure theory and set theory.

For general point processes, sometimes a subscript on the point symbol, for example x, is included so one writes (with set notation) $x_i \in N$ instead of $x \in N$, and x can be used for the dummy variable in integral expressions such as Campbell's theorem, instead of denoting random points. Sometimes an uppercase letter denotes the point process, while a lowercase denotes a point from the processs, so, for example, the point x or x_i belongs to or is a point of the point process X, and be written with set notation as $x \in X$ or $x_i \in X$.

Furthermore, the set theory and integral or measure theory notation can be used interchangeably. For example, for a point process N defined on the Euclidean state space $\mathbf{R}^d$ and a (measurable) function f on $\mathbf{R}^d$, the expression

$$\int_{\mathbf{R}^d} f(x)N(dx) = \sum_{x_i \in N} f(x_i),$$

demonstrates two different ways to write a summation over a point process. More specifically, the integral notation on the left-hand side is interpreting the point process as a random counting measure while the sum on the right-hand side suggests a random set interpretation.

Functionals and Moment Measures

In probability theory, operations are applied to random variables for different purposes. Sometimes these operations are regular expectations that produce the average or variance of a random variable. Others, such as characteristic functions (or Laplace transforms) of a random variable can be used to uniquely identify or characterize random variables and prove results like the central limit theorem. In the theory of point processes there exist analogous mathematical tools which usually exist in the forms of measures and functionals instead of moments and functions respectively. For measures, often their densities (or Radon-Nikodym derivatives), if they exist, are also expressed with respect to the Lebesgue measure.

Laplace Functionals

For a Poisson point process N with intensity measure Λ, the Laplace functional is given by:

$$L_N(f) = e^{-\int_{\mathbf{R}^d}(1-e^{f(x)})\Lambda(dx)},$$

which for the homogeneous case is:

$$L_N(f) = e^{-\lambda \int_{\mathbf{R}^d}(1-e^{f(x)})dx}.$$

One version of Campbell's theorem involves the Laplace functional of the Poisson point process.

Probability Generating Functionals

The probability generating function of non-negative integer-valued random variable leads to the probability generating functional being defined analogously with respect to any non-negative bounded function v on $\mathbf{R}^d$ such that $0 \le v(x) \le 1$. For a point process N the probability generating functional is defined as:

$$G(v) = E\left[\prod_{x \in N} v(x)\right]$$

where the product is performed for all the points in N. If the intensity measure Λ of N is locally finite, then the G is well-defined for any measurable function u on $\mathbf{R}^d$. For a Poisson point process with intensity measure Λ the generating functional is given by:

$$G(v) = e^{-\int_{\mathbf{R}^d}[1-v(x)]\Lambda(dx)},$$

which in the homogeneous case reduces to

$$G(v) = e^{-\lambda\int_{\mathbf{R}^d}[1-v(x)]dx}.$$

Moment Measure

For a general Poisson point process with intensity measure Λ the first moment measure is its intensity measure:

$$M^1(B) = \Lambda(B),$$

which for a homogeneous Poisson point process with constant intensity λ means:

$$M^1(B) = \lambda\,|\,B\,|,$$

where $|\,B\,|$ is the length, area or volume (or more generally, the Lebesgue measure) of B.

For the Poisson case with measure Λ the second moment measure is:

$$M^2(B) = \Lambda(B) + \Lambda(B)^2.$$

which in the homogeneous case reduces to

$$M^2(B) = \lambda\,|\,B\,| + (\lambda\,|\,B\,|)^2.$$

Factorial Moment Measure

For a general Poisson point process with intensity measure Λ the n-th factorial moment measure is given by the expression:

$$M^{(n)}(B_1 \times, \ldots, \times B_n) = \prod_{i=1}^{n}[\Lambda(B_i)],$$

where Λ is the intensity measure or first moment measure of N, which for some Borel set B is given by:

$$\Lambda(B) = M^1(B) = E[N(B)].$$

For a homogeneous Poisson point process the n-th factorial moment measure is simply:

$$M^{(n)}(B_1 \times, \ldots, \times B_n) = \lambda^n \prod_{i=1}^{n}|B_i\,|,$$

where $|B_i|$ is the length, area, or volume (or more generally, the Lebesgue measure) of B_i. Furthermore, the n-th factorial moment density is:

$$\mu^{(n)}(x_1,\ldots,x_n) = \lambda^n.$$

Avoidance Function

The avoidance function or void probability v of a point process N is defined in relation to some set B, which is a subset of the underlying space $\mathbf{R}^d$, as the probability of no points of N existing in B. More precisely, for a test set B, the avoidance function is given by:

$$v(B) = P(N(B) = 0).$$

For a general Poisson point process N with intensity measure Λ, its avoidance function is give by:

$$v(B) = e^{-\Lambda(B)}$$

Rényi's Theorem

It can be shown that simple point processes are completely characterized by their void probabilities. In other words, complete information of a simple point process is captured entirely in its void probabilities. The case for Poisson process is sometimes known as Rényi's theorem, which is named after Alfréd Rényi who discovered the result for the case of a homogeneous point process in one-dimension.

In one form, the Rényi's theorem says for a diffuse (or non-atomic) Radon measure Ë on $\mathbf{R}^d$ and a set A is a finite union of rectangles (so not Borel[d]) that if N is a countable subset of $\mathbf{R}^d$ such that:

$$P(N(A) = 0) = v(A) = e^{-\Lambda(A)}$$

then N is a Poisson point process with intensity measure

Point Process Operations

Mathematical operations can be performed on point processes in order to develop suitable mathematical models. One example of an operation is known as thinning which entails deleting or removing the points of some point process according to a rule, hence creating a new process with the remaining points (the deleted points also form a point process). Another example of a point process operation is superimposing (or combining) point processes into one point process.

One of the reasons why the Poisson point process is often used as model is that, under suitable conditions, when performed on a Poisson point process these operations often produce another (usually different) Poisson point process, demonstrating an aspect of mathematical closure. The operations can also be used to create new point processes, which are then also used as mathematical models for the random placement of certain objects.

Thinning

For the Poisson process, the independent $p(x)$-thinning operations results in another Poisson point process. More specifically, a $p(x)$-thinning operation applied to a Poisson point process with intensity measure Ë gives a point process of removed points that is also Poisson point process N_p with intensity measure Λ_p, which for a bounded Borel set B is given by:

$$\Lambda_p(B) = \int_B p(x)\Lambda(dx)$$

Furthermore, after randomly thinning a Poisson point process, the kept or remaining points also form a Poisson point process, which has the intensity measure

$$\Lambda_p(B) = \int_B (1 - p(x))\Lambda(dx).$$

The two separate Poisson point processes formed respectively from the removed and kept points are stochastically independent of each other. In other words, if a region is known to contain n kept points (from the original Poisson point process), then this will have no influence on the random number of removed points in the same region. This ability to randomly create two independent Poisson point processes from one is sometimes known as *splitting* the Poisson point process.

Superposition

If there is a countable collection of point processes $N_1, N_2 \ldots$, then their superposition, or, in set theory language, their union

$$N = \bigcup_{i=1}^{\infty} N_i,$$

also forms a point process. In other words, any points located in any of the point processes $N_1, N_2 \ldots$ will also be located in the superposition of these point processes

Superposition Theorem

The Superposition theorem of the Poisson point process, which stems directly from the complete independence property, says that the superposition of independent Poisson point processes $N_1, N_2 \ldots$ with mean measures $\Lambda_1, \Lambda_2, \ldots$ will also be a Poisson point process with mean measure

$$\Lambda = \sum_{i=1}^{\infty} \Lambda_i.$$

In other words, the union of two (or countably more) Poisson processes is another Poisson process. If a point x is sampled from a countable n union of Poisson processes, then the probability that the point x belongs to the jth Poisson process N_j is given by:

$$P(x \in N_j) = \frac{\Lambda_j}{\sum_{i=1}^{n} \Lambda_i}.$$

Homogeneous Case

In the homogeneous case with constant $\lambda_1, \lambda_2 \ldots$, the two previous expressions reduce to

$$\lambda = \sum_{i=1}^{\infty} \lambda_i,$$

and

$$P(x \in N_j) = \frac{\lambda_j}{\sum_{i=1}^{n} \lambda_i}.$$

Clustering

The operation clustering is performed when each point x of some point process N is replaced by another (possibly different) point process. If the original process N is a Poisson point process, then the resulting process N_c is called a Poisson cluster point process.

Random Displacement

A mathematical model may require randomly moving points of a point process to other locations on the underlying mathematical space, which gives rise to a point process operation known as displacement or translation. The Poisson point process has been used to model, for example, the movement of plants between generations, owing to the displacement theorem, which loosely says that the random independent displacement of points of a Poisson point process (on the same underlying space) forms another Poisson point process.

Displacement Theorem

One version of the displacement theorem entails first considering a Poisson point process N on $\mathbf{R}^d$ with intensity function $\lambda(x)$. It is then assumed the points of N are randomly displaced somewhere else in $\mathbf{R}^d$ so that each point's displacement is independent and that the displacement of a point formerly at x is a random vector with a probability density $\rho(x, \cdot)$. [e] Then the new point process N_D is also a Poisson point process with intensity function

$$\lambda_D(y) = \int_{\mathbf{R}^d} \lambda(x)\rho(x, y)dx,$$

which for the homogeneous case with a constant $\lambda > 0$ means

$$\lambda_D(y) = \lambda.$$

In other words, after each random and independent displacement of points, the original Poisson point process still exists.

The displacement theorem can be extended such that the Poisson points are randomly displaced from one Euclidean space $\mathbf{R}^d$ to another Euclidean space $\mathbf{R}^{d'}$, where $d' \geq 1$ is not necessarily equal to d.

Mapping

Another property that is considered useful is the ability to map a Poisson point process from one underlying space to another space.

Mapping Theorem

If the mapping (or transformation) adheres to some conditions, then the resulting mapped (or transformed) collection of points also form a Poisson point process, and this result is sometimes referred to as the Mapping theorem. The theorem involves some Poisson point process with mean measure on some underlying space. If the locations of the points are mapped (that is, the point process is transformed) according to some function to another underlying space, then the resulting point process is also a Poisson point process but with a different mean measure Λ'.

More specifically, one can consider a (Borel measurable) function f that maps a point process N with intensity measure Λ from one space S, to another space T in such a manner so that the new point process N' has the intensity measure:

$$\Lambda(B)' = \Lambda(f^{-1}(B))$$

with no atoms, where B is a Borel set and f^{-1} denotes the inverse of the function f. If N is a Poisson point process, then the new process N' is also a Poisson point process with the intensity measure Λ'.

Approximations with Poisson Point Processes

The tractability of the Poisson process means that sometimes it is convenient to approximate a non-Poisson point process with a Poisson one. The overall aim is to approximate the both number of points of some point process and the location of each point by a Poisson point process. There a number of methods that can be used to justify, informally or rigorously, approximating the occurrence of random events or phenomena with suitable Poisson point processes. The more rigorous methods involve deriving upper bounds on the probability metrics between the Poisson and non-Poisson point processes, while other methods can be justified by less formal heuristics.

Clumping Heuristic

One method for approximating random events or phenomena with Poisson processes is called the clumping heuristic. The general heuristic or principle involves using the Poisson point process (or Poisson distribution) to approximate events, which are considered rare or unlikely, of some stochastic process. In some cases these rare events are close to independent, hence a Poisson point process can be used. When the events are not independent, but tend to occur in clusters or *clumps*, then if these clumps are suitably defined such that they are approximately independent of each other, then the number of clumps occurring will be close to a Poisson random variable and the locations of the clumps will be close to a Poisson process.

Stein's Method

Stein's method, a rigorous mathematical technique originally developed for approximating random variables such as Gaussian and Poisson variables, has also been developed and applied to point processes. Stein's method can be used to derive upper bounds on probability metrics, which give way to quantify how different two random mathematical objects vary stochastically, of the Poisson and other point processes. Upperbounds on probability metrics such as total variation and Wasserstein distance have been derived.

Researchers have applied Stein's method to Poisson point processes in a number of ways, such as using Palm calculus. Techniques based on Stein's method have been developed to factor into the upper bounds the effects of certain point process operations such as thinning and superposition. Stein's method has also been used to derive upper bounds on metrics of Poisson and other processes such as the Cox point process, which is a Poisson process with a random intensity measure.

Convergence to a Poisson Point Process

In general, when an operation is applied to a general point process the resulting process is usually not a Poisson point process. For example, if a point process, other than a Poisson, has its points randomly and independently displaced, then the process would not necessarily be a Poisson point process. However, under certain mathematical conditions for both the original point process and the random displacement, it has been shown via limit theorems that if the points of a point process are repeatedly displaced in a random and independent manner, then the finite-distribution of the point process will converge (weakly) to that of a Poisson point process.

Similar convergence results have been developed for thinning and superposition operations that show that such repeated operations on point processes can, under certain conditions, result in the process converging to a Poisson point processes, provided a suitable rescaling of the intensity measure (otherwise values of the intensity measure of the resulting point processes would approach zero or infinity). Such convergence work is directly related to the results known as the Palm–Khinchin[f] equations, which has its origins in the work of Conny Palm and Aleksandr Khinchin, and help explains why the Poisson process can often be used as a mathematical model of various random phenomena.

Generalizations of Poisson Point Processes

The Poisson point process can be generalized by, for example, changing its intensity measure or defining on more general mathematical spaces. These generalizations can be studied mathematically as well as used to mathematically model or represent physical phenomena.

Poisson Point Processes on More General Spaces

For mathematical models the Poisson point process is often defined in Euclidean space, but has been generalized to more abstract spaces and plays a fundamental role in the study of random measures, which requires an understanding of certain mathematical fields such as probability theory, measure theory, topology and functional analysis.

In general, the concept of distance is of practical interest for applications while topological structure is needed for Palm distributions, hence point processes are often defined on mathematical spaces equipped with metrics. The necessity of convergence of sequences requires the space to be complete, which has inspired point processes to be studied on specific complete metric spaces.

Furthermore, every realization of a point process in general can be regarded as a counting measure, which has motivated point processes being considered as random measures. Using the techniques of random measures, the Poisson and other point processes has been defined and studied on a locally compact second countable Hausdorff space.

Cox Point Process

A Poisson point process can be generalized by letting its intensity measure Λ to be also random and independent of the underlying Poisson process, which gives rise to the Cox process or doubly stochastic Poisson process, introduced by David Cox in 1955 under the latter name. The intensity measure may be a realization of random variable or a random field. For example, if the logarithm of the intensity measure is a Gaussian random field, then the resulting process is known as a *log Gaussian Cox process*. More generally, the intensity measures is a realization of a non-negative locally finite random measure. Cox point processes exhibit a *clustering* of points, which can be shown mathematically to be larger than those of Poisson point processes. The generality and tractability of Cox processes has resulted in them being used as models in fields such as spatial statistics and wireless networks.

Marked Poisson Point Process

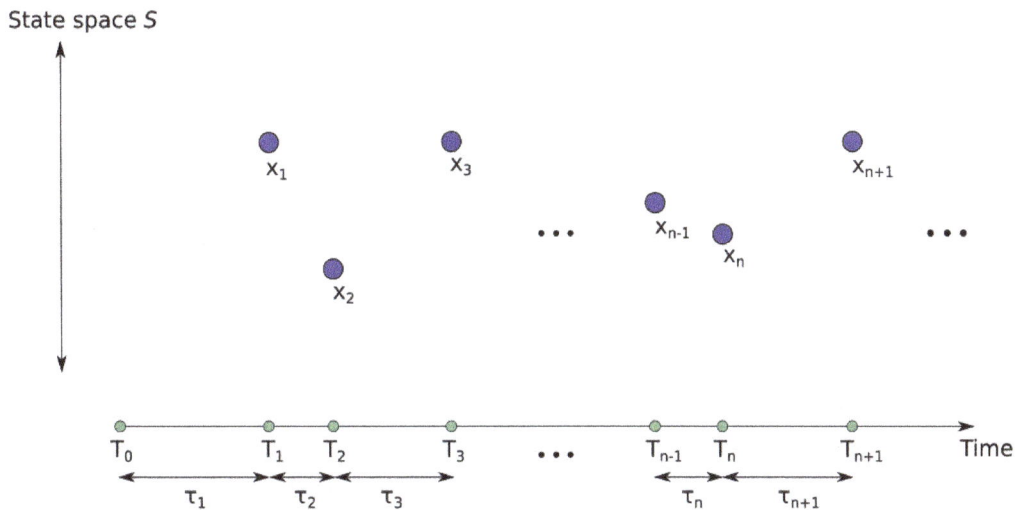

An illustration of a marked point process. If the Poisson process is homogeneous, then the gaps τ_i in the diagram are drawn from an exponential distribution.

For a given point process, each random point of a point process can have a random mathematical object, known as a *mark*, assigned to it. These marks can be as diverse as integers, real numbers, lines, geometrical objects or other point processes. The pair consisting of a point of the point process and its corresponding mark is called a marked point, and all the marked points form a marked point process. It is often assumed that the random marks are independent of each other and identically distributed, which makes the process easier to work with, yet the mark of a point can still depend on the location of its corresponding point in the underlying (state) space. If the underlying point process is a Poisson point process, then one obtains a marked Poisson point process.

Marking Theorem

If a general point process is defined on some mathematical space and the random marks are defined on another mathematical space, then the marked point process is defined on the Cartesian product of these two spaces. For a marked Poisson point process with independent and identically distributed marks, the Marking theorem states that this marked point process is also a (non-marked) Poisson point process defined on the aforementioned Cartesian product of the two mathematical spaces, which is not true for general point processes.

Compound Poisson point process

The compound Poisson point process is formed by adding random values or weights to each point of Poisson point process defined on some underlying state space, so the process is constructed from a marked Poisson point process, where the marks form a collection of independent and identically distributed non-negative random variables. In other words, for each point of the original Poisson process, there is an independent and identically distributed non-negative random variable, and then the compound Poisson process is then formed from the sum of all the random variables corresponding to points of the Poisson process located in a some region of the underlying mathematical space.

IF there is a marked Poisson point processes formed from a Poisson point process N (defined on, for example, $\mathbf{R}^d$) and a collection of independent and identically distributed non-negative marks $\{M_i\}$ such that for each point x_i of the Poisson process N, then there is a non-negative random variable M_i. The resulting compound Poisson process is then:

$$C(B) = \sum_{i=1}^{N(B)} M_i,$$

where $B \subset \mathbf{R}^d$ is a Borel measurable set. If the collection of random variables or marks $\{M_i\}$ are non-negative integer-valued random variables, then the resulting process is called a compound Poisson counting process.

For general random variables, if the compound Poisson point process is formed from a homogeneous Point point process defined on the real line, often representing time, then the resulting compound Poisson process is an example of a Lévy process.

References

- Dodge, Yadolah (2006). The Oxford Dictionary of Statistical Terms. Oxford, England: Oxford University Press. p. 335. ISBN 9780199206131.

- Lindsey, J. K. (2004). Statistical Analysis of Stochastic Processes in Time. Cambridge, England: Cambridge University Press. p. 3. ISBN 9780521837415.

- Durrett, Rick (2010). Probability: Theory and Examples (Fourth ed.). Cambridge: Cambridge University Press. ISBN 978-0-521-76539-8.

- Allen, Linda J. S., An Introduction to Stochastic Processes with Applications to Biology, 2nd Edition, Chapman and Hall, 2010, ISBN 1-4398-1882-7

- Gardiner, C. Handbook of Stochastic Methods: for Physics, Chemistry and the Natural Sciences, 3rd ed., Springer, 2004, ISBN 3540208828

- Kleinert, Hagen (2004). Path Integrals in Quantum Mechanics, Statistics, Polymer Physics, and Financial Markets (4th ed.). Singapore: World Scientific. ISBN 981-238-107-4. (also available online: PDF-files)

- Stark, Henry; Woods, John (2002). Probability and Random Processes with Applications to Signal Processing (3rd ed.). New Jersey: Prentice Hall. ISBN 0-13-020071-9.

- Risken, H. (1989). The Fokker–Planck Equation: Method of Solution and Applications. New York: Springer-Verlag. ISBN 0387504982.

- Weiss, George H. (1994). Aspects and Applications of the Random Walk. Random Materials and Processes. North-Holland Publishing Co., Amsterdam. ISBN 0-444-81606-2. MR 1280031.

- Stoyan, Dietrich; Kendall, Wilfred S.; Mecke, Joseph (1995). Stochastic geometry and its applications. Wiley. ISBN 0471950998.

- Moller, Jesper; Waagepetersen, Rasmus P. (2003). Statistical Inference and Simulation for Spatial Point Processes. CRC Press. ISBN 1584882654.

- Daley, Daryl J.; Vere-Jones, David (2003). An Introduction to the Theory of Point Processes: Volume I: Elementary Theory and Methods. Springer. ISBN 1475781091.

- Streit, Streit (2010). Poisson Point Processes: Imaging, Tracking, and Sensing. Springer Science& Business Media. ISBN 1441969225.

Statistical Models: An Integrated Study

A statistical model is a mathematical model that helps in the demonstration of a set of assumptions. These assumptions are concerned with the generation of sample data from a larger population. The aspects elucidated in this section are of vital importance and provide a better understanding of statistical models.

Statistical Model

A statistical model is a class of mathematical model, which embodies a set of assumptions concerning the generation of some sample data, and similar data from a larger population. A statistical model represents, often in considerably idealized form, the data-generating process.

The assumptions embodied by a statistical model describe a set of probability distributions, some of which are assumed to adequately approximate the distribution from which a particular data set is sampled. The probability distributions inherent in statistical models are what distinguishes statistical models from other, non-statistical, mathematical models.

A statistical model is usually specified by mathematical equations that relate one or more random variables and possibly other non-random variables. As such, "a model is a formal representation of a theory" (Herman Adèr quoting Kenneth Bollen).

All statistical hypothesis tests and all statistical estimators are derived from statistical models. More generally, statistical models are part of the foundation of statistical inference.

Formal Definition

In mathematical terms, a statistical model is usually thought of as a pair $(S, \mathcal{P})$, where S is the set of possible observations, i.e. the sample space, and $\mathcal{P}$ is a set of probability distributions on S.

The intuition behind this definition is as follows. It is assumed that there is a "true" probability distribution induced by the process that generates the observed data. We choose $\mathcal{P}$ to represent a set (of distributions) which contains a distribution that adequately approximates the true distribution. Note that we do not require that $\mathcal{P}$ contains the true distribution, and in practice that is rarely the case. Indeed, as Burnham & Anderson state, "A model is a simplification or approximation of reality and hence will not reflect all of reality"—whence the saying "all models are wrong".

The set $\mathcal{P}$ is almost always parameterized: $\mathcal{P} = \{P_\theta : \theta \in \Theta\}$. The set Θ defines the parameters of the model. A parameterization is generally required to have distinct parameter values give rise to distinct distributions, i.e. $P_{\theta_1} = P_{\theta_2} \Rightarrow \theta_1 = \theta_2$ must hold (in other words, it must be injective). A parameterization that meets the condition is said to be identifiable.

An Example

Height and age are each probabilistically distributed over humans. They are stochastically related: when we know that a person is of age 10, this influences the chance of the person being 5 feet tall. We could formalize that relationship in a linear regression model with the following form: $\text{height}_i = b_0 + b_1 \text{age}_i + \varepsilon_i$, where b_0 is the intercept, b_1 is a parameter that age is multiplied by to get a prediction of height, ε is the error term, and i identifies the person. This implies that height is predicted by age, with some error.

An admissible model must be consistent with all the data points. Thus, the straight line ($\text{height}_i = b_0 + b_1 \text{age}_i$) is *not* a model of the data. The line cannot be a model, unless it exactly fits all the data points—i.e. all the data points lie perfectly on a straight line. The error term, ε_i, must be included in the model, so that the model is consistent with all the data points.

To do statistical inference, we would first need to assume some probability distributions for the ε_i. For instance, we might assume that the ε_i distributions are i.i.d. Gaussian, with zero mean. In this instance, the model would have 3 parameters: b_0, b_1, and the variance of the Gaussian distribution.

We can formally specify the model in the form $(S, \mathcal{P})$ as follows. The sample space, S, of our model comprises the set of all possible pairs (age, height). Each possible value of $\theta = (b_0, b_1, \sigma^2)$ determines a distribution on S; denote that distribution by P_θ. If Θ is the set of all possible values of θ, then $\mathcal{P} = \{P_\theta : \theta \in \Theta\}$. (The parameterization is identifiable, and this is easy to check.)

In this example, the model is determined by (1) specifying S and (2) making some assumptions relevant to $\mathcal{P}$. There are two assumptions: that height can be approximated by a linear function of age; that errors in the approximation are distributed as i.i.d. Gaussian. The assumptions are sufficient to specify $\mathcal{P}$ —as they are required to do.

General Remarks

A statistical model is a special class of mathematical model. What distinguishes a statistical model from other mathematical models is that a statistical model is non-deterministic. Thus, in a statistical model specified via mathematical equations, some of the variables do not have specific values, but instead have probability distributions; i.e. some of the variables are stochastic. In the example above, ε is a stochastic variable; without that variable, the model would be deterministic.

Statistical models are often used even when the physical process being modeled is deterministic. For instance, coin tossing is, in principle, a deterministic process; yet it is commonly modeled as stochastic (via a Bernoulli process).

There are three purposes for a statistical model, according to Konishi & Kitagawa.

- Predictions

- Extraction of information

- Description of stochastic structures

Dimension of a Model

Suppose that we have a statistical model $(S, \mathcal{P})$ with $\mathcal{P} = \{P_\theta : \theta \in \Theta\}$ The model is said to be parametric if $\grave{E}$ has a finite dimension. In notation, we write that $\Theta \subseteq \mathbb{R}^d$ where d is a positive integer ($\mathbb{R}$ denotes the real numbers; other sets can be used, in principle). Here, d is called the dimension of the model.

As an example, if we assume that data arise from a univariate Gaussian distribution, then we are assuming that

$$\mathcal{P} = \{P_{\mu,\sigma}(x) \equiv \frac{1}{\sqrt{2\pi}\sigma} \exp\left(-\frac{(x-\mu)^2}{2\sigma^2}\right) : \mu \in \mathbb{R}, \sigma > 0\} \ .$$

In this example, the dimension, d, equals 2.

As another example, suppose that the data consists of points (x, y) that we assume are distributed according to a straight line with i.i.d. Gaussian residuals (with zero mean). Then the dimension of the statistical model is 3: the intercept of the line, the slope of the line, and the variance of the distribution of the residuals. (Note that in geometry, a straight line has dimension 1.)

A statistical model is nonparametric if the parameter set $\grave{E}$ is infinite dimensional. A statistical model is semiparametric if it has both finite-dimensional and infinite-dimensional parameters. Formally, if d is the dimension of Θ and n is the number of samples, both semiparametric and nonparametric models have $d \to \infty$ as $n \to \infty$.. If $d/n \to 0$ as $n \to \infty$, then the model is semi-parametric; otherwise, the model is nonparametric.

Parametric models are by far the most commonly used statistical models. Regarding semiparametric and nonparametric models, Sir David Cox has said: "These typically involve fewer assumptions of structure and distributional form but usually contain strong assumptions about independencies".

Nested Models

Two statistical models are nested if the first model can be transformed into the second model by imposing constraints on the parameters of the first model. For example, the set of all Gaussian distributions has, nested within it, the set of zero-mean Gaussian distributions: we constrain the mean in the set of all Gaussian distributions to get the zero-mean distributions.

In that example, the first model has a higher dimension than the second model (the zero-mean model has dimension 1). Such is usually, but not always, the case. As a different example, the set of positive-mean Gaussian distributions, which has dimension 2, is nested within the set of all Gaussian distributions.

Comparing Models

It is assumed that there is a "true" probability distribution underlying the observed data, induced by the process that generated the data. The main goal of model selection is to make statements about which elements of $\mathcal{P}$ are most likely to adequately approximate the true distribution.

Models can be compared to each other by exploratory data analysis or confirmatory data analysis.

In exploratory analysis, a variety of models are formulated and an assessment is performed of how well each one describes the data. In confirmatory analysis, a previously formulated model or models are compared to the data. Common criteria for comparing models include R^2, Bayes factor, and the likelihood-ratio test together with its generalization relative likelihood.

Konishi & Kitagawa state: "The majority of the problems in statistical inference can be considered to be problems related to statistical modeling. They are typically formulated as comparisons of several statistical models." Relatedly, Sir David Cox has said, "How [the] translation from subject-matter problem to statistical model is done is often the most critical part of an analysis".

Regression Analysis

In statistical modeling, regression analysis is a statistical process for estimating the relationships among variables. It includes many techniques for modeling and analyzing several variables, when the focus is on the relationship between a dependent variable and one or more independent variables (or 'predictors'). More specifically, regression analysis helps one understand how the typical value of the dependent variable (or 'criterion variable') changes when any one of the independent variables is varied, while the other independent variables are held fixed. Most commonly, regression analysis estimates the conditional expectation of the dependent variable given the independent variables – that is, the average value of the dependent variable when the independent variables are fixed. Less commonly, the focus is on a quantile, or other location parameter of the conditional distribution of the dependent variable given the independent variables. In all cases, the estimation target is a function of the independent variables called the regression function. In regression analysis, it is also of interest to characterize the variation of the dependent variable around the regression function which can be described by a probability distribution. A related but distinct approach is necessary condition analysis (NCA), which estimates the maximum (rather than average) value of the dependent variable for a given value of the independent variable (ceiling line rather than central line) in order to identify what value of the independent variable is necessary but not sufficient for a given value of the dependent variable.

Regression analysis is widely used for prediction and forecasting, where its use has substantial overlap with the field of machine learning. Regression analysis is also used to understand which among the independent variables are related to the dependent variable, and to explore the forms of these relationships. In restricted circumstances, regression analysis can be used to infer causal relationships between the independent and dependent variables. However this can lead to illusions or false relationships, so caution is advisable; for example, correlation does not imply causation.

Many techniques for carrying out regression analysis have been developed. Familiar methods such as linear regression and ordinary least squares regression are parametric, in that the regression function is defined in terms of a finite number of unknown parameters that are estimated from the data. Nonparametric regression refers to techniques that allow the regression function to lie in a specified set of functions, which may be infinite-dimensional.

The performance of regression analysis methods in practice depends on the form of the data generating process, and how it relates to the regression approach being used. Since the true form of the data-generating process is generally not known, regression analysis often depends to some extent on making assumptions about this process. These assumptions are sometimes testable if a sufficient quantity of data is available. Regression models for prediction are often useful even when the assumptions are moderately violated, although they may not perform optimally. However, in many applications, especially with small effects or questions of causality based on observational data, regression methods can give misleading results.

In a narrower sense, regression may refer specifically to the estimation of continuous response variables, as opposed to the discrete response variables used in classification. The case of a continuous output variable may be more specifically referred to as metric regression to distinguish it from related problems.

History

The earliest form of regression was the method of least squares, which was published by Legendre in 1805, and by Gauss in 1809. Legendre and Gauss both applied the method to the problem of determining, from astronomical observations, the orbits of bodies about the Sun (mostly comets, but also later the then newly discovered minor planets). Gauss published a further development of the theory of least squares in 1821, including a version of the Gauss–Markov theorem.

The term "regression" was coined by Francis Galton in the nineteenth century to describe a biological phenomenon. The phenomenon was that the heights of descendants of tall ancestors tend to regress down towards a normal average (a phenomenon also known as regression toward the mean). For Galton, regression had only this biological meaning, but his work was later extended by Udny Yule and Karl Pearson to a more general statistical context. In the work of Yule and Pearson, the joint distribution of the response and explanatory variables is assumed to be Gaussian. This assumption was weakened by R.A. Fisher in his works of 1922 and 1925. Fisher assumed that the conditional distribution of the response variable is Gaussian, but the joint distribution need not be. In this respect, Fisher's assumption is closer to Gauss's formulation of 1821.

In the 1950s and 1960s, economists used electromechanical desk calculators to calculate regressions. Before 1970, it sometimes took up to 24 hours to receive the result from one regression.

Regression methods continue to be an area of active research. In recent decades, new methods have been developed for robust regression, regression involving correlated responses such as time series and growth curves, regression in which the predictor (independent variable) or response variables are curves, images, graphs, or other complex data objects, regression methods accommodating various types of missing data, nonparametric regression, Bayesian methods for regression, regression in which the predictor variables are measured with error, regression with more predictor variables than observations, and causal inference with regression.

Regression Models

Regression models involve the following variables:

- The unknown parameters, denoted as β, which may represent a scalar or a vector.
- The independent variables, X.
- The dependent variable, Y.

In various fields of application, different terminologies are used in place of dependent and independent variables.

A regression model relates Y to a function of X and β.

$$Y \approx f(x, \beta)$$

The approximation is usually formalized as $E(Y \mid X) = f(X, \beta)$. To carry out regression analysis, the form of the function f must be specified. Sometimes the form of this function is based on knowledge about the relationship between Y and X that does not rely on the data. If no such knowledge is available, a flexible or convenient form for f is chosen.

Assume now that the vector of unknown parameters β is of length k. In order to perform a regression analysis the user must provide information about the dependent variable Y:

- If N data points of the form (Y, X) are observed, where $N < k$, most classical approaches to regression analysis cannot be performed: since the system of equations defining the regression model is underdetermined, there are not enough data to recover β.

- If exactly $N = k$ data points are observed, and the function f is linear, the equations $Y = f(X, \beta)$ can be solved exactly rather than approximately. This reduces to solving a set of N equations with N unknowns (the elements of β), which has a unique solution as long as the X are linearly independent. If f is nonlinear, a solution may not exist, or many solutions may exist.

- The most common situation is where $N > k$ data points are observed. In this case, there is enough information in the data to estimate a unique value for β that best fits the data in some sense, and the regression model when applied to the data can be viewed as an overdetermined system in β.

In the last case, the regression analysis provides the tools for:

1. Finding a solution for unknown parameters β that will, for example, minimize the distance between the measured and predicted values of the dependent variable Y (also known as method of least squares).

2. Under certain statistical assumptions, the regression analysis uses the surplus of information to provide statistical information about the unknown parameters β and predicted values of the dependent variable Y.

Necessary Number of Independent Measurements

Consider a regression model which has three unknown parameters, β_0, β_1, and β_2. Suppose an experimenter performs 10 measurements all at exactly the same value of independent variable vector X (which contains the independent variables X_1, X_2, and X_3). In this case, regression analysis fails to give a unique set of estimated values for the three unknown parameters; the experimenter did not provide enough information. The best one can do is to estimate the average value and the standard deviation of the dependent variable Y. Similarly, measuring at two different values of X would give enough data for a regression with two unknowns, but not for three or more unknowns.

If the experimenter had performed measurements at three different values of the independent variable vector X, then regression analysis would provide a unique set of estimates for the three unknown parameters in β.

In the case of general linear regression, the above statement is equivalent to the requirement that the matrix X^TX is invertible.

Statistical Assumptions

When the number of measurements, N, is larger than the number of unknown parameters, k, and the measurement errors ε_i are normally distributed then *the excess of information* contained in $(N - k)$ measurements is used to make statistical predictions about the unknown parameters. This excess of information is referred to as the degrees of freedom of the regression.

Underlying Assumptions

Classical assumptions for regression analysis include:

- The sample is representative of the population for the inference prediction.

- The error is a random variable with a mean of zero conditional on the explanatory variables.

- The independent variables are measured with no error. (Note: If this is not so, modeling may be done instead using errors-in-variables model techniques).

- The independent variables (predictors) are linearly independent, i.e. it is not possible to express any predictor as a linear combination of the others.

- The errors are uncorrelated, that is, the variance–covariance matrix of the errors is diagonal and each non-zero element is the variance of the error.

- The variance of the error is constant across observations (homoscedasticity). If not, weighted least squares or other methods might instead be used.

These are sufficient conditions for the least-squares estimator to possess desirable properties; in particular, these assumptions imply that the parameter estimates will be unbiased, consistent, and efficient in the class of linear unbiased estimators. It is important to note that actual data rarely satisfies the assumptions. That is, the method is used even though the assumptions are not true. Variation from the assumptions can sometimes be used as a measure of how far the model is from being useful. Many of these assumptions may be relaxed in more advanced treatments. Reports of statistical analyses usually include analyses of tests on the sample data and methodology for the fit and usefulness of the model.

Assumptions include the geometrical support of the variables. Independent and dependent variables often refer to values measured at point locations. There may be spatial trends and spatial autocorrelation in the variables that violate statistical assumptions of regression. Geographic weighted regression is one technique to deal with such data. Also, variables may include values aggregated by areas. With aggregated data the modifiable areal unit problem can cause extreme variation in regression parameters. When analyzing data aggregated by political boundaries, postal codes or census areas results may be very distinct with a different choice of units.

Linear Regression

In linear regression, the model specification is that the dependent variable, y_i is a linear combination of the *parameters* (but need not be linear in the *independent variables*). For example, in simple linear regression for modeling x_i data points there is one independent variable: β_0, and two parameters, β_0 and β_1:

$$\text{straight line: } y_i = \beta_0 + \beta_1 x_i + \varepsilon_i, \quad i = 1, \ldots, n.$$

In multiple linear regression, there are several independent variables or functions of independent variables.

Adding a term in x_i^2 to the preceding regression gives:

$$\text{parabola: } y_i = \beta_0 + \beta_1 x_i + \beta_2 x_i^2 + \varepsilon_i, \ i = 1, \ldots, n.$$

This is still linear regression; although the expression on the right hand side is quadratic in the independent variable x_i, it is linear in the parameters β_0, β_1 and β_2.

In both cases, ε_i is an error term and the subscript i indexes a particular observation.

Returning our attention to the straight line case: Given a random sample from the population, we estimate the population parameters and obtain the sample linear regression model:

$$\widehat{y_i} = \hat{\beta}_0 + \hat{\beta}_1 x_i.$$

The residual, $e_i = y_i - \hat{y}_i$, is the difference between the value of the dependent variable predicted by the model, $\hat{y}_i$, and the true value of the dependent variable, y_i. One method of estimation is ordinary least squares. This method obtains parameter estimates that minimize the sum of squared residuals, SSE, also sometimes denoted RSS:

$$SSE = \sum_{i=1}^{n} e_i^2.$$

Minimization of this function results in a set of normal equations, a set of simultaneous linear equations in the parameters, which are solved to yield the parameter estimators, $\hat{\beta}_0, \hat{\beta}_1$.

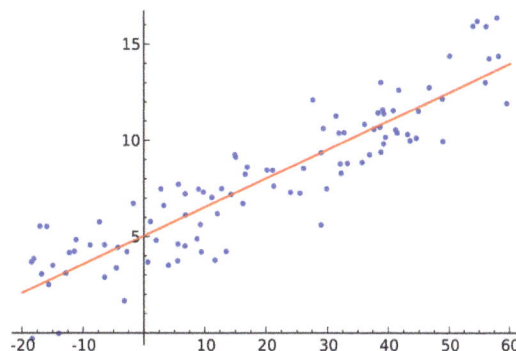

Illustration of linear regression on a data set.

In the case of simple regression, the formulas for the least squares estimates are

$$\widehat{\beta_1} = \frac{\sum (x_i - \bar{x})(y_i - \bar{y})}{\sum (x_i - \bar{x})^2} \text{ and } \widehat{\beta_0} = \bar{y} - \widehat{\beta_1}\bar{x}$$

where $\bar{x}$ is the mean (average) of the x values and $\bar{y}$ is the mean of the y values.

Under the assumption that the population error term has a constant variance, the estimate of that variance is given by:

$$\hat{\sigma}_\varepsilon^2 = \frac{SSE}{n-2}.$$

This is called the mean square error (MSE) of the regression. The denominator is the sample size reduced by the number of model parameters estimated from the same data, (n-p) for p regressors or (n-p-1) if an intercept is used. In this case, p=1 so the denominator is n-2.

The standard errors of the parameter estimates are given by

$$\hat{\sigma}_{\beta_0} = \hat{\sigma}_\varepsilon \sqrt{\frac{1}{n} + \frac{\bar{x}^2}{\sum (x_i - \bar{x})^2}}$$

$$\hat{\sigma}_{\beta_1} = \hat{\sigma}_\varepsilon \sqrt{\frac{1}{\sum (x_i - \bar{x})^2}}.$$

Under the further assumption that the population error term is normally distributed, the researcher can use these estimated standard errors to create confidence intervals and conduct hypothesis tests about the population parameters.

General Linear Model

In the more general multiple regression model, there are p independent variables:

$$y_i = \beta_1 x_{i1} + \beta_2 x_{i2} + \cdots + \beta_p x_{ip} + \varepsilon_i,$$

where x_{ij} is the ith observation on the jth independent variable. If the first independent variable takes the value 1 for all i, $x_{i1} = 1$, then β_1 is called the regression intercept.

The least squares parameter estimates are obtained from p normal equations. The residual can be written as

$$\varepsilon_i = y_i - \hat{\beta}_1 x_{i1} - \cdots - \hat{\beta}_p x_{ip}.$$

The normal equations are

$$\sum_{i=1}^{n} \sum_{k=1}^{p} X_{ij} X_{ik} \hat{\beta}_k = \sum_{i=1}^{n} X_{ij} y_i, \; j = 1, \dots, p.$$

In matrix notation, the normal equations are written as

$$(X^\top X)\hat{\beta} = X^\top Y,$$

where the ij element of X is x_{ij}, the i element of the column vector Y is y_i, and the j element of $\hat{\beta}$ is $\hat{\beta}_j$. Thus X is $n{\times}p$, Y is $n{\times}1$, and $\hat{\beta}$ is $p{\times}1$. The solution is

$$\hat{\beta} = (X^\top X)^{-1} X^\top Y.$$

Diagnostics

Once a regression model has been constructed, it may be important to confirm the goodness of fit of the model and the statistical significance of the estimated parameters. Commonly used checks of goodness of fit include the R-squared, analyses of the pattern of residuals and hypothesis testing. Statistical significance can be checked by an F-test of the overall fit, followed by t-tests of individual parameters.

Interpretations of these diagnostic tests rest heavily on the model assumptions. Although examination of the residuals can be used to invalidate a model, the results of a t-test or F-test are sometimes more difficult to interpret if the model's assumptions are violated. For example, if the error term does not have a normal distribution, in small samples the estimated parameters will not follow normal distributions and complicate inference. With relatively large samples, however, a central limit theorem can be invoked such that hypothesis testing may proceed using asymptotic approximations.

"Limited Dependent" Variables

The phrase "limited dependent" is used in econometric statistics for categorical and constrained variables.

The response variable may be non-continuous ("limited" to lie on some subset of the real line). For binary (zero or one) variables, if analysis proceeds with least-squares linear regression, the model is called the linear probability model. Nonlinear models for binary dependent variables include the probit and logit model. The multivariate probit model is a standard method of estimating a joint relationship between several binary dependent variables and some independent variables. For categorical variables with more than two values there is the multinomial logit. For ordinal variables with more than two values, there are the ordered logit and ordered probit models. Censored regression models may be used when the dependent variable is only sometimes observed, and Heckman correction type models may be used when the sample is not randomly selected from the population of interest. An alternative to such procedures is linear regression based on polychoric correlation (or polyserial correlations) between the categorical variables. Such procedures differ in the assumptions made about the distribution of the variables in the population. If the variable is positive with low values and represents the repetition of the occurrence of an event, then count models like the Poisson regression or the negative binomial model may be used instead.

Interpolation and Extrapolation

Regression models predict a value of the Y variable given known values of the X variables. Prediction *within* the range of values in the dataset used for model-fitting is known informally as interpolation. Prediction *outside* this range of the data is known as extrapolation. Performing extrapolation relies strongly on the regression assumptions. The further the extrapolation goes outside the data, the more room there is for the model to fail due to differences between the assumptions and the sample data or the true values.

It is generally advised that when performing extrapolation, one should accompany the estimated value of the dependent variable with a prediction interval that represents the uncertainty. Such intervals tend to expand rapidly as the values of the independent variable(s) moved outside the range covered by the observed data.

For such reasons and others, some tend to say that it might be unwise to undertake extrapolation.

However, this does not cover the full set of modelling errors that may be being made: in particular, the assumption of a particular form for the relation between Y and X. A properly conducted regression analysis will include an assessment of how well the assumed form is matched by the observed data, but it can only do so within the range of values of the independent variables actually available. This means that any extrapolation is particularly reliant on the assumptions being made about the structural form of the regression relationship. Best-practice advice here is that a linear-in-variables and linear-in-parameters relationship should not be chosen simply for computational convenience, but that all available knowledge should be deployed in constructing a regression model. If this knowledge includes the fact that the dependent variable cannot go outside a certain range of values, this can be made use of in selecting the model – even if the observed dataset has no values particularly near such bounds. The implications of this step of choosing an appropriate functional form for the regression can be great when extrapolation is considered. At a minimum, it can ensure that any extrapolation arising from a fitted model is "realistic" (or in accord with what is known).

Nonlinear Regression

When the model function is not linear in the parameters, the sum of squares must be minimized by an iterative procedure. This introduces many complications which are summarized in Differences between linear and non-linear least squares

Power and Sample Size Calculations

There are no generally agreed methods for relating the number of observations versus the number of independent variables in the model. One rule of thumb suggested by Good and Hardin is $N = m^n$, where N is the sample size,   is the number of independent variables and m is the number of observations needed to reach the desired precision if the model had only one independent variable. For example, a researcher is building a linear regression model using a dataset that contains 1000 patients (N). If the researcher decides that five observations are needed to precisely define a straight line (m), then the maximum number of independent variables the model can support is 4, because

$$\frac{\log 1000}{\log 5} = 4.29.$$

Other Methods

Although the parameters of a regression model are usually estimated using the method of least squares, other methods which have been used include:

- Bayesian methods, e.g. Bayesian linear regression

- Percentage regression, for situations where reducing *percentage* errors is deemed more appropriate.

- Least absolute deviations, which is more robust in the presence of outliers, leading to quantile regression

- Nonparametric regression, requires a large number of observations and is computationally intensive

- Distance metric learning, which is learned by the search of a meaningful distance metric in a given input space.

Software

All major statistical software packages perform least squares regression analysis and inference. Simple linear regression and multiple regression using least squares can be done in some spreadsheet applications and on some calculators. While many statistical software packages can perform various types of nonparametric and robust regression, these methods are less standardized; different software packages implement different methods, and a method with a given name may be implemented differently in different packages. Specialized regression software has been developed for use in fields such as survey analysis and neuroimaging.

Bayesian Hierarchical Modeling

Bayesian hierarchical modelling is a statistical model written in multiple levels (hierarchical form) that estimates the parameters of the posterior distribution using the Bayesian method. The sub-models combine to form the hierarchical model, and the Bayes' theorem is used to integrate them with the observed data, and account for all the uncertainty that is present. The result of this integration is the posterior distribution, also known as the updated probability estimate, as additional evidence on the prior distribution is acquired.

Frequentist statistics, the more popular foundation of statistics, has been known to contradict Bayesian statistics due to its treatment of the parameters as a random variable, and its use of subjective information in establishing assumptions on these parameters. However, Bayesians argue that relevant information regarding decision making and updating beliefs cannot be ignored and that hierarchical modeling has the potential to overrule classical methods in applications where respondents give multiple observational data. Moreover, the model has proven to be robust, with the posterior distribution less sensitive to the more flexible hierarchical priors.

Hierarchical modeling is used when information is available on several different levels of observational units. The hierarchical form of analysis and organization helps in the understanding of multiparameter problems and also plays an important role in developing computational strategies.

Philosophy

Numerous statistical applications involve multiple parameters that can be regarded as related or connected in such a way that the problem implies dependence of the joint probability model for these parameters. Individual degrees of belief, expressed in the form of probabilities, come with uncertainty. Amidst this is the change of the degrees of belief over time. As was stated by Professor José M. Bernardo and Professor Adrian F. Smith, "The actuality of the learning process consists in the evolution of individual and subjective beliefs about the reality." These subjective probabilities are more directly involved in the mind rather than the physical probabilities. Hence, it is with this need of updating beliefs that Bayesians have formulated an alternative statistical model which takes into account the prior occurrence of a particular event.

Bayes' Theorem

The assumed occurrence of a real-world event will typically modify preferences between certain options. This is done by modifying the degrees of belief attached, by an individual, to the events defining the options.

Suppose in a study of the effectiveness of cardiac treatments, with the patients in hospital j having survival probability θ_j, the survival probability will be updated with the occurrence of y, the event in which a hypothetical controversial serum is created which, as believed by some, increases survival in cardiac patients.

In order to make updated probability statements about θ_j, given the occurrence of event y, we must begin with a model providing a joint probability distribution for θ_j and y. This can be written as a product of the two distributions that are often referred to as the prior distribution $P(\theta)$ and the sampling distribution $P(y \mid \theta)$ respectively:

$$P(\theta, y) = P(\theta)P(y \mid \theta)$$

Using the basic property of conditional probability, the posterior distribution will yield:

$$P(\theta \mid y) = \frac{P(\theta, y)}{P(y)} = \frac{P(y \mid \theta)P(\theta)}{P(y)}$$

This equation, showing the relationship between the conditional probability and the individual events, is known as Bayes' theorem. This simple expression encapsulates the technical core of Bayesian inference which aims to incorporate the updated belief, $P(\theta \mid y)$, in appropriate and solvable ways.

Exchangeability

The usual starting point of a statistical analysis is the assumption that the n values y_n are exchangeable. If no information – other than data y – is available to distinguish any of the θ_j's from any others, and no ordering or grouping of the parameters can be made, one must assume symmetry among the parameters in their prior distribution. This symmetry is represented probabilistically by exchangeability. Generally, it is useful and appropriate to model data from an exchangeable distribution, as independently and identically distributed, given some unknown parameter vector θ, with distribution $P(\theta)$.

Finite Exchangeability

For a fixed number n, the set $y_1, y_2, \ldots, y_n$ is exchangeable if the joint probability $P(y_1, y_2, \ldots, y_n)$ is invariant under permutations of the indices. That is, for every permutation π or $(\pi_1, \pi_2, \ldots, \pi_n)$ of (1, 2, ..., n),

To visualize this is an exchangeable but not independent and identical (iid) example: Consider an urn with one red ball and one blue ball with probability $\frac{1}{2}$ of drawing either. Balls are drawn without replacement, i.e. after one ball is drawn from the n balls, there will be $n - 1$ remaining balls left for the next draw.

$$\text{Let } Y_i = \begin{cases} 1, & \text{if the } i\text{th ball is red,} \\ 0, & \text{otherwise.} \end{cases}$$

Since the probability of selecting a red ball in the first draw and a blue ball in the second draw is equal to the probability of selecting a blue ball on the first draw and a red on the second draw, both of which are equal to 1/2 (i.e. $[P(y_1 = 1, y_2 = 0) = P(y_1 = 0, y_2 = 1) = \frac{1}{2}]$), then y_1 and   are exchangeable.

But the probability of selecting a red ball on the second draw given that the red ball has already been selected in the first draw is 0, and is not equal to the probability that the red ball is selected in the second draw which is equal to 1/2 (i.e. $[P(y_2 = 1 | y_1 = 1) = 0 \neq P(y_2 = 1) = \frac{1}{2}]$). Thus, y_1 and y_2 are not independent.

If $x_1, \ldots, x_n$ are independent and identically distributed, then they are exchangeable, but not conversely true.

Infinite Exchangeability

An infinite exchangeability implies that every finite subset of an infinite sequence $y_1, y_2, \ldots$ is exchangeable. That is, for any n, the sequence $y_1, y_2, \ldots, y_n$ is exchangeable.

Hierarchical Models

Components

Bayesian hierarchical modeling makes use of two important concepts in deriving the posterior distribution, namely:

1. Hyperparameter: parameter of the prior distribution

2. Hyperprior: distribution of a Hyperparameter

Say a random variable Y follows a normal distribution with parameters θ as the mean and 1 as the variance, that is $Y|\theta \sim N(\theta,1)$.. The parameter θ has a prior distribution given by a normal distribution with mean $\theta \mid \mu \sim N(\mu,1)$ and variance 1, i.e. μ. Furthermore, $N(0,1)$. follows another distribution given, for example, by the standard normal distribution, μ. The parameter $N(0,1)$ is called the hyperparameter, while its distribution given by $N(0,1)$ is an example of a hyperprior distribution. The notation of the distribution of Y changes as another parameter is added, i.e. μ. If there is another stage, say, β follows another normal distribution with mean ϵ and variance $\mu \sim N(\beta, \epsilon)$, meaning β, and ϵ can also be called hyperparameters while their distributions are hyperprior distributions as well.

Framework

Let y_j be an observation and θ_j a parameter governing the data generating process for y_j. Assume further that the parameters $\theta_1, \theta_2, \ldots, \theta_j$ are generated exchangeably from a common population, with distribution governed by a hyperparameter ϕ. The Bayesian hierarchical model contains the following stages:

Stage I: $y_j \mid \theta_j, \phi \sim P(y_j \mid \theta_j, \phi)$

Stage II: $\theta_j \mid \phi \sim P(\theta_j \mid \phi)$

Stage III: $\phi \sim P(\phi)$

The likelihood, as seen in stage I is $P(y_j \mid \theta_j, \phi)$ with $P(\theta_j, \phi)$ as its prior distribution. Note that the likelihood depends on ϕ only through θ_j.

The prior distribution from stage I can be broken down into:

$P(\theta_j, \phi) = P(\theta_j \mid \phi)P(\phi)$ *[using Bayes' Theorem]*

With ϕ as its hyperparameter with hyperprior distribution, $P(\phi)$.

Thus, the posterior distribution is proportional to:

$P(\phi, \theta_j \mid y) \propto P(y_j \mid \theta_j, \phi)P(\theta_j \mid \phi)$ *[using Bayes' Theorem]*

$P(\phi, \theta_j \mid y) \propto P(y_j \mid \theta_j)P(\theta_j, \phi)$

Example

To further illustrate this, consider the example: A teacher wants to estimate how well a male student did in his SAT. He uses information on the student's high school grades and his current grade point average (GPA) to come up with an estimate. His current GPA, denoted by Y, has a likelihood given by some probability function with parameter θ, i.e. $Y \mid \theta \sim P(Y \mid \theta)$.. This parameter θ is the SAT score of the student. The SAT score is viewed as a sample coming from a common population distribution indexed by another parameter ϕ, which is the high school grade of the student. That is, $\theta \mid \phi \sim P(\theta \mid \phi)$ Moreover, the hyperparameter ϕ follows its own distribution given by $P(\phi)$, a hyperprior. To solve for the SAT score given information on the GPA,

$P(\theta, \phi \mid Y) \propto P(Y \mid \theta, \phi)P(\theta, \phi)$

$P(\theta, \phi \mid Y) \propto P(Y \mid \theta)P(\theta \mid \phi)P(\phi)$

All information in the problem will be used to solve for the posterior distribution. Instead of solving only using the prior distribution and the likelihood function, the use of hyperpriors gives more information to make more accurate beliefs in the behavior of a parameter.

2-stage Hierarchical Model

In general, the joint posterior distribution of interest in 2-stage hierarchical models is:

$$P(\theta,\phi \mid Y) = \frac{P(Y \mid \theta,\phi)P(\theta,\phi)}{P(Y)} = \frac{P(Y \mid \theta)P(\theta \mid \phi)P(\phi)}{P(Y)}$$

$$P(\theta,\phi \mid Y) \propto P(Y \mid \theta)P(\theta \mid \phi)P(\phi)$$

3-stage Hierarchical Model

For 3-stage hierarchical models, the posterior distribution is given by:

$$P(\theta,\phi,X \mid Y) = \frac{P(Y \mid \theta)P(\theta \mid \phi)P(\phi \mid X)P(X)}{P(Y)}$$

$$P(\theta,\phi,X \mid Y) \propto P(Y \mid \theta)P(\theta \mid \phi)P(\phi \mid X)P(X)$$

Errors-in-Variables Models

In statistics, errors-in-variables models or measurement error models are regression models that account for measurement errors in the independent variables. In contrast, standard regression models assume that those regressors have been measured exactly, or observed without error; as such, those models account only for errors in the dependent variables, or responses.

In the case when some regressors have been measured with errors, estimation based on the standard assumption leads to inconsistent estimates, meaning that the parameter estimates do not tend to the true values even in very large samples. For simple linear regression the effect is an underestimate of the coefficient, known as the *attenuation bias*. In non-linear models the direction of the bias is likely to be more complicated.

Motivational Example

Consider a simple linear regression model of the form

$$y_t = \alpha + \beta x_t^* + \varepsilon_t, \quad t = 1,\dots,T,$$

where x_t^* denotes the *true* but unobserved regressor. Instead we observe this value with an error:

$$x_t = x_t^* + \eta_t$$

where the measurement error η_t is assumed to be independent from the true value x_t^*.

If the y_t's are simply regressed on the x_t's, then the estimator for the slope coefficient is

$$\hat{\beta} = \frac{\frac{1}{T}\sum_{t=1}^{T}(x_t - \bar{x})(y_t - \bar{y})}{\frac{1}{T}\sum_{t=1}^{T}(x_t - \bar{x})^2},$$

which converges as the sample size T increases without bound:

$$\hat{\beta} \xrightarrow{p} \frac{\text{Cov}[x_t, y_t]}{\text{Var}[x_t]} = \frac{\beta\sigma_{x^*}^2}{\sigma_{x^*}^2 + \sigma_\eta^2} = \frac{\beta}{1 + \sigma_\eta^2 / \sigma_{x^*}^2}.$$

Variances are non-negative, so that in the limit the estimate is smaller in magnitude than the true value of β, an effect which statisticians call *attenuation* or regression dilution. Thus the 'naïve' least squares estimator is inconsistent in this setting. However, the estimator is a consistent estimator of the parameter required for a best linear predictor of y given x: in some applications this may be what is required, rather than an estimate of the 'true' regression coefficient, although that would assume that the variance of the errors in observing x^* remains fixed. This follows directly from the result quoted immediately above, and the fact that the regression coefficient relating the x_t's to the actually observed y_t's, in a simple linear regression, is given by

$$\beta_x = \frac{\text{Cov}[x_t, y_t]}{\text{Var}[x_t]}.$$

It is this coefficient, rather than β, that would be required for constructing a predictor of y based on an observed x which is subject to noise.

It can be argued that almost all existing data sets contain errors of different nature and magnitude, so that attenuation bias is extremely frequent (although in multivariate regression the direction of bias is ambiguous. Jerry Hausman sees this as an *iron law of econometrics*: "The magnitude of the estimate is usually smaller than expected."

Specification

Usually measurement error models are described using the latent variables approach. If y is the response variable and x are observed values of the regressors, then it is assumed there exist some latent variables y^* and x^* which follow the model's "true" functional relationship $g(\cdot)$, and such that the observed quantities are their noisy observations:

$$\begin{cases} x = x^* + \eta, \\ y = y^* + \varepsilon, \\ y^* = g(x^*, w|\theta), \end{cases}$$

where θ is the model's parameter and w are those regressors which are assumed to be error-free (for example when linear regression contains an intercept, the regressor which corresponds to the constant certainly has no "measurement errors"). Depending on the specification these error-free regressors may or may not be treated separately; in the latter case it is simply assumed that corresponding entries in the variance matrix of η's are zero.

The variables y, x, w are all *observed*, meaning that the statistician possesses a data set of n statistical units $x\{y_i, x_i, w_i\}_{i=1,\dots,n}$ which follow the data generating process described above; the latent variables x^*, y^*, ε, and η are not observed however.

This specification does not encompass all the existing errors-in-variables models. For example in some of them function $g(\cdot)$ may be non-parametric or semi-parametric. Other approaches model the relationship between y^* and x^* as distributional instead of functional, that is they assume that y^* conditionally on x^* follows a certain (usually parametric) distribution.

Terminology and Assumptions

- The observed variable x may be called the *manifest*, *indicator*, or *proxy* variable.

- The unobserved variable x^* may be called the *latent* or *true* variable. It may be regarded either as an unknown constant (in which case the model is called a *functional model*), or as a random variable (correspondingly a *structural model*).

- The relationship between the measurement error η and the latent variable x^* can be modeled in different ways:

  o *Classical errors*: $\eta \perp x^*$ the errors are independent from the latent variable. This is the most common assumption, it implies that the errors are introduced by the measuring device and their magnitude does not depend on the value being measured.

  o *Mean-independence*: $E[\eta \mid x^*]=0$, the errors are mean-zero for every value of the latent regressor. This is a less restrictive assumption than the classical one, as it allows for the presence of heteroscedasticity or other effects in the measurement errors.

  o *Berkson's errors*: $\eta \perp x$, the errors are independent from the *observed* regressor x. This assumption has very limited applicability. One example is round-off errors: for example if a person's AGE* is a continuous random variable, whereas the observed AGE is truncated to the next smallest integer, then the truncation error is approximately independent from the observed AGE. Another possibility is with the fixed design experiment: for example if a scientist decides to make a measurement at a certain predetermined moment of time x, say at $x=10s$, then the real measurement may occur at some other value of x^* (for example due to her finite reaction time) and such measurement error will be generally independent from the "observed" value of the regressor.

  o *Misclassification errors*: special case used for the dummy regressors. If x^* is an indicator of a certain event or condition (such as person is male/female, some medical treatment given/not, etc.), then the measurement error in such regressor will correspond to the incorrect classification similar to type I and type II errors in statistical testing. In this case the error η may take only 3 possible values, and its distribution conditional on x^* is modeled with two parameters: $\alpha = \Pr[\eta = -1 \mid x^* = 1]$, and $\beta = \Pr[\eta = 1 \mid x^* = 0]$. The necessary condition for identification is that $\alpha + \beta < 1$, that is misclassification should not happen "too often". (This idea can be generalized to discrete variables with more than two possible values.)

Linear Model

Linear errors-in-variables models were studied first, probably because linear models were so widely used and they are easier than non-linear ones. Unlike standard least squares regression (OLS), extending errors in variables regression (EiV) from the simple to the multivariable case is not straightforward.

Simple Linear Model

The simple linear errors-in-variables model was already presented in the "motivation" section:

$$\begin{cases} y_t = \alpha + \beta x_t^* + \varepsilon_t, \\ x_t = x_t^* + \eta_t, \end{cases}$$

where all variables are scalar. Here α and β are the parameters of interest, whereas σ_ε and σ_η—standard deviations of the error terms—are the nuisance parameters. The "true" regressor x^* is treated as a random variable (*structural* model), independent from the measurement error η (*classic* assumption).

This model is identifiable in two cases: (1) either the latent regressor x^* is *not* normally distributed, (2) or x^* has normal distribution, but neither ε_t nor η_t are divisible by a normal distribution. That is, the parameters α, β can be consistently estimated from the data set without any additional information, provided the latent regressor is not Gaussian.

Before this identifiability result was established, statisticians attempted to apply the maximum likelihood technique by assuming that all variables are normal, and then concluded that the model is not identified. The suggested remedy was to *assume* that some of the parameters of the model are known or can be estimated from the outside source. Such estimation methods include

- Deming regression — assumes that the ratio $\delta = \sigma_\varepsilon^2 / \sigma_\eta^2$ is known. This could be appropriate for example when errors in y and x are both caused by measurements, and the accuracy of measuring devices or procedures are known. The case when $\delta = 1$ is also known as the orthogonal regression.

- Regression with known reliability ratio $\lambda = \sigma_\square^2 / (\sigma_\eta^2 + \sigma_\square^2)$, where $\sigma_\square^2$ is the variance of the latent regressor. Such approach may be applicable for example when repeating measurements of the same unit are available, or when the reliability ratio has been known from the independent study. In this case the consistent estimate of slope is equal to the least-squares estimate divided by λ.

- Regression with known σ_η^2 may occur when the source of the errors in x's is known and their variance can be calculated. This could include rounding errors, or errors introduced by the measuring device. When σ^2_η is known we can compute the reliability ratio as $\lambda = (\sigma^2_x - \sigma^2_\eta) / \sigma^2_x$ and reduce the problem to the previous case.

Newer estimation methods that do not assume knowledge of some of the parameters of the model, include

- Method of moments — the GMM estimator based on the third- (or higher-) order joint cumulants of observable variables. The slope coefficient can be estimated from

$$\hat{\beta} = \frac{\hat{K}(n_1, n_2 + 1)}{\hat{K}(n_1 + 1, n_2)}, \quad n_1, n_2 > 0,$$

where (n_1, n_2) are such that $K(n_1+1, n_2)$ — the joint cumulant of (x,y) — is not zero. In the case when the third central moment of the latent regressor x^* is non-zero, the formula reduces to

$$\hat{\beta} = \frac{\frac{1}{T}\sum_{t=1}^{T}(x_t - \overline{x})(y_t - \overline{y})^2}{\frac{1}{T}\sum_{t=1}^{T}(x_t - \overline{x})^2(y_t - \overline{y})}.$$

- Instrumental variables — a regression which requires that certain additional data variables z, called *instruments*, were available. These variables should be uncorrelated with the errors in the equation for the dependent variable (*valid*), and they should also be correlated (*relevant*) with the true regressors x^*. If such variables can be found then the estimator takes form

$$\hat{\beta} = \frac{\frac{1}{T}\sum_{t=1}^{T}(z_t - \overline{z})(y_t - \overline{y})}{\frac{1}{T}\sum_{t=1}^{T}(z_t - \overline{z})(x_t - \overline{x})}.$$

Multivariable Linear Model

Multivariable model looks exactly like the simple linear model, only this time β, η_t, x_t and x^*_t are $k\times1$ vectors.

$$\begin{cases} y_t = \alpha + \beta'x^*_t + \varepsilon_t, \\ x_t = x^*_t + \eta_t. \end{cases}$$

The general identifiability condition for this model remains an open question. It is known however that in the case when (ε,η) are independent and jointly normal, the parameter β is identified if and only if it is impossible to find a non-singular $k\times k$ block matrix $[a\ A]$ (where a is a $k\times1$ vector) such that $a'x^*$ is distributed normally and independently from $A'x^*$.

Some of the estimation methods for multivariable linear models are

- Total least squares is an extension of Deming regression to the multivariable setting. When all the $k+1$ components of the vector (ε, η) have equal variances and are independent, this is equivalent to running the orthogonal regression of y on the vector x — that is, the regression which minimizes the sum of squared distances between points (y_t, x_t) and the k-dimensional hyperplane of "best fit".

- The method of moments estimator can be constructed based on the moment conditions $E[z_t \cdot (y_t - \alpha - \beta \cdot x_t)] = 0$, where the $(5k+3)$-dimensional vector of instruments z_t is defined as

$$z_t = \left(1\ z_{t1'}\ z_{t2'}\ z_{t3'}\ z_{t4'}\ z_{t5'}\ z_{t6'}\ z_{t7'}\right)',\quad \text{where}$$

$$z_{t1} = x_t {}^\circ x_t$$

$$z_{t2} = x_t y_t$$

$$z_{t3} = y_t^2$$

$$z_{t4} = x_t {}^\circ x_t {}^\circ x_t - 3\left(E[x_t x_{t'}]{}^\circ I_k\right)x_t$$

$$z_{t5} = x_t {}^\circ x_t y_t - 2\left(E[y_t x_{t'}]{}^\circ I_k\right)x_t - y_t\left(E[x_t x_{t'}]{}^\circ I_k\right)\iota_k$$

$$z_{t6} = x_t y_t^2 - E[y_t^2]x_t - 2y_t\ E[x_t y_t]$$

$$z_{t7} = y_t^3 - 3y_t\ E[y_t^2]$$

where $\circ$ designates the Hadamard product of matrices, and variables x_t, y_t have been preliminarily de-meaned. The authors of the method suggest to use Fuller's modified IV estimator.

This method can be extended to use moments higher than the third order, if necessary, and to accommodate variables measured without error.

- The instrumental variables approach requires to find additional data variables z_t which would serve as *instruments* for the mismeasured regressors x_t. This method is the simplest from the implementation point of view, however its disadvantage is that it requires to collect additional data, which may be costly or even impossible. When the instruments can be found, the estimator takes standard form

$$\hat{\beta} = \left(X'Z(Z'Z)^{-1}Z'X\right)^{-1} X'Z(Z'Z)^{-1}Z'y.$$

Non-linear Models

A generic non-linear measurement error model takes form

$$\begin{cases} y_t = g(x_t^*) + \varepsilon_t, \\ x_t = x_t^* + \eta_t. \end{cases}$$

Here function g can be either parametric or non-parametric. When function g is parametric it will be written as $g(x^*, \beta)$.

For a general vector-valued regressor x^* the conditions for model identifiability are not known. However in the case of scalar x^* the model is identified unless the function g is of the "log-exponential" form

$$g(x^*) = a + b \ln\left(e^{cx^*} + d\right)$$

and the latent regressor x^* has density

$$f_{x^*}(x) = \begin{cases} A e^{-Be^{Cx} + CDx} (e^{Cx} + E)^{-F}, & \text{if } d > 0 \\ A e^{-Bx^2 + Cx} & \text{if } d = 0 \end{cases}$$

where constants A,B,C,D,E,F may depend on a,b,c,d.

Despite this optimistic result, as of now no methods exist for estimating non-linear errors-in-variables models without any extraneous information. However there are several techniques which make use of some additional data: either the instrumental variables, or repeated observations.

Instrumental Variables Methods

- Newey's simulated moments method for parametric models — requires that there is an additional set of observed *predictor variabels* z_t, such that the true regressor can be expressed as

$$x_t^* = \pi_{0'} z_t + \sigma_0 \zeta_t,$$

where π_0 and σ_0 are (unknown) constant matrices, and $\zeta_t \perp z_t$. The coefficient π_0 can be estimated using standard least squares regression of x on z. The distribution of ζ_t is unknown, however we can model it as belonging to a flexible parametric family — the Edgeworth series:

$$f_\zeta(v;\gamma) = \phi(v) \sum_{j=1}^{J} \gamma_j v^j$$

where ϕ is the standard normal distribution.

Simulated moments can be computed using the importance sampling algorithm: first we generate several random variables $\{v_{ts} \sim \phi, s = 1,...,S, t = 1,...,T\}$ from the standard normal distribution, then we compute the moments at t-th observation as

$$m_t(\theta) = A(z_t) \frac{1}{S} \sum_{s=1}^{S} H(x_t, y_t, z_t, v_{ts}; \theta) \sum_{j=1}^{J} \gamma_j v_{ts}^j,$$

where $\theta = (\beta, \sigma, \gamma)$, A is just some function of the instrumental variables z, and H is a two-component vector of moments

$$H_1(x_t, y_t, z_t, v_{ts}; \theta) = y_t - g(\hat{\pi}' z_t + \sigma v_{ts}, \beta),$$
$$H_2(x_t, y_t, z_t, v_{ts}; \theta) = z_t y_t - (\hat{\pi}' z_t + \sigma v_{ts}) g(\hat{\pi}' z_t + \sigma v_{ts}, \beta)$$

With moment functions m_t one can apply standard GMM technique to estimate the unknown parameter θ.

Repeated Observations

In this approach two (or maybe more) repeated observations of the regressor x^* are available. Both observations contain their own measurement errors, however those errors are required to be independent:

$$\begin{cases} x_{1t} = x_t^* + \eta_{1t}, \\ x_{2t} = x_t^* + \eta_{2t}, \end{cases}$$

where $x^* \perp \eta_1 \perp \eta_2$. Variables η_1, η_2 need not be identically distributed (although if they are efficiency of the estimator can be slightly improved). With only these two observations it is possible to consistently estimate the density function of x^* using Kotlarski's deconvolution technique.

- Li's conditional density method for parametric models. The regression equation can be written in terms of the observable variables as

$$\mathrm{E}[y_t \mid x_t] = \int g(x_t^*, \beta) f_{x^*|x}(x_t^* \mid x_t) dx_t^*,$$

where it would be possible to compute the integral if we knew the conditional density function $f_{x^*|x}$. If this function could be known or estimated, then the problem turns into standard non-linear regression, which can be estimated for example using the NLLS method. Assuming for simplicity that η_1, η_2 are identically distributed, this conditional density can be computed as

$$\hat{f}_{x^*|x}(x^* \mid x) = \frac{\hat{f}_{x^*}(x^*)}{\hat{f}_x(x)} \prod_{j=1}^{k} \hat{f}_{\eta_j}\left(x_j - x_j^*\right),$$

where with slight abuse of notation x_j denotes the j-th component of a vector. All densities in this formula can be estimated using inversion of the empirical characteristic functions. In particular,

$$\hat{\varphi}_{\eta_j}(v) = \frac{\hat{\varphi}_{x_j}(v,0)}{\hat{\varphi}_{x_j^*}(v)}, \quad \text{where } \hat{\varphi}_{x_j}(v_1, v_2) = \frac{1}{T} \sum_{t=1}^{T} e^{iv_1 x_{1tj} + iv_2 x_{2tj}},$$

$$\hat{\varphi}_{x_j^*}(v) = \exp \int_0^v \frac{\partial \hat{\varphi}_{x_j}(0, v_2)/\partial v_1}{\hat{\varphi}_{x_j}(0, v_2)} dv_2,$$

$$\hat{\varphi}_x(u) = \frac{1}{2T} \sum_{t=1}^{T} \left(e^{iu'x_{1t}} + e^{iu'x_{2t}}\right), \quad \hat{\varphi}_{x^*}(u) = \frac{\hat{\varphi}_x(u)}{\prod_{j=1}^{k} \hat{\varphi}_{\eta_j}(u_j)}.$$

In order to invert these characteristic function one has to apply the inverse Fourier transform, with a trimming parameter C needed to ensure the numerical stability. For example:

$$\hat{f}_x(x) = \frac{1}{(2\pi)^k} \int_{-C}^{C} \cdots \int_{-C}^{C} e^{-iu'x} \hat{\varphi}_x(u) du.$$

- Schennach's estimator for a parametric linear-in-parameters nonlinear-in-variables model. This is a model of the form

$$\begin{cases} y_t = \sum_{j=1}^{k} \beta_j g_j(x_t^*) + \sum_{j=1}^{\ell} \beta_{k+j} w_{jt} + \varepsilon_t, \\ x_{1t} = x_t^* + \eta_{1t}, \\ x_{2t} = x_t^* + \eta_{2t}, \end{cases}$$

where w_t represents variables measured without errors. The regressor x^* here is scalar (the method can be extended to the case of vector x^* as well). If not for the measurement errors, this would have been a standard linear model with the estimator

$$\hat{\beta} = \left(\hat{E}[\xi_t \xi_{t'}] \right)^{-1} \hat{E}[\xi_t y_t],$$

where

$$\xi_{t'} = (g_1(x_t^*), \cdots, g_k(x_t^*), w_{1,t}, \cdots, w_{l,t}).$$

It turns out that all the expected values in this formula are estimable using the same deconvolution trick. In particular, for a generic observable w_t (which could be 1, w_{1t}, ..., $w_{t\,t'}$, or y_t) and some function h (which could represent any g_j or $g_i g_j$) we have

$$E[w_t h(x_t^*)] = \frac{1}{2\pi} \int_{-\infty}^{\infty} \varphi_h(-u) \psi_w(u) du,$$

where φ_h is the Fourier transform of $h(x^*)$, but using the same convention as for the characteristic functions,

$$\varphi_h(u) = \int e^{iux} h(x) dx, ,$$

and

$$\psi_w(u) = E[w_t e^{iux^*}] = \frac{E[w_t e^{iux_{1t}}]}{E[e^{iux_{1t}}]} \exp \int_0^u i \frac{E[x_{2t} e^{ivx_{1t}}]}{E[e^{ivx_{1t}}]} dv$$

The resulting estimator $\hat{\beta}$ is consistent and asymptotically normal.

- Schennach's estimator for a nonparametric model. The standard Nadaraya–Watson estimator for a nonparametric model takes form

$$\hat{g}(x) = \frac{\hat{E}[y_t K_h(x_t^* - x)]}{\hat{E}[K_h(x_t^* - x)]},$$

for a suitable choice of the kernel K and the bandwidth h. Both expectations here can be estimated using the same technique as in the previous method.

Generalized Linear Model

In statistics, the generalized linear model (GLM) is a flexible generalization of ordinary linear regression that allows for response variables that have error distribution models other than a normal distribution. The GLM generalizes linear regression by allowing the linear model to be related to the response variable via a *link function* and by allowing the magnitude of the variance of each measurement to be a function of its predicted value.

Generalized linear models were formulated by John Nelder and Robert Wedderburn as a way of unifying various other statistical models, including linear regression, logistic regression and Poisson regression. They proposed an iteratively reweighted least squares method for maximum likelihood estimation of the model parameters. Maximum-likelihood estimation remains popular and is the default method on many statistical computing packages. Other approaches, including Bayesian approaches and least squares fits to variance stabilized responses, have been developed.

Intuition

Ordinary linear regression predicts the expected value of a given unknown quantity (the *response variable*, a random variable) as a linear combination of a set of observed values (*predictors*). This implies that a constant change in a predictor leads to a constant change in the response variable (i.e. a *linear-response model*). This is appropriate when the response variable has a normal distribution (intuitively, when a response variable can vary essentially indefinitely in either direction with no fixed "zero value", or more generally for any quantity that only varies by a relatively small amount, e.g. human heights).

However, these assumptions are inappropriate for some types of response variables. For example, in cases where the response variable is expected to be always positive and varying over a wide range, constant input changes lead to geometrically varying, rather than constantly varying, output changes. As an example, a prediction model might predict that 10 degree temperature decrease would lead to 1,000 fewer people visiting the beach is unlikely to generalize well over both small beaches (e.g. those where the expected attendance was 50 at a particular temperature) and large beaches (e.g. those where the expected attendance was 10,000 at a low temperature). The problem with this kind of prediction model would imply a temperature drop of 10 degrees would lead to 1,000 fewer people visiting the beach, a beach whose expected attendance was 50 at a higher temperature would now be predicted to have the impossible attendance value of −950. Logically, a more realistic model would instead predict a constant *rate* of increased beach attendance (e.g. an increase in 10 degrees leads to a doubling in beach attendance, and a drop in 10 degrees leads to a halving in attendance). Such a model is termed an *exponential-response model* (or *log-linear model*, since the logarithm of the response is predicted to vary linearly).

Similarly, a model that predicts a probability of making a yes/no choice (a Bernoulli variable) is even less suitable as a linear-response model, since probabilities are bounded on both ends (they must be between 0 and 1). Imagine, for example, a model that predicts the likelihood of a given person going to the beach as a function of temperature. A reasonable model might predict, for example, that a change in 10 degrees makes a person two times more or less likely to go to the beach. But what does "twice as likely" mean in terms of a probability? It cannot literally mean to double the probability value (e.g. 50% becomes 100%, 75% becomes 150%, etc.). Rather, it is the *odds* that are doubling: from 2:1 odds, to 4:1 odds, to 8:1 odds, etc. Such a model is a *log-odds model*.

Generalized linear models cover all these situations by allowing for response variables that have arbitrary distributions (rather than simply normal distributions), and for an arbitrary function of the response variable (the *link function*) to vary linearly with the predicted values (rather than assuming that the response itself must vary linearly). For example, the case above of predicted number of beach attendees would typically be modeled with a Poisson distribution and a log link, while the case of predicted probability of beach attendance would typically be modeled with a Bernoulli distribution (or binomial distribution, depending on exactly how the problem is phrased) and a log-odds (or *logit*) link function.

Overview

In a generalized linear model (GLM), each outcome Y of the dependent variables is assumed to be generated from a particular distribution in the exponential family, a large range of probability distributions that includes the normal, binomial, Poisson and gamma distributions, among others. The mean, μ, of the distribution depends on the independent variables, X, through:

$$E(Y) = \mu = g^{-1}(X\beta)$$

where E(Y) is the expected value of Y; $X\beta$ is the *linear predictor*, a linear combination of unknown parameters β; g is the link function.

In this framework, the variance is typically a function, V, of the mean:

$$Var(Y) = V(\mu) = V(g^{-1}(X\beta)).$$

It is convenient if V follows from the exponential family distribution, but it may simply be that the variance is a function of the predicted value.

The unknown parameters, β, are typically estimated with maximum likelihood, maximum quasi-likelihood, or Bayesian techniques.

Model Components

The GLM consists of three elements:

1. A probability distribution from the exponential family.

2. A linear predictor $\eta = X\beta$.

3. A link function g such that $E(Y) = \mu = g^{-1}(\eta)$.

Probability Distribution

The overdispersed exponential family of distributions is a generalization of the exponential family and exponential dispersion model of distributions and includes those probability distributions, parameterized by θ and τ , whose density functions f (or probability mass function, for the case of a discrete distribution) can be expressed in the form

$$f_Y(y \mid \theta, \tau) = h(y, \tau) \exp\left(\frac{b(\theta)^T T(y) - A(\theta)}{d(\tau)} \right).$$

τ, called the *dispersion parameter*, typically is known and is usually related to the variance of the distribution. The functions $h(y, \tau)$, $b(\theta)$, $T(y)$, $A(\theta)$ and $d(\tau)$ are known. Many common distributions are in this family.

For scalar Y and θ, this reduces to

$$f_Y(y \mid \theta, \tau) = h(y, \tau) \exp\left(\frac{b(\theta) T(y) - A(\theta)}{d(\tau)} \right).$$

θ is related to the mean of the distribution. If $b(\theta)$ is the identity function, then the distribution is said to be in canonical form (or *natural form*). Note that any distribution can be converted to canonical form by rewriting θ as θ' and then applying the transformation $\theta = b(\theta')$. It is always possible to convert $A(\theta)$ in terms of the new parametrization, even if $b(\theta)$ is not a one-to-one function. If, in addition, $T(y)$ is the identity and τ is known, then θ is called the *canonical parameter* (or *natural parameter*) and is related to the mean through

$$\mu = E(Y) = \nabla A(\theta).$$

For scalar Y and θ, this reduces to

$$\mu = E(Y) = \Lambda'(\theta).$$

Under this scenario, the variance of the distribution can be shown to be

$$\mathrm{Var}(Y) = \nabla \nabla^T A(\theta) d(\tau).$$

For scalar Y and θ, this reduces to

$$\mathrm{Var}(Y) = A''(\theta) d(\tau).$$

Linear Predictor

The linear predictor is the quantity which incorporates the information about the independent variables into the model. The symbol η (Greek "eta") denotes a linear predictor. It is related to the expected value of the data (thus, "predictor") through the link function.

η is expressed as linear combinations (thus, "linear") of unknown parameters β. The coefficients of the linear combination are represented as the matrix of independent variables X. η can thus be expressed as

$$\eta = X\beta.$$

Link Function

The link function provides the relationship between the linear predictor and the mean of the distribution function. There are many commonly used link functions, and their choice is informed by several considerations. There is always a well-defined *canonical* link function which is derived from exponential of the response's density function. However in some cases it makes sense to try to match the domain of the link function to the range of the distribution function's mean, or use a non-canonical link function for algorithmic purposes, for example Bayesian probit regression.

When using a distribution function with a canonical parameter θ, the canonical link function is the function that expresses θ in terms of μ, i.e. $\theta = b(\mu)$. For the most common distributions, the mean μ is one of the parameters in the standard form of the distribution's density function, and then $b(\mu)$ is the function as defined above that maps the density function into its canonical form. When using the canonical link function, $b(\mu) = \theta = X\beta$, which allows $\mathbf{X}^T\mathbf{Y}$ to be a sufficient statistic for β.

Following is a table of several exponential-family distributions in common use and the data they are typically used for, along with the canonical link functions and their inverses (sometimes referred to as the mean function, as done here).

Common distributions with typical uses and canonical link functions					
Distribution	**Support of distribution**	**Typical uses**	**Link name**	**Link function**	**Mean function**
Normal	real: $(-\infty, +\infty)$	Linear-response data	Identity	$X\beta = \mu$	$\mu = X\beta$
Exponential	real: $(0, +\infty)$	Exponential-response data, scale parameters	Inverse	$X\beta = \mu^{-1}$	$\mu = (X\beta)^{-1}$
Gamma					
Inverse Gaussian	real: $(0, +\infty)$		Inverse squared	$X\beta = \mu^{-2}$	$\mu = (X\beta)^{-1/2}$
Poisson	integer: $0, 1, 2, \ldots$	count of occurrences in fixed amount of time/space	Log	$X\beta = \ln(\mu)$	$\mu = \exp(X\beta)$
Bernoulli	integer: $\{0, 1\}$	outcome of single yes/no occurrence	Logit	$X\beta = \ln\left(\dfrac{\mu}{1-\mu}\right)$	$\mu = \dfrac{\exp(X\beta)}{1+\exp(X\beta)} = \dfrac{1}{1+\exp(-X\beta)}$
Binomial	integer: $0, 1, \ldots, N$	count of # of "yes" occurrences out of N yes/no occurrences			
Categorical	integer: $[0, K)$				
	K-vector of integer: $[0, 1]$, where exactly one element in the vector has the value 1	outcome of single K-way occurrence			
Multinomial	K-vector of integer: $[0, N]$	count of occurrences of different types $(1 .. K)$ out of N total K-way occurrences			

In the cases of the exponential and gamma distributions, the domain of the canonical link function is not the same as the permitted range of the mean. In particular, the linear predictor may be negative, which would give an impossible negative mean. When maximizing the likelihood, precautions must be taken to avoid this. An alternative is to use a noncanonical link function.

Note also that in the case of the Bernoulli, binomial, categorical and multinomial distributions, the support of the distributions is not the same type of data as the parameter being predicted. In all of these cases, the predicted parameter is one or more probabilities, i.e. real numbers in the range [0,1]. The resulting model is known as *logistic regression* (or *multinomial logistic regression* in the case that K-way rather than binary values are being predicted).

For the Bernoulli and binomial distributions, the parameter is a single probability, indicating the likelihood of occurrence of a single event. The Bernoulli still satisfies the basic condition of the generalized linear model in that, even though a single outcome will always be either 0 or 1, the *expected value* will nonetheless be a real-valued probability, i.e. the probability of occurrence of a "yes" (or 1) outcome. Similarly, in a binomial distribution, the expected value is Np, i.e. the expected proportion of "yes" outcomes will be the probability to be predicted.

For categorical and multinomial distributions, the parameter to be predicted is a K-vector of probabilities, with the further restriction that all probabilities must add up to 1. Each probability indicates the likelihood of occurrence of one of the K possible values. For the multinomial distribution, and for the vector form of the categorical distribution, the expected values of the elements of the vector can be related to the predicted probabilities similarly to the binomial and Bernoulli distributions.

Fitting

Maximum Likelihood

Biologist and statistician Ronald Fisher

The maximum likelihood estimates can be found using an iteratively reweighted least squares algorithm using either a Newton–Raphson method with updates of the form:

$$\beta^{(t+1)} = \beta^{(t)} + \mathcal{J}^{-1}(\beta^{(t)})\mathbf{u}(\beta^{(t)}),$$

where $\mathcal{J}(\beta^{(t)})$ is the observed information matrix (the negative of the Hessian matrix) and $u(\beta^{(t)})$ is the score function; or a Fisher's scoring method:

$$\beta^{(t+1)} = \beta^{(t)} + \mathcal{I}^{-1}(\beta^{(t)})u(\beta^{(t)}),$$

where $\mathcal{I}(\beta^{(t)})$ is the Fisher information matrix. Note that if the canonical link function is used, then they are the same.

Bayesian Methods

In general, the posterior distribution cannot be found in closed form and so must be approximated, usually using Laplace approximations or some type of Markov chain Monte Carlo method such as Gibbs sampling.

Examples

General Linear Models

A possible point of confusion has to do with the distinction between generalized linear models and the general linear model, two broad statistical models. The general linear model may be viewed as a special case of the generalized linear model with identity link and responses normally distributed. As most exact results of interest are obtained only for the general linear model, the general linear model has undergone a somewhat longer historical development. Results for the generalized linear model with non-identity link are asymptotic (tending to work well with large samples).

Linear Regression

A simple, very important example of a generalized linear model (also an example of a general linear model) is linear regression. In linear regression, the use of the least-squares estimator is justified by the Gauss-Markov theorem, which does not assume that the distribution is normal.

From the perspective of generalized linear models, however, it is useful to suppose that the distribution function is the normal distribution with constant variance and the link function is the identity, which is the canonical link if the variance is known.

For the normal distribution, the generalized linear model has a closed form expression for the maximum-likelihood estimates, which is convenient. Most other GLMs lack closed form estimates.

Binary Data

When the response data, Y, are binary (taking on only values 0 and 1), the distribution function is generally chosen to be the Bernoulli distribution and the interpretation of μ_i is then the probability, p, of Y_i taking on the value one.

There are several popular link functions for binomial functions; the most typical is the canonical logit link:

$$g(p) = \ln\left(\frac{p}{1-p}\right).$$

GLMs with this setup are logistic regression models (or *logit models*).

In addition, the inverse of any continuous cumulative distribution function (CDF) can be used for the link since the CDF's range is $[0,1]$, the range of the binomial mean. The normal CDF Φ is a popular choice and yields the probit model. Its link is

$$g(p) = \Phi^{-1}(p).$$

The reason for the use of the probit model is that a constant scaling of the input variable to a normal CDF (which can be absorbed through equivalent scaling of all of the parameters) yields a function that is practically identical to the logit function, but probit models are more tractable in some situations than logit models. (In a Bayesian setting in which normally distributed prior distributions are placed on the parameters, the relationship between the normal priors and the normal CDF link function means that a probit model can be computed using Gibbs sampling, while a logit model generally cannot.)

Complementary Log-log (Cloglog)

The complementary log-log function $\log(-\log(1-p))$ may also be used. This link function is asymmetric and will often produce different results from the probit and logit link functions. The cloglog model corresponds to applications where we observe either zero events (e.g., defects) or one or more, where the number of events is assumed to follow the Poisson distribution. The Poisson assumption means that

$$\Pr(0) = \exp(-\mu),$$

and we assume that $\log(\mu)$ is a linear model. If p represents the proportion of observations with at least one event, its complement

$$(1-p) = \Pr(0) = \exp(-\mu),$$

and then

$$(-\log(1-p)) = \mu.$$

Since μ must be positive, we can enforce that by throwing zero to negative infinity by taking the logarithm. This produces the "cloglog" transformation

$$\log(-\log(1-p)) = \log(\mu).$$

The identity link is also sometimes used for binomial data to yield a linear probability model. However, the identity link can predict nonsense "probabilities" less than zero or greater than one. This can be avoided by using a transformation like cloglog, probit or logit (or any inverse cumulative distribution function). A primary merit of the identity link is that it can be estimated using linear math—and other standard link functions are approximately linear matching the identity link near p = 0.5.

The variance function for "quasibinomial" data is:

$$\text{Var}(Y_i) = \tau\mu_i(1-\mu_i)$$

where the dispersion parameter τ is exactly one for the binomial distribution. Indeed, the standard binomial likelihood omits τ. When it is present, the model is called "quasibinomial", and the modified likelihood is called a quasi-likelihood, since it is not generally the likelihood corresponding to any real probability distribution. If τ exceeds one, the model is said to exhibit overdispersion.

Multinomial Regression

The binomial case may be easily extended to allow for a multinomial distribution as the response (also, a Generalized Linear Model for counts, with a constrained total). There are two ways in which this is usually done:

Ordered Response

If the response variable is an ordinal measurement, then one may fit a model function of the form:

$$g(\mu_m) = \eta_m = \beta_0 + X_1\beta_1 + \cdots + X_p\beta_p + \gamma_2 + \cdots + \gamma_m = \eta_1 + \gamma_2 + \cdots + \gamma_m \text{ where } \mu_m = P(Y \leq m).$$

for $m > 2$. Different links g lead to proportional odds models or ordered probit models.

Unordered Response

If the response variable is a nominal measurement, or the data do not satisfy the assumptions of an ordered model, one may fit a model of the following form:

$$g(\mu_m) = \eta_m = \beta_{m,0} + X_1\beta_{m,1} + \cdots + X_p\beta_{m,p} \text{ where } \mu_m = P(Y = m \mid Y \in \{1, m\}).$$

for $m > 2$. Different links g lead to multinomial logit or multinomial probit models. These are more general than the ordered response models, and more parameters are estimated.

Count Data

Another example of generalized linear models includes Poisson regression which models count data using the Poisson distribution. The link is typically the logarithm, the canonical link.

The variance function is proportional to the mean

$$\text{var}(Y_i) = \tau\mu_i,$$

where the dispersion parameter τ is typically fixed at exactly one. When it is not, the resulting quasi-likelihood model is often described as poisson with overdispersion or *quasipoisson*.

Extensions

Correlated or Clustered Data

The standard GLM assumes that the observations are uncorrelated. Extensions have been developed to allow for correlation between observations, as occurs for example in longitudinal studies and clustered designs:

- Generalized estimating equations (GEEs) allow for the correlation between observations without the use of an explicit probability model for the origin of the correlations, so there is no explicit likelihood. They are suitable when the random effects and their variances are not of inherent interest, as they allow for the correlation without explaining its origin. The focus is on estimating the average response over the population ("population-averaged" effects) rather than the regression parameters that would enable prediction of the effect of changing one or more components of X on a given individual. GEEs are usually used in conjunction with Huber-White standard errors.

- Generalized linear mixed models (GLMMs) are an extension to GLMs that includes random effects in the linear predictor, giving an explicit probability model that explains the origin of the correlations. The resulting "subject-specific" parameter estimates are suitable when the focus is on estimating the effect of changing one or more components of X on a given individual. GLMMs are also referred to as multilevel models and as mixed model. In general, fitting GLMMs is more computationally complex and intensive than fitting GEEs.

Generalized Additive Models

Generalized additive models (GAMs) are another extension to GLMs in which the linear predictor η is not restricted to be linear in the covariates X but is the sum of smoothing functions applied to the x_is:

$$\eta = \beta_0 + f_1(x_1) + f_2(x_2) + \cdots$$

The smoothing functions f_i are estimated from the data. In general this requires a large number of data points and is computationally intensive.

Vector Generalized Linear Model

In statistics, the class of vector generalized linear models (VGLMs) was proposed to enlarge the scope of models catered for by generalized linear models (GLMs). In particular, VGLMs allow for response variables outside the classical exponential family and for more than one parameter. Each parameter (not necessarily a mean) can be transformed by a *link function*. The VGLM framework is also large enough to naturally accommodate multiple responses; these are several independent responses each coming from a particular statistical distribution with possibly different parameter values.

Vector generalized linear models are described in detail in Yee (2015). The central algorithm adopted is the iteratively reweighted least squares method, for maximum likelihood estimation of usually all the model parameters. In particular, Fisher scoring is implemented by such, which, for most models, uses the first and expected second derivatives of the log-likelihood function.

Motivation

GLMs essentially cover one-parameter models from the classical exponential family, and include 3 of the most important statistical regression models: the linear model, Poisson regression for counts, and logistic regression for binary responses. However, the exponential family is far too limiting for regular data analysis. For example, for counts, zero-inflation, zero-truncation and overdispersion are regularly encountered, and the makeshift adaptations made to the binomial and Poisson models in the form of quasi-binomial and quasi-Poisson can be argued as being ad hoc and unsatisfactory. But the VGLM framework readily handles models such as zero-inflated Poisson regression, zero-altered Poisson (hurdle) regression, positive-Poisson regression, and negative binomial regression. As another example, for the linear model, the variance of a normal distribution is relegated as a scale parameter and it is treated often as a nuisance parameter (if it is considered as a parameter at all). But the VGLM framework allows the variance to be modelled using covariates.

As a whole, one can loosely think of VGLMs as GLMs that handle many models outside the classical exponential family and are not restricted to estimating a single mean. During estimation, rather than using weighted least squares during IRLS, one uses generalized least squares to handle the correlation between the M linear predictors.

Data and Notation

We suppose that the response or outcome or the dependent variable(s), $\mathbf{y} = (y_1, \ldots, y_{Q_1})^T$, are assumed to be generated from a particular distribution. Most distributions are univariate, so that $Q_1 = 1$, and an example of $Q_1 = 2$ is the bivariate normal distribution.

Sometimes we write our data as $(\mathbf{x}_i, w_i, \mathbf{y}_i)$ for $i = 1, \ldots, n$. Each of the n observations are considered to be independent. Then $\mathbf{y}_i = (y_{i1}, \ldots, y_{iQ_1})^T$. The w_i are known positive prior weights, and often $w_i = 1$.

The explanatory or independent variables are written $\mathbf{x} = (x_1, \ldots, x_p)^T$, or when i is needed, as $\mathbf{x}_i = (x_{i1}, \ldots, x_{ip})^T$. Usually there is an *intercept*, in which case $x_1 = 1$ or $x_{i1} = 1$.

Actually, the VGLM framework allows for S responses, each of dimension Q_1. In the above $S = 1$. Hence the dimension of $\mathbf{y}_i$ is more generally $Q = S \times Q_1$. One handles S responses by code such as vglm(cbind(y1, y2, y3) ~ x2 + x3, ..., data = mydata) for $S = 3$. To simplify things, most of this article has $S = 1$.

Model Components

The VGLM usually consists of four elements:

>   1. A probability density function or probability mass function from some statistical distribution which has a log-likelihood ℓ, first derivatives $\partial \ell / \partial \theta_j$ and expected information matrix that can be computed. The model is required to satisfy the usual MLE regularity conditions.

>   2. Linear predictors η_j described below to model each parameter θ_j, $j = 1, \ldots, M$.

3. Link functions g_j such that $\theta_j = g_j^{-1}(\eta_j)$.

4. *Constraint matrices* $\mathbf{H}_k$ for $k = 1, \ldots, p$, each of full column-rank and known.

Linear Predictors

Each *linear predictor* is a quantity which incorporates information about the independent variables into the model. The symbol η_j (Greek "eta") denotes a linear predictor and a subscript j is used to denote the jth one. It relates the jth parameter to the explanatory variables, and η_j is expressed as linear combinations (thus, "linear") of unknown parameters β_j, .e., of regression coefficients $\beta_{(j)k}$.

The jth parameter, θ_j, of the distribution depends on the independent variables, $\mathbf{x}$, through

$$g_j(\theta_j) = \eta_j = \beta_j^T \mathbf{x}.$$

Let $\varsigma = (\eta_1, \ldots, \eta_M)^T$ be the vector of all the linear predictors. (For convenience we always let ς be of dimension M). Thus *all* the covariates comprising $\mathbf{x}$ potentially affect *all* the parameters through the linear predictors η_j. Later, we will allow the linear predictors to be generalized to additive predictors, which is the sum of smooth functions of each x_k and each function is estimated from the data.

Link Functions

Each link function provides the relationship between a linear predictor and a parameter of the distribution. There are many commonly used link functions, and their choice can be somewhat arbitrary. It makes sense to try to match the domain of the link function to the range of the distribution's parameter value. Notice above that the g_j allows a different link function for each parameter. They have similar properties as with generalized linear models, for example, common link functions include the *logit* link for parameters in $(0,1)$, and the log link for positive parameters. The VGAM package has function identitylink() for parameters that can assume both positive and negative values.

Constraint Matrices

More generally, the VGLM framework allows for any linear constraints between the regression coefficients of each linear predictors. For example, we may want to set some to be equal to 0, or constraint some of them to be equal. We have

$$\eta = \sum_{k=1}^{p} \beta_{(k)} x_k = \sum_{k=1}^{p} \mathbf{H}_k \, \beta_{(k)}^* x_k$$

where the $\mathbf{H}_k$ are the *constraint matrices*. Each constraint matrix is known and prespecified, and has M rows, and between 1 and M columns. The elements of constraint matrices are finite-valued, and often they are just 0 or 1. For example, the value 0 effectively omits that element while a 1 includes it. It is common for some models to have a *parallelism* assumption, which means that $\mathbf{H}_k = \mathbf{1}_M$ for $k = 2, \ldots, p$, and for some models, for $k = 1$ too. The special case when $\mathbf{H}_k = \mathbf{I}_M$ for all $k = 1, \ldots, p$ is known as *trivial constraints*; all the regression coefficients are estimated and are unrelated. And θ_j is known as an *intercept-only* parameter if the jth row of all the $\mathbf{H}_k =$ are equal to $\mathbf{0}^T$ for $k = 2, \ldots, p$, i.e., $\eta_j = \beta^*_{(j)1}$ equals an intercept only. Intercept-only parameters are thus modelled as simply as possible, as a scalar.

The unknown parameters, $\beta^* = (\beta^{*T}_{(1)}, \ldots, \beta^{*T}_{(p)})^T$ are typically estimated by the method of maximum likelihood. All the regression coefficients may be put into a matrix as follows:

$$\eta_i = \mathbf{b}^T \mathbf{x}_i = \begin{pmatrix} \beta^T_1 \mathbf{x}_i \\ \vdots \\ \beta^T_M \mathbf{x}_i \end{pmatrix} = \left(\beta_{(1)}, \ldots, \beta_{(p)} \right) \mathbf{x}_i.$$

The xij Facility

With even more generally, one can allow the value of a variable x_k to have a different value for each η_j. For example, if each linear predictor is for a different time point then one might have a time-varying covariate. For example, in discrete choice models, one has *conditional* logit models, *nested* logit models, *generalized* logit models, and the like, to distinguish between certain variants and fit a multinomial logit model to, e.g., transport choices. A variable such as cost differs depending on the choice, for example, taxi is more expensive than bus, which is more expensive than walking. The xij facility of VGAM allows one to generalize $\eta_j(\mathbf{x}_i)$ to $\eta_j(\mathbf{x}_{ij})$

The most general formula is

$$\eta_i = \mathbf{o}_i + \sum_{k=1}^{p} \mathrm{diag}(x_{ik1}, \ldots, x_{ikM}) \mathbf{H}_k \beta^*_{(k)}.$$

Here the $\mathbf{o}_i$ is an optional *offset*; which translates to be a $n \times M$ matrix in practice. The VGAM package has an xij argument that allows the successive elements of the diagonal matrix to be inputted.

Software

Yee (2015) describes an R package implementation in the called VGAM. Currently this software fits approximately 150 models/distributions. The central modelling functions are vglm() and vgam(). The family argument is assigned a *VGAM family function*, e.g., family = negbinomial for negative binomial regression, family = poissonff for Poisson regression, family = propodds for the *proportional odd model* or *cumulative logit model* for ordinal categorical regression.

Fitting

Maximum Likelihood

We are maximizing a log-likelihood

$$\ell = \sum_{i=1}^{n} w_i \ell_i,$$

where the w_i are positive and known *prior weights*. The maximum likelihood estimates can be found using an iteratively reweighted least squares algorithm using Fisher's scoring method, with updates of the form:

$$\beta^{(a+1)} = \beta^{(a)} + \mathcal{I}^{-1}(\beta^{(a)}) \mathbf{u}(\beta^{(a)}),$$

where $\mathcal{I}(\beta^{(a)})$ is the Fisher information matrix at iteration a. It is also called the *expected information matrix*, or *EIM*.

VLM

For the computation, the (small) *model matrix* constructed from the RHS of the formula in vglm() and the constraint matrices are combined to form a *big* model matrix. The IRLS is applied to this big X. This matrix is known as the VLM matrix, since the *vector linear model* is the underlying least squares problem being solved. A VLM is a weighted multivariate regression where the variance-covariance matrix for each row of the response matrix is not necessarily the same, and is known. (In classical multivariate regression, all the errors have the same variance-covariance matrix, and it is unknown). In particular, the VLM minimizes the weighted sum of squares

$$ResSS = \sum_{i=1}^{n} \mathbf{w}_i \left\{ \mathbf{z}_i^{(a-1)} - \eta_i^{(a-1)} \right\}^{\mathrm{T}} \mathbf{W}_i^{(a-1)} \left\{ \mathbf{z}_i^{(a-1)} - \eta_i^{(a-1)} \right\}$$

This quantity is minimized at each IRLS iteration. The *working responses* (also known as *pseudo-response* and *adjusted* dependent vectors) are

$$\mathbf{z}_i = \eta_i + \mathbf{W}_i^{-1} \mathbf{u}_i,$$

where the $\mathbf{W}_i$ are known as *working weights* or *working weight matrices*. They are symmetric and positive-definite. Using the EIM helps ensure that they are all positive-definite (and not just the sum of them) over much of the parameter space. In contrast, using Newton–Raphson would mean the observed information matrices would be used, and these tend to be positive-definite in a smaller subset of the parameter space.

Computationally, the Cholesky decomposition is used to invert the working weight matrices and to convert the overall generalized least squares problem into an ordinary least squares problem.

Examples

Generalized Linear Models

Of course, all generalized linear models are a special cases of VGLMs. But we often estimate all parameters by full maximum likelihood estimation rather than using the method of moments for the scale parameter.

Ordered Categorical Response

If the response variable is an ordinal measurement with $M + 1$ *levels*, then one may fit a model function of the form:

$$g(\theta_j) = \eta_j \quad \text{where } \theta_j = \Pr(Y \le j),$$

for $j = 1, \ldots, M$. Different links g lead to proportional odds models or ordered probit models, e.g., the VGAM family function cumulative(link = probit) assigns a probit link to the cumulative probabilities, therefore this model is also called the *cumulative probit model*. In general they are called *cumulative link models*.

For categorical and multinomial distributions, the fitted values are an $(M + 1)$-vector of probabilities, with the property that all probabilities add up to 1. Each probability indicates the likelihood of occurrence of one of the $M + 1$ possible values.

Unordered Categorical Response

If the response variable is a nominal measurement, or the data do not satisfy the assumptions of an ordered model, then one may fit a model of the following form:

$$\log\left[\frac{Pr(Y = j)}{\Pr(Y = M + 1)} \right] = \eta_j,$$

for $j = 1, \ldots, M$. The above link is sometimes called the *multilogit* link, and the model is called the multinomial logit model. It is common to choose the first or the last level of the response as the *reference* or *baseline* group; the above uses the last level. The VGAM family function multinomial() fits the above model, and it has an argument called refLevel that can be assigned the level used for as the reference group.

Count Data

Classical GLM theory performs Poisson regression for count data. The link is typically the logarithm, which is known as the *canonical link*. The variance function is proportional to the mean:

$$\text{Var}(Y_i) = \tau \mu_i,$$

where the dispersion parameter τ is typically fixed at exactly one. When it is not, the resulting quasi-likelihood model is often described as Poisson with overdispersion, or *quasi-Poisson*; then τ is commonly estimated by the method-of-moments and as such, confidence intervals for τ are difficult to obtain.

In contrast, VGLMs offer a much richer set of models to handle overdispersion with respect to the Poisson, e.g., the negative binomial distribution and several variants thereof. Another count regression model is the *generalized Poisson distribution*. Other possible models are the *zeta distribution* and the *Zipf distribution*.

Extensions

Reduced-rank Vector Generalized Linear Models

RR-VGLMs are VGLMs where a subset of the B matrix is of a lower rank. Without loss of generality, suppose that $\mathbf{x} = (\mathbf{x}_1^T, \mathbf{x}_2^T)^T$ is a partition of the covariate vector. Then the part of the B matrix corresponding to x_2 is of the form AC^T where A and C are thin matrices (i.e., with R columns), e.g., vectors if the rank $R = 1$. RR-VGLMs potentially offer several advantages when applied to certain models and data sets. Firstly, if M and p are large then the number of regression coefficients that are estimated by VGLMs is large ($M \times p$). Then RR-VGLMs can reduce the number of estimated regression coefficients enormously if R is low, e.g., $R = 1$ or $R = 2$. An example of a model where this is particularly useful is the RR-multinomial logit model, also known as the *stereotype model*. Secondly, $v = C^T x_2 = (v_1, \ldots, v_R)^T$ is an R-vector of latent variables, and often these can be usefully interpreted. If $R = 1$ then we can write $v = c^T x_2$ so that the latent variable comprises loadings on the explanatory variables. It may be seen that RR-VGLMs take optimal linear combinations of the x_2 and then a VGLM is fitted to the explanatory variables (x_1, v) Thirdly, a biplot can be produced if $R' = 2$, *and this allows the model to be visualized.*

It can be shown that RR-VGLMs are simply VGLMs where the constraint matrices for the variables in x_2 are unknown and to be estimated. It then transpires that $H_k = A$ for such variables. RR-VGLMs can be estimated by an *alternating* algorithm which fixes A and estimates C, and then fixes C and estimates A, etc.

In practice, some uniqueness constraints are needed for A and/or C. In VGAM, the rrvglm() function uses *corner constraints* by default, which means that the top R rows of A is set to I_R. RR-VGLMs were proposed in 2003.

Two to One

A special case of RR-VGLMs is when $R = 1$ and $M = 2$. This is *dimension reduction* from 2 parameters to 1 parameter. Then it can be shown that

$$\theta_2 = g_2^{-1}\left(t_1 + a_{21} \cdot g_1(\theta_1)\right),$$

where elements t_1 and a_{21} are estimated. Equivalently,

$$\eta_2 = t_1 + a_{21} \cdot \eta_1.$$

This formula provides a coupling of η_1 and η_2. It induces a relationship between two parameters of a model that can be useful, e.g., for modelling a mean-variance relationship. Sometimes there is some choice of link functions, therefore it offers a little flexibility when coupling the two parameters, e.g., a logit, probit, cauchit or cloglog link for parameters in the unit interval. The above formula is particularly useful for the negative binomial distribution, so that the RR-NB has variance function

$$\mathrm{Var}(Y \mid x) = \mu(x) + \delta_1 \mu(x)^{\delta_2}.$$

This has been called the *NB-P* variant by some authors. The δ_1 and δ_2 are estimated, and it is also possible to obtain approximate confidence intervals for them too.

Incidentally, several other useful NB variants can also be fitted, with the help of selecting the right combination of constraint matrices. For example, *NB − 1*, *NB − 2* (negbinomial() default), *NB − H*.

RCIMs

The subclass of *row-column interaction models* (RCIMs) has also been proposed; these are a special type of RR-VGLM. RCIMs apply only to a matrix Y response and there are no explicit explanatory variables **x**. Instead, indicator variables for each row and column are explicitly set up, and an order-*R* interaction of the form $\mathbf{AC}^T$ is allowed. Special cases of this type of model include the *Goodman RC association model* and the quasi-variances methodology as implemented by the qvcalc R package.

RCIMs can be defined as a RR-VGLM applied to Y with

$$g_1(\theta_1) \equiv \eta_{1ij} = \beta_0 + \alpha_i + \gamma_j + \sum_{r=1}^{R} c_{ir}\, a_{jr}.$$

For the Goodman RC association model, we have $\eta_{1ij} = \log \mu_{ij}$, so that if $R = 0$ then it is a Poisson regression fitted to a matrix of counts with row effects and column effects; this has a similar idea to a no-interaction two-way ANOVA model.

Another example of a RCIM is if g_1 is the identity link and the parameter is the median and the model corresponds to an asymmetric Laplace distribution; then a no-interaction RCIM is similar to a technique called *median polish*.

In VGAM, rcim() and grc() functions fit the above models. And also Yee and Hadi (2014) show that RCIMs can be used to fit unconstrained quadratic ordination models to species data; this is an example of indirect gradient analysis in ordination (a topic in statistical ecology).

Vector Generalized Additive Models

Vector generalized additive models (VGAMs) are a major extension to VGLMs in which the linear predictor η_j is not restricted to be linear in the covariates x_k but is the sum of smoothing functions applied to the x_k:

$$\eta(\mathbf{x}) = H_1 \boldsymbol{\beta}_{(1)}^* + H_2 \mathbf{f}_{(2)}^*(\mathbf{x}_2) + H_3 F_{(3)}^*(\mathbf{x}_3) + \cdots$$

where $\mathbf{f}_{(k)}^*(x_k) = (f_{(1)k}^*(x_k), f_{(2)k}^*(x_k), \ldots)^T$. These are *M additive predictors*. Each smooth function $f_{(j)k}^*$ is estimated from the data. Thus VGLMs are *model-driven* while VGAMs are *data-driven*. Currently, only smoothing splines are implemented in the VGAM package. For $M > 1$ they are actually *vector splines*, which estimate the component functions in $f_{(j)k}^*(x_k)$ simultaneously. Of course, one could use regression splines with VGLMs. The motivation behind VGAMs is similar to that of Hastie and Tibshirani (1990) and Wood (2006). VGAMs were proposed in 1996 .

Currently, work is being done to estimate VGAMs using *P-splines* of Eilers and Marx (1996) . This allows for several advantages over using smoothing splines and vector backfitting, such as the ability to perform automatic smoothing parameter selection easier.

Quadratic Reduced-rank Vector Generalized Linear Models

These add on a quadratic in the latent variable to the RR-VGLM class. The result is a bell-shaped curve can be fitted to each response, as a function of the latent variable. For $R = 2$, one has bell-shaped surfaces as a function of the 2 latent variables---somewhat similar to a bivariate normal distribution. Particular applications of QRR-VGLMs can be found in ecology, in a field of multivariate analysis called ordination.

As a specific rank-1 example of a QRR-VGLM, consider Poisson data with S species. The model for Species s is the Poisson regression

$$\log \mu_s(\nu) = \eta_s(\nu) = \beta_{(s)1} + \beta_{(s)2}\nu + \beta_{(s)3}\nu^2 = \alpha_s - \frac{1}{2}\left(\frac{\nu - u_s}{t_s}\right)^2,$$

for $s = 1, \ldots, S$. The right-most parameterization which uses the symbols α_s, u_s, t_s, has particular ecological meaning, because they relate to the species *abundance*, *optimum* and *tolerance* respectively. For example, the tolerance is a measure of niche width, and a large value means that that species can live in a wide range of environments. In the above equation, one would need $\beta_{(s)3} < 0$ in order to obtain a bell-shaped curve.

QRR-VGLMs fit Gaussian ordination models by maximum likelihood estimation, and they are an example of direct gradient analysis. The cqo() function in the VGAM package currently calls optim() to search for the optimal **C**, and given that, it is easy to calculate the site scores and fit a suitable generalized linear model to that. The function is named after the acronym CQO, which stands for *constrained quadratic ordination*: the *constrained* is for direct gradient analysis (there are environmental variables, and a linear combination of these is taken as the latent variable) and the *quadratic* is for the quadratic form in the latent variables i on the ς scale. Unfortunately QRR-VGLMs are sensitive to outliers in both the response and explanatory variables, as well as being computationally expensive, and may give a local solution rather than a global solution. QRR-VGLMs were proposed in 2004.

References

- Burnham, K. P.; Anderson, D. R. (2002), Model Selection and Multimodel Inference (2nd ed.), Springer-Verlag, ISBN 0-387-95364-7.

- Ronald A. Fisher (1954). Statistical Methods for Research Workers (Twelfth ed.). Edinburgh: Oliver and Boyd. ISBN 0-05-002170-2.

- Chiang, C.L, (2003) Statistical methods of analysis, World Scientific. ISBN 981-238-310-7 - page 274 section 9.7.4 "interpolation vs extrapolation"

- Good, P. I.; Hardin, J. W. (2009). Common Errors in Statistics (And How to Avoid Them) (3rd ed.). Hoboken, New Jersey: Wiley. p. 211. ISBN 978-0-470-45798-6.

- Gelman, Andrew; Carlin, John B.; Stern, Hal S. & Rubin, Donald B. (2004). Bayesian Data Analysis (second ed.). Boca Raton, Florida: CRC Press. pp. 4–5. ISBN 1-58488-388-X.

- Bernardo, Degroot, Lindley (September 1983). "Proceedings of the Second Valencia International Meeting". Bayesian Statistics 2. Amsterdam: Elsevier Science Publishers B.V, ISBN 0-444-87746-0, pp. 371–372

- Carroll, Raymond J.; Ruppert, David; Stefanski, Leonard A.; Crainiceanu, Ciprian (2006). Measurement Error in Nonlinear Models: A Modern Perspective (Second ed.). ISBN 1-58488-633-1.

- Fuller, Wayne A. (1987). "A Single Explanatory Variable". Measurement Error Models. John Wiley & Sons. pp. 1–99. ISBN 0-471-86187-1.

- Madsen, Henrik; Thyregod, Poul (2011). Introduction to General and Generalized Linear Models. Chapman & Hall/CRC. ISBN 978-1-4200-9155-7.

- McCullagh, Peter; Nelder, John (1989). Generalized Linear Models, Second Edition. Boca Raton: Chapman and Hall/CRC. ISBN 0-412-31760-5.

- Wood, Simon (2006). Generalized Additive Models: An Introduction with R. Chapman & Hall/CRC. ISBN 1-58488-474-6.

- Yee, T. W. (2015). Vector Generalized Linear and Additive Models: With an Implementation in R. New York, USA: Springer. ISBN 978-1-4939-2817-0.

Mathematical Statistics

The application of mathematics to statistics is known as mathematical statistics. The techniques used in this are mathematical analysis, linear algebra, stochastic analysis and differential equations. This chapter helps the readers in developing an in-depth understanding of mathematical statistics.

Mathematical Statistics

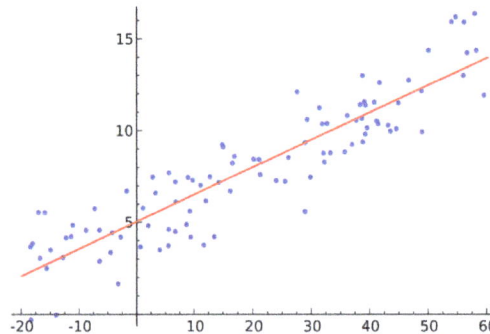

Illustration of linear regression on a data set. Regression analysis is an important part of mathematical statistics.

Mathematical statistics is the application of mathematics to statistics, which was originally conceived as the science of the state — the collection and analysis of facts about a country: its economy, land, military, population, and so forth. Mathematical techniques which are used for this include mathematical analysis, linear algebra, stochastic analysis, differential equations, and measure-theoretic probability theory.

Introduction

Statistical science is concerned with the planning of studies, especially with the design of randomized experiments and with the planning of surveys using random sampling. The initial analysis of the data from properly randomized studies often follows the study protocol.

Of course, the data from a randomized study can be analyzed to consider secondary hypotheses or to suggest new ideas. A secondary analysis of the data from a planned study uses tools from data analysis.

Data analysis is divided into:

- descriptive statistics - the part of statistics that describes data, i.e. summarises the data and their typical properties.

- inferential statistics - the part of statistics that draws conclusions from data (using some model for the data): For example, inferential statistics involves selecting a model for the data, checking whether the data fulfill the conditions of a particular model, and with quantifying the involved uncertainty (e.g. using confidence intervals).

While the tools of data analysis work best on data from randomized studies, they are also applied to other kinds of data --- for example, from natural experiments and observational studies, in which case the inference is dependent on the model chosen by the statistician, and so subjective.

Mathematical statistics has been inspired by and has extended many options in applied statistics.

Topics

The following are some of the important topics in mathematical statistics:

Probability Distributions

A probability distribution assigns a probability to each measurable subset of the possible outcomes of a random experiment, survey, or procedure of statistical inference. Examples are found in experiments whose sample space is non-numerical, where the distribution would be a categorical distribution; experiments whose sample space is encoded by discrete random variables, where the distribution can be specified by a probability mass function; and experiments with sample spaces encoded by continuous random variables, where the distribution can be specified by a probability density function. More complex experiments, such as those involving stochastic processes defined in continuous time, may demand the use of more general probability measures.

A probability distribution can either be univariate or multivariate. A univariate distribution gives the probabilities of a single random variable taking on various alternative values; a multivariate distribution (a joint probability distribution) gives the probabilities of a random vector—a set of two or more random variables—taking on various combinations of values. Important and commonly encountered univariate probability distributions include the binomial distribution, the hypergeometric distribution, and the normal distribution. The multivariate normal distribution is a commonly encountered multivariate distribution.

Special Distributions

- Normal distribution (Gaussian distribution), the most common continuous distribution

- Bernoulli distribution, for the outcome of a single Bernoulli trial (e.g. success/failure, yes/no)

- Binomial distribution, for the number of "positive occurrences" (e.g. successes, yes votes, etc.) given a fixed total number of independent occurrences

- Negative binomial distribution, for binomial-type observations but where the quantity of interest is the number of failures before a given number of successes occurs

- Geometric distribution, for binomial-type observations but where the quantity of interest is the number of failures before the first success; a special c*Discrete uniform distribution, for a finite set of values (e.g. the outcome of a fair die)

- Continuous uniform distribution, for continuously distributed values

- Poisson distribution, for the number of occurrences of a Poisson-type event in a given period of time

- Exponential distribution, for the time before the next Poisson-type event occurs

- Gamma distribution, for the time before the next k Poisson-type events occur

- Chi-squared distribution, the distribution of a sum of squared standard normal variables; useful e.g. for inference regarding the sample variance of normally distributed samples

- Student's t distribution, the distribution of the ratio of a standard normal variable and the square root of a scaled chi squared variable; useful for inference regarding the mean of normally distributed samples with unknown variance

- Beta distribution, for a single probability (real number between 0 and 1); conjugate to the Bernoulli distribution and binomial distribution

Statistical Inferences

Statistical inference is the process of drawing conclusions from data that are subject to random variation, for example, observational errors or sampling variation. Initial requirements of such a system of procedures for inference and induction are that the system should produce reasonable answers when applied to well-defined situations and that it should be general enough to be applied across a range of situations. Inferential statistics are used to test hypotheses and make estimations using sample data. Whereas descriptive statistics describe a sample, inferential statistics infer predictions about a larger population that the sample represents.

The outcome of statistical inference may be an answer to the question "what should be done next?", where this might be a decision about making further experiments or surveys, or about drawing a conclusion before implementing some organizational or governmental policy. For the most part, statistical inference makes propositions about populations, using data drawn from the population of interest via some form of random sampling. More generally, data about a random process is obtained from its observed behavior during a finite period of time. Given a parameter or hypothesis about which one wishes to make inference, statistical inference most often uses:

- a statistical model of the random process that is supposed to generate the data, which is known when randomization has been used, and

- a particular realization of the random process; i.e., a set of data.

Regression

In statistics, regression analysis is a statistical process for estimating the relationships among variables. It includes many techniques for modeling and analyzing several variables, when the focus is on the relationship between a dependent variable and one or more independent variables. More specifically, regression analysis helps one understand how the typical value of the dependent variable (or 'criterion variable') changes when any one of the independent variables is varied, while the other independent variables are held fixed. Most commonly, regression analysis estimates the conditional expectation of the dependent variable given the independent variables – that is, the average value of the dependent variable when the independent variables are fixed. Less commonly, the focus is on a quantile, or other location parameter of the conditional distribution of the dependent variable given the independent variables. In all cases, the estimation target is a function of the independent variables called the regression function. In regression analysis, it is also of interest to characterize the variation of the dependent variable around the regression function which can be described by a probability distribution.

Many techniques for carrying out regression analysis have been developed. Familiar methods such as linear regression and ordinary least squares regression are parametric, in that the regression function is defined in terms of a finite number of unknown parameters that are estimated from the data. Nonparametric regression refers to techniques that allow the regression function to lie in a specified set of functions, which may be infinite-dimensional.

Nonparametric Statistics

Nonparametric statistics are statistics not based on parameterized families of probability distributions. They include both descriptive and inferential statistics. The typical parameters are the mean, variance, etc. Unlike parametric statistics, nonparametric statistics make no assumptions about the probability distributions of the variables being assessed.

Non-parametric methods are widely used for studying populations that take on a ranked order (such as movie reviews receiving one to four stars). The use of non-parametric methods may be necessary when data have a ranking but no clear numerical interpretation, such as when assessing preferences. In terms of levels of measurement, non-parametric methods result in "ordinal" data.

As non-parametric methods make fewer assumptions, their applicability is much wider than the corresponding parametric methods. In particular, they may be applied in situations where less is known about the application in question. Also, due to the reliance on fewer assumptions, non-parametric methods are more robust.

Another justification for the use of non-parametric methods is simplicity. In certain cases, even when the use of parametric methods is justified, non-parametric methods may be easier to use. Due both to this simplicity and to their greater robustness, non-parametric methods are seen by some statisticians as leaving less room for improper use and misunderstanding.

Statistics, Mathematics, and Mathematical Statistics

Mathematical statistics has substantial overlap with the discipline of statistics. Statistical theorists study and improve statistical procedures with mathematics, and statistical research often raises mathematical questions. Statistical theory relies on probability and decision theory.

Mathematicians and statisticians like Gauss, Laplace, and C. S. Peirce used decision theory with probability distributions and loss functions (or utility functions). The decision-theoretic approach to statistical inference was reinvigorated by Abraham Wald and his successors, and makes extensive use of scientific computing, analysis, and optimization; for the design of experiments, statisticians use algebra and combinatorics.

Descriptive Statistics

Descriptive statistics are statistics that quantitatively describe or summarise features of a collection of information. Descriptive statistics are distinguished from inferential statistics (or inductive statistics), in that descriptive statistics aim to summarize a sample, rather than use the data to learn about the population that the sample of data is thought to represent. This generally means that descriptive statistics, unlike inferential statistics, are not developed on the basis of probability theory. Even when a data analysis draws its main conclusions using inferential statistics, descriptive statistics are generally also presented. For example in papers reporting on human subjects, typically a table is included giving the overall sample size, sample sizes in important subgroups (e.g., for each treatment or exposure group), and demographic or clinical characteristics such as the average age, the proportion of subjects of each sex, the proportion of subjects with related comorbidities etc.

Some measures that are commonly used to describe a data set are measures of central tendency and measures of variability or dispersion. Measures of central tendency include the mean, median and mode, while measures of variability include the standard deviation (or variance), the minimum and maximum values of the variables, kurtosis and skewness.

Use in Statistical Analysis

Descriptive statistics provide simple summaries about the sample and about the observations that have been made. Such summaries may be either quantitative, i.e. summary statistics, or visual, i.e. simple-to-understand graphs. These summaries may either form the basis of the initial description of the data as part of a more extensive statistical analysis, or they may be sufficient in and of themselves for a particular investigation.

For example, the shooting percentage in basketball is a descriptive statistic that summarizes the performance of a player or a team. This number is the number of shots made divided by the number of shots taken. For example, a player who shoots 33% is making approximately one shot in every three. The percentage summarizes or describes multiple discrete events. Consider also the grade point average. This single number describes the general performance of a student across the range of their course experiences.

The use of descriptive and summary statistics has an extensive history and, indeed, the simple tabulation of populations and of economic data was the first way the topic of statistics appeared. More recently, a collection of summarisation techniques has been formulated under the heading of exploratory data analysis: an example of such a technique is the box plot.

In the business world, descriptive statistics provides a useful summary of many types of data. For example, investors and brokers may use a historical account of return behavior by performing empirical and analytical analyses on their investments in order to make better investing decisions in the future.

Univariate Analysis

Univariate analysis involves describing the distribution of a single variable, including its central tendency (including the mean, median, and mode) and dispersion (including the range and quantiles of the data-set, and measures of spread such as the variance and standard deviation). The shape of the distribution may also be described via indices such as skewness and kurtosis. Characteristics of a variable's distribution may also be depicted in graphical or tabular format, including histograms and stem-and-leaf display.

Bivariate and Multivariate Analysis

When a sample consists of more than one variable, descriptive statistics may be used to describe the relationship between pairs of variables. In this case, descriptive statistics include:

- Cross-tabulations and contingency tables

- Graphical representation via scatterplots

- Quantitative measures of dependence

- Descriptions of conditional distributions

The main reason for differentiating univariate and bivariate analysis is that bivariate analysis is not only simple descriptive analysis, but also it describes the relationship between two different variables. Quantitative measures of dependence include correlation (such as Pearson's r when both variables are continuous, or Spearman's rho if one or both are not) and covariance (which reflects the scale variables are measured on). The slope, in regression analysis, also reflects the relationship between variables. The unstandardised slope indicates the unit change in the criterion variable for a one unit change in the predictor. The standardised slope indicates this change in standardised (z-score) units. Highly skewed data are often transformed by taking logarithms. Use of logarithms makes graphs more symmetrical and look more similar to the normal distribution, making them easier to interpret intuitively.

Nonparametric Statistics

Nonparametric statistics are statistics not based on parameterized families of probability distributions. They include both descriptive and inferential statistics. The typical parameters are the mean, variance, etc. Unlike parametric statistics, nonparametric statistics make no assumptions about the probability distributions of the variables being assessed. The difference between parametric models and non-parametric models is that the former has a fixed number of parameters, while the latter grows the number of parameters with the amount of training data. Note that the *non*-parametric model does, counterintuitively, contain parameters: the distinction is that parameters are determined by the training data in the case of non-parametric statistics, not the model.

Definitions

In statistics, the term "non-parametric statistics" has at least two different meanings:

1. The first meaning of *non-parametric* covers techniques that do not rely on data belonging to any particular distribution. These include, among others:

   ○ *distribution free* methods, which do not rely on assumptions that the data are drawn from a given probability distribution. As such it is the opposite of parametric statistics. It includes non-parametric descriptive statistics, statistical models, inference and statistical tests.

   ○ *non-parametric statistics* (in the sense of a statistic over data, which is defined to be a function on a sample that has no dependency on a parameter), whose interpretation does not depend on the population fitting any parameterised distributions. Order statistics, which are based on the ranks of observations, are one example of such statistics and these play a central role in many non-parametric approaches.

The following discussion is taken from *Kendall's*.

Statistical hypotheses concern the behavior of observable random variables.... For example, the hypothesis (a) that a normal distribution has a specified mean and variance is statistical; so is the hypothesis (b) that it has a given mean but unspecified variance; so is the hypothesis (c) that a distribution is of normal form with both mean and variance unspecified; finally, so is the hypothesis (d) that two unspecified continuous distributions are identical.

It will have been noticed that in the examples (a) and (b) the distribution underlying the observations was taken to be of a certain form (the normal) and the hypothesis was concerned entirely with the value of one or both of its parameters. Such a hypothesis, for obvious reasons, is called *parametric*.

Hypothesis (c) was of a different nature, as no parameter values are specified in the statement of the hypothesis; we might reasonably call such a hypothesis *non-parametric*. Hypothesis (d) is also non-parametric but, in addition, it does not even specify the underlying form of the distribution and may now be reasonably termed *distribution-free*. Notwithstanding these distinctions, the statistical literature now commonly applies the label "non-parametric" to test procedures that we have just termed "distribution-free", thereby losing a useful classification.

2. The second meaning of *non-parametric* covers techniques that do not assume that the *structure* of a model is fixed. Typically, the model grows in size to accommodate the complexity of the data. In these techniques, individual variables *are* typically assumed to belong to parametric distributions, and assumptions about the types of connections among variables are also made. These techniques include, among others:

   ○ *non-parametric regression*, which refers to modeling where the structure of the relationship between variables is treated non-parametrically, but where nevertheless there may be parametric assumptions about the distribution of model residuals.

   ○ *non-parametric hierarchical Bayesian models*, such as models based on the Dirichlet process, which allow the number of latent variables to grow as necessary to fit the data, but where individual variables still follow parametric distributions and even the process controlling the rate of growth of latent variables follows a parametric distribution.

Applications and Purpose

Non-parametric methods are widely used for studying populations that take on a ranked order (such as movie reviews receiving one to four stars). The use of non-parametric methods may be necessary when data have a ranking but no clear numerical interpretation, such as when assessing preferences. In terms of levels of measurement, non-parametric methods result in "ordinal" data.

As non-parametric methods make fewer assumptions, their applicability is much wider than the corresponding parametric methods. In particular, they may be applied in situations where less is known about the application in question. Also, due to the reliance on fewer assumptions, non-parametric methods are more robust.

Another justification for the use of non-parametric methods is simplicity. In certain cases, even when the use of parametric methods is justified, non-parametric methods may be easier to use. Due both to this simplicity and to their greater robustness, non-parametric methods are seen by some statisticians as leaving less room for improper use and misunderstanding.

The wider applicability and increased robustness of non-parametric tests comes at a cost: in cases where a parametric test would be appropriate, non-parametric tests have less power. In other words, a larger sample size can be required to draw conclusions with the same degree of confidence.

Non-parametric Models

Non-parametric models differ from parametric models in that the model structure is not specified *a priori* but is instead determined from data. The term *non-parametric* is not meant to imply that such models completely lack parameters but that the number and nature of the parameters are flexible and not fixed in advance.

- A histogram is a simple nonparametric estimate of a probability distribution.

- Kernel density estimation provides better estimates of the density than histograms.

- Nonparametric regression and semiparametric regression methods have been developed based on kernels, splines, and wavelets.

- Data envelopment analysis provides efficiency coefficients similar to those obtained by multivariate analysis without any distributional assumption.

- KNNs classify the unseen instance based on the K points in the training set which are nearest to it.

- A support vector machine (with a Gaussian kernel) is a nonparametric large-margin classifier.

- Non-parametrics models can be extended to artificial neural networks

Methods

Non-parametric (or distribution-free) inferential statistical methods are mathematical procedures for statistical hypothesis testing which, unlike parametric statistics, make no assumptions about the probability distributions of the variables being assessed. The most frequently used tests include

- Analysis of similarities

- Anderson–Darling test: tests whether a sample is drawn from a given distribution

- Statistical bootstrap methods: estimates the accuracy/sampling distribution of a statistic

- Cochran's Q: tests whether k treatments in randomized block designs with 0/1 outcomes have identical effects

- Cohen's kappa: measures inter-rater agreement for categorical items

- Friedman two-way analysis of variance by ranks: tests whether k treatments in randomized block designs have identical effects

- Kaplan–Meier: estimates the survival function from lifetime data, modeling censoring

- Kendall's tau: measures statistical dependence between two variables

- Kendall's W: a measure between 0 and 1 of inter-rater agreement

- Kolmogorov–Smirnov test: tests whether a sample is drawn from a given distribution, or whether two samples are drawn from the same distribution

- Kruskal–Wallis one-way analysis of variance by ranks: tests whether > 2 independent samples are drawn from the same distribution

- Kuiper's test: tests whether a sample is drawn from a given distribution, sensitive to cyclic variations such as day of the week

- Logrank test: compares survival distributions of two right-skewed, censored samples

- Mann–Whitney U or Wilcoxon rank sum test: tests whether two samples are drawn from the same distribution, as compared to a given alternative hypothesis.

- McNemar's test: tests whether, in 2×2 contingency tables with a dichotomous trait and matched pairs of subjects, row and column marginal frequencies are equal

- Median test: tests whether two samples are drawn from distributions with equal medians

- Pitman's permutation test: a statistical significance test that yields exact p values by examining all possible rearrangements of labels

- Rank products: detects differentially expressed genes in replicated microarray experiments

- Siegel–Tukey test: tests for differences in scale between two groups

- Sign test: tests whether matched pair samples are drawn from distributions with equal medians

- Spearman's rank correlation coefficient: measures statistical dependence between two variables using a monotonic function

- Squared ranks test: tests equality of variances in two or more samples

- Tukey–Duckworth test: tests equality of two distributions by using ranks

- Wald–Wolfowitz runs test: tests whether the elements of a sequence are mutually independent/random

- Wilcoxon signed-rank test: tests whether matched pair samples are drawn from populations with different mean ranks

Probability Distribution

In probability and statistics, a probability distribution is a mathematical description of a random phenomenon in terms of the probabilities of events. Examples of random phenomena include the results of an experiment or survey. A probability distribution is defined in terms of an underlying sample space, which is the set of all possible outcomes of the random phenomenon being observed. The sample space may be the set of real numbers or a higher-dimensional vector space, or it may be a list of non-numerical values; for example, the sample space of a coin flip would be {Heads, Tails}. Probability distributions are generally divided into two classes. A discrete probability distribution can be encoded by a list of the probabilities of the outcomes, known as a probability mass function. On the other hand, in a continuous probability distribution, the probability of any individual outcome is 0. Continuous probability distributions can often be described by probability density functions; however, more complex experiments, such as those involving stochastic processes defined in continuous time, may demand the use of more general probability measures.

In applied probability, a probability distribution can be specified in a number of different ways, often chosen for mathematical convenience:

- by supplying a valid probability mass function or probability density function

- by supplying a valid cumulative distribution function or survival function

- by supplying a valid hazard function

- by supplying a valid characteristic function

- by supplying a rule for constructing a new random variable from other random variables whose joint probability distribution is known.

A probability distribution whose sample space is the set of real numbers is called univariate, while a distribution whose sample space is a vector space is called multivariate. A univariate distribution gives the probabilities of a single random variable taking on various alternative values; a multivariate distribution (a joint probability distribution) gives the probabilities of a random vector—a list of two or more random variables—taking on various combinations of values. Important and commonly encountered univariate probability distributions include the binomial distribution, the hypergeometric distribution, and the normal distribution. The multivariate normal distribution is a commonly encountered multivariate distribution.

Introduction

To define probability distributions for the simplest cases, one needs to distinguish between discrete and continuous random variables. In the discrete case, it is sufficient to specify a probability mass function assigning a probability to each possible outcome: for example, when throwing a fair dice, each of the six values *1* to *6* has the probability 1/6. The probability of an event is then defined to be the sum of the probabilities of the outcomes that satisfy the event

In contrast, when a random variable takes values from a continuum then typically, any individual outcome has probability zero and only events that include infinitely many outcomes, such as intervals, can have positive probability. For example, the probability that a given object weighs *exactly* 500 g is zero, because the probability of measuring exactly 500 g tends to zero as the accuracy of our measuring instruments increases. Nevertheless, in quality control one might demand that the probability of a "500 g" package containing between 490 g and 510 g should be no less than 98%, and this demand is less sensitive to the accuracy of our instruments.

Continuous probability distributions can be described in several ways. The probability density function describes the infinitesimal probability of any given value, and the probability that the outcome lies in a given interval can be computed by integrating the probability density function over that interval. On the other hand, the cumulative distribution function describes the probability that the random variable is no larger than a given value; the probability that the outcome lies in a given interval can be computed by taking the difference between the values of the cumulative distribution function at the endpoints of the interval. The cumulative distribution function is the antiderivative of the probability density function provided that the latter function exists.

Terminology

As probability theory is used in quite diverse applications, terminology is not uniform and sometimes confusing. The following terms are used for non-cumulative probability distribution functions:

- Probability mass, Probability mass function, p.m.f.: for discrete random variables.

- Categorical distribution: for discrete random variables with a finite set of values.

- Probability density, Probability density function, p.d.f.: most often reserved for continuous random variables.

The following terms are somewhat ambiguous as they can refer to non-cumulative or cumulative distributions, depending on authors' preferences:

- Probability distribution function: continuous or discrete, non-cumulative or cumulative.

- Probability function: even more ambiguous, can mean any of the above or other things.

Finally,

- Probability distribution: sometimes the same as *probability distribution function*, but usually refers to the more complete assignment of probabilities to all measurable subsets of outcomes (i.e. the corresponding probability measure), not just to specific outcomes or ranges of outcomes.

Basic Terms

- Mode: for a discrete random variable, the value with highest probability (the location at which the probability mass function has its peak); for a continuous random variable, the location at which the probability density function has its peak.

- Support: the smallest closed set whose complement has probability zero.

- Head: the range of values where the pmf or pdf is relatively high.

- Tail: the complement of the head within the support; the large set of values where the pmf or pdf is relatively low.

- Expected value or mean: the weighted average of the possible values, using their probabilities as their weights; or the continuous analog thereof.

- Median: the value such that the set of values less than the median has a probability of one-half.

- Variance: the second moment of the pmf or pdf about the mean; an important measure of the dispersion of the distribution.

- Standard deviation: the square root of the variance, and hence another measure of dispersion.

- Symmetry: a property of some distributions in which the portion of the distribution to the left of a specific value is a mirror image of the portion to its right.

- Skewness: a measure of the extent to which a pmf or pdf "leans" to one side of its mean. The third standardized moment of the distribution.

- Kurtosis: a measure of the "fatness" of the tails of a pmf or pdf. The fourth standardized moment of the distribution.

Cumulative Distribution Function

Because a probability distribution Pr on the real line is determined by the probability of a scalar random variable X being in a half-open interval $(-\infty, x]$, the probability distribution is completely characterized by its cumulative distribution function:

$$F(x) = \Pr[X \le x] \qquad \text{for all } x \in \mathbb{R}.$$

Discrete Probability Distribution

A discrete probability distribution is a probability distribution characterized by a probability mass function. Thus, the distribution of a random variable X is discrete, and X is called a discrete random variable, if

$$\sum_u \Pr(X = u) = 1$$

as u runs through the set of all possible values of X. A discrete random variable can assume only a finite or countably infinite number of values. For the number of potential values to be countably infinite, even though their probabilities sum to 1, the probabilities have to decline to zero fast enough. For example, if $\Pr(X = n) = \frac{1}{2^n}$ for $n = 1, 2, ...$, we have the sum of probabilities $1/2 + 1/4 + 1/8 + ... = 1$.

Well-known discrete probability distributions used in statistical modeling include the Poisson distribution, the Bernoulli distribution, the binomial distribution, the geometric distribution, and the negative binomial distribution. Additionally, the discrete uniform distribution is commonly used in computer programs that make equal-probability random selections between a number of choices.

Measure Theoretic Formulation

A measurable function $X : A \to B$ between a probability space $(A, \mathcal{A}, P)$ and a measurable space $(B, \mathcal{B})$ is called a discrete random variable provided its image is a countable set and the pre-image of singleton sets are measurable, i.e., $X^{-1}(b) \in \mathcal{A}$ for all $b \in B$. The latter requirement induces a probability mass function $f_X : X(A) \to \mathbb{R}$ via $f_X(b) := P(X^{-1}(b))$. Since the pre-images of disjoint sets are disjoint

$$\sum_{b \in X(A)} f_X(b) = \sum_{b \in X(A)} P(X^{-1}(b)) = P\left(\bigcup_{b \in X(A)} X^{-1}(b) \right) = P(A) = 1.$$

Cumulative Density

Equivalently to the above, a discrete random variable can be defined as a random variable whose cumulative distribution function (cdf) increases only by jump discontinuities—that is, its cdf increases only where it "jumps" to a higher value, and is constant between those jumps. The points where jumps occur are precisely the values which the random variable may take.

Delta-function Representation

Consequently, a discrete probability distribution is often represented as a generalized probability density function involving Dirac delta functions, which substantially unifies the treatment of continuous and discrete distributions. This is especially useful when dealing with probability distributions involving both a continuous and a discrete part.

Indicator-function Representation

For a discrete random variable X, let u_0, u_1, ... be the values it can take with non-zero probability. Denote

$$\Omega_i = X^{-1}(u_i) = \{\omega : X(\omega) = u_i\}, i = 0,1,2,\dots$$

These are disjoint sets, and by formula (1)

$$\Pr\left(\bigcup_i \Omega_i\right) = \sum_i \Pr(\Omega_i) = \sum_i \Pr(X = u_i) = 1.$$

It follows that the probability that X takes any value except for u_0, u_1, ... is zero, and thus one can write X as

$$X(\omega) = \sum_i u_i 1_{\Omega_i}(\omega)$$

except on a set of probability zero, where 1_A is the indicator function of A. This may serve as an alternative definition of discrete random variables.

Continuous Probability Distribution

A continuous probability distribution is a *probability distribution* that has a cumulative distribution function that is continuous. Most often they are generated by having a probability density function. Mathematicians call distributions with probability density functions absolutely continuous, since their cumulative distribution function is absolutely continuous with respect to the Lebesgue measure λ. If the distribution of X is continuous, then X is called a continuous random variable. There are many examples of continuous probability distributions: normal, uniform, chi-squared, and others.

Intuitively, a continuous random variable is the one which can take a continuous range of values—as opposed to a discrete distribution, where the set of possible values for the random variable is at most countable. While for a discrete distribution an event with probability zero is impossible (e.g., rolling 31/2 on a standard dice is impossible, and has probability zero), this is not so in the case of a continuous random variable. For example, if one measures the width of an oak leaf, the result of 3½ cm is possible; however, it has probability zero because uncountably many other potential values exist even between 3 cm and 4 cm. Each of these individual outcomes has probability zero, yet the probability that the outcome will fall into the interval (3 cm, 4 cm) is nonzero. This apparent paradox is resolved by the fact that the probability that X attains some value within an infinite set, such as an interval, cannot be found by naively adding the probabilities for individual values. Formally, each value has an infinitesimally small probability, which statistically is equivalent to zero.

Formally, if X is a continuous random variable, then it has a probability density function $f(x)$, and therefore its probability of falling into a given interval, say $[a, b]$ is given by the integral

$$\Pr[a \leq X \leq b] = \int f(x)dx$$

In particular, the probability for X to take any single value a (that is $a \leq X \leq a$) is zero, because an integral with coinciding upper and lower limits is always equal to zero.

The definition states that a continuous probability distribution must possess a density, or equivalently, its cumulative distribution function be absolutely continuous. This requirement is stronger than simple continuity of the cumulative distribution function, and there is a special class of distributions, singular distributions, which are neither continuous nor discrete nor a mixture of those. An example is given by the Cantor distribution. Such singular distributions however are never encountered in practice.

Note on terminology: some authors use the term "continuous distribution" to denote the distribution with continuous cumulative distribution function. Thus, their definition includes both the (absolutely) continuous and singular distributions.

By one convention, a probability distribution μ is called *continuous* if its cumulative distribution function $F(x) = \mu(-\infty, x]$ is continuous and, therefore, the probability measure of singletons $\mu\{x\}=0$ for all x.

Another convention reserves the term *continuous probability distribution* for absolutely continuous distributions. These distributions can be characterized by a probability density function: a non-negative Lebesgue integrable function f defined on the real numbers such that

$$F(x) = \mu(-\infty, x] = \int_{-\infty}^{x} f(t)dt.$$

Discrete distributions and some continuous distributions (like the Cantor distribution) do not admit such a density.

Some Properties

- The probability distribution of the sum of two independent random variables is the convolution of each of their distributions.

- Probability distributions are not a vector space—they are not closed under linear combinations, as these do not preserve non-negativity or total integral 1—but they are closed under convex combination, thus forming a convex subset of the space of functions (or measures).

Kolmogorov Definition

In the measure-theoretic formalization of probability theory, a random variable is defined as a measurable function X from a probability space $(\Omega, \mathcal{F}, P)$ to measurable space $(\mathcal{X}, \mathcal{A})$. A probability distribution of X is the pushforward measure X_*P of X, which is a probability measure on $(\mathcal{X}, \mathcal{A})$ satisfying $X_*P = PX^{-1}$.

Random Number Generation

A frequent problem in statistical simulations (the Monte Carlo method) is the generation of pseudo-random numbers that are distributed in a given way. Most algorithms are based on a pseudo-random number generator that produces numbers X that are uniformly distributed in the interval [0,1). These random variates X are then transformed via some algorithm to create a new random variate having the required probability distribution.

Applications

The concept of the probability distribution and the random variables which they describe underlies the mathematical discipline of probability theory, and the science of statistics. There is spread or variability in almost any value that can be measured in a population (e.g. height of people, durability of a metal, sales growth, traffic flow, etc.); almost all measurements are made with some intrinsic error; in physics many processes are described probabilistically,from the kinetic properties of gases to the quantum mechanical description of fundamental particles. For these and many other reasons, simple numbers are often inadequate for describing a quantity, while probability distributions are often more appropriate.

As a more specific example of an application, the cache language models and other statistical language models used in natural language processing to assign probabilities to the occurrence of particular words and word sequences do so by means of probability distributions.

Common Probability Distributions

The following is a list of some of the most common probability distributions, grouped by the type of process that they are related to. For a more complete, there is list which groups by the nature of the outcome being considered (discrete, continuous, multivariate, etc.)

Note also that all of the univariate distributions below are singly peaked; that is, it is assumed that the values cluster around a single point. In practice, actually observed quantities may cluster around multiple values. Such quantities can be modeled using a mixture distribution.

Related to Real-valued Quantities that Grow Linearly (e.g. Errors, Offsets)

- Normal distribution (Gaussian distribution), for a single such quantity; the most common continuous distribution

Related to Positive Real-valued Quantities that Grow Exponentially (e.g. Prices, Incomes, Populations)

- Log-normal distribution, for a single such quantity whose log is normally distributed

- Pareto distribution, for a single such quantity whose log is exponentially distributed; the prototypical power law distribution

Related to Real-valued Quantities that are Assumed to Be Uniformly Distributed Over a (Possibly Unknown) Region

- Discrete uniform distribution, for a finite set of values (e.g. the outcome of a fair die)

- Continuous uniform distribution, for continuously distributed values

Related to Bernoulli Trials (Yes/No Events, With a Given Probability)

- Basic distributions:

- o   Bernoulli distribution, for the outcome of a single Bernoulli trial (e.g. success/failure, yes/no)

- o   Binomial distribution, for the number of "positive occurrences" (e.g. successes, yes votes, etc.) given a fixed total number of independent occurrences

- o   Negative binomial distribution, for binomial-type observations but where the quantity of interest is the number of failures before a given number of successes occurs

- o   Geometric distribution, for binomial-type observations but where the quantity of interest is the number of failures before the first success; a special case of the negative binomial distribution

- •   Related to sampling schemes over a finite population:

  - o   Hypergeometric distribution, for the number of "positive occurrences" (e.g. successes, yes votes, etc.) given a fixed number of total occurrences, using sampling without replacement

  - o   Beta-binomial distribution, for the number of "positive occurrences" (e.g. successes, yes votes, etc.) given a fixed number of total occurrences, sampling using a Polya urn scheme (in some sense, the "opposite" of sampling without replacement)

Related to Categorical Outcomes (Events with K Possible Outcomes, With a Given Probability for Each Outcome)

- •   Categorical distribution, for a single categorical outcome (e.g. yes/no/maybe in a survey); a generalization of the Bernoulli distribution

- •   Multinomial distribution, for the number of each type of categorical outcome, given a fixed number of total outcomes; a generalization of the binomial distribution

- •   Multivariate hypergeometric distribution, similar to the multinomial distribution, but using sampling without replacement; a generalization of the hypergeometric distribution

Related to Events in a Poisson Process (Events that Occur Independently With a Given Rate)

- •   Poisson distribution, for the number of occurrences of a Poisson-type event in a given period of time

- •   Exponential distribution, for the time before the next Poisson-type event occurs

- •   Gamma distribution, for the time before the next k Poisson-type events occur

Related to the Absolute Values of Vectors With Normally Distributed Components

- •   Rayleigh distribution, for the distribution of vector magnitudes with Gaussian distributed orthogonal components. Rayleigh distributions are found in RF signals with Gaussian real and imaginary components.

- Rice distribution, a generalization of the Rayleigh distributions for where there is a stationary background signal component. Found in Rician fading of radio signals due to multipath propagation and in MR images with noise corruption on non-zero NMR signals.

Related to Normally Distributed Quantities Operated With Sum of Squares (for Hypothesis Testing)

- Chi-squared distribution, the distribution of a sum of squared standard normal variables; useful e.g. for inference regarding the sample variance of normally distributed samples

- Student's t distribution, the distribution of the ratio of a standard normal variable and the square root of a scaled chi squared variable; useful for inference regarding the mean of normally distributed samples with unknown variance

- F-distribution, the distribution of the ratio of two scaled chi squared variables; useful e.g. for inferences that involve comparing variances or involving R-squared (the squared correlation coefficient)

Useful as Conjugate Prior Distributions in Bayesian Inference

- Beta distribution, for a single probability (real number between 0 and 1); conjugate to the Bernoulli distribution and binomial distribution

- Gamma distribution, for a non-negative scaling parameter; conjugate to the rate parameter of a Poisson distribution or exponential distribution, the precision (inverse variance) of a normal distribution, etc.

- Dirichlet distribution, for a vector of probabilities that must sum to 1; conjugate to the categorical distribution and multinomial distribution; generalization of the beta distribution

- Wishart distribution, for a symmetric non-negative definite matrix; conjugate to the inverse of the covariance matrix of a multivariate normal distribution; generalization of the gamma distribution

References

- Lakshmikantham,, ed. by D. Kannan,... V. (2002). Handbook of stochastic analysis and applications. New York: M. Dekker. ISBN 0824706609.

- Wald, Abraham (1947). Sequential analysis. New York: John Wiley and Sons. ISBN 0-471-91806-7. See Dover reprint, 2004: ISBN 0-486-43912-7

- Nick, Todd G. (2007). "Descriptive Statistics". Topics in Biostatistics. Methods in Molecular Biology. 404. New York: Springer. pp. 33–52. doi:10.1007/978-1-59745-530-5_3. ISBN 978-1-58829-531-6.

- Bagdonavicius, V., Kruopis, J., Nikulin, M.S. (2011). "Non-parametric tests for complete data", ISTE & WILEY: London & Hoboken. ISBN 978-1-84821-269-5.

- Gibbons, Jean Dickinson; Chakraborti, Subhabrata (2003). Nonparametric Statistical Inference, 4th Ed. CRC Press. ISBN 0-8247-4052-1.

- B. S. Everitt: The Cambridge Dictionary of Statistics, Cambridge University Press, Cambridge (3rd edition, 2006). ISBN 0-521-69027-7

Statistical Inference and Hypothesis Testing

The technique of analyzing properties of an underlying distribution by analysis of data is termed as statistical inference. The topics elucidated in the section are Bayesian inference, asymptotic theory, estimation theory and statistical hypothesis theory. The chapter serves as a source to understand the major categories related to statistical inference and hypothesis testing.

Statistical Inference

Statistical inference is the process of deducing properties of an underlying distribution by analysis of data. Inferential statistical analysis infers properties about a population: this includes testing hypotheses and deriving estimates. The population is assumed to be larger than the observed data set; in other words, the observed data is assumed to be sampled from a larger population.

Inferential statistics can be contrasted with descriptive statistics. Descriptive statistics is solely concerned with properties of the observed data, and does not assume that the data came from a larger population.

Introduction

Statistical inference makes propositions about a population, using data drawn from the population with some form of sampling. Given a hypothesis about a population, for which we wish to draw inferences, statistical inference consists of (firstly) selecting a statistical model of the process that generates the data and (secondly) deducing propositions from the model.

Konishi & Kitagawa state, "The majority of the problems in statistical inference can be considered to be problems related to statistical modeling". Relatedly, Sir David Cox has said, "How [the] translation from subject-matter problem to statistical model is done is often the most critical part of an analysis".

The conclusion of a statistical inference is a statistical proposition.Some common forms of statistical proposition are the following:

- a point estimate, i.e. a particular value that best approximates some parameter of interest;

- an interval estimate, e.g. a confidence interval (or set estimate), i.e. an interval constructed using a dataset drawn from a population so that, under repeated sampling of such datasets, such intervals would contain the true parameter value with the probability at the stated confidence level;

- a credible interval, i.e. a set of values containing, for example, 95% of posterior belief;

- rejection of a hypothesis;

- clustering or classification of data points into groups.

Models and Assumptions

Any statistical inference requires some assumptions. A statistical model is a set of assumptions concerning the generation of the observed data and similar data. Descriptions of statistical models usually emphasize the role of population quantities of interest, about which we wish to draw inference. Descriptive statistics are typically used as a preliminary step before more formal inferences are drawn.

Degree of Models/Assumptions

Statisticians distinguish between three levels of modeling assumptions;

- Fully parametric: The probability distributions describing the data-generation process are assumed to be fully described by a family of probability distributions involving only a finite number of unknown parameters. For example, one may assume that the distribution of population values is truly Normal, with unknown mean and variance, and that datasets are generated by 'simple' random sampling. The family of generalized linear models is a widely used and flexible class of parametric models.

- Non-parametric: The assumptions made about the process generating the data are much less than in parametric statistics and may be minimal. For example, every continuous probability distribution has a median, which may be estimated using the sample median or the Hodges–Lehmann–Sen estimator, which has good properties when the data arise from simple random sampling.

- Semi-parametric: This term typically implies assumptions 'in between' fully and non-parametric approaches. For example, one may assume that a population distribution has a finite mean. Furthermore, one may assume that the mean response level in the population depends in a truly linear manner on some covariate (a parametric assumption) but not make any parametric assumption describing the variance around that mean (i.e. about the presence or possible form of any heteroscedasticity). More generally, semi-parametric models can often be separated into 'structural' and 'random variation' components. One component is treated parametrically and the other non-parametrically. The well-known Cox model is a set of semi-parametric assumptions.

Importance of Valid Models/Assumptions

Whatever level of assumption is made, correctly calibrated inference in general requires these assumptions to be correct; i.e. that the data-generating mechanisms really have been correctly specified.

Incorrect assumptions of 'simple' random sampling can invalidate statistical inference. More complex semi- and fully parametric assumptions are also cause for concern. For example, incorrectly assuming the Cox model can in some cases lead to faulty conclusions. Incorrect assumptions of

Normality in the population also invalidates some forms of regression-based inference. The use of any parametric model is viewed skeptically by most experts in sampling human populations: "most sampling statisticians, when they deal with confidence intervals at all, limit themselves to statements about [estimators] based on very large samples, where the central limit theorem ensures that these [estimators] will have distributions that are nearly normal." In particular, a normal distribution "would be a totally unrealistic and catastrophically unwise assumption to make if we were dealing with any kind of economic population." Here, the central limit theorem states that the distribution of the sample mean "for very large samples" is approximately normally distributed, if the distribution is not heavy tailed.

Approximate Distributions

Given the difficulty in specifying exact distributions of sample statistics, many methods have been developed for approximating these.

With finite samples, approximation results measure how close a limiting distribution approaches the statistic's sample distribution: For example, with 10,000 independent samples the normal distribution approximates (to two digits of accuracy) the distribution of the sample mean for many population distributions, by the Berry–Esseen theorem. Yet for many practical purposes, the normal approximation provides a good approximation to the sample-mean's distribution when there are 10 (or more) independent samples, according to simulation studies and statisticians' experience. Following Kolmogorov's work in the 1950s, advanced statistics uses approximation theory and functional analysis to quantify the error of approximation. In this approach, the metric geometry of probability distributions is studied; this approach quantifies approximation error with, for example, the Kullback–Leibler divergence, Bregman divergence, and the Hellinger distance.

With indefinitely large samples, limiting results like the central limit theorem describe the sample statistic's limiting distribution, if one exists. Limiting results are not statements about finite samples, and indeed are irrelevant to finite samples. However, the asymptotic theory of limiting distributions is often invoked for work with finite samples. For example, limiting results are often invoked to justify the generalized method of moments and the use of generalized estimating equations, which are popular in econometrics and biostatistics. The magnitude of the difference between the limiting distribution and the true distribution (formally, the 'error' of the approximation) can be assessed using simulation. The heuristic application of limiting results to finite samples is common practice in many applications, especially with low-dimensional models with log-concave likelihoods (such as with one-parameter exponential families).

Randomization-based Models

For a given dataset that was produced by a randomization design, the randomization distribution of a statistic (under the null-hypothesis) is defined by evaluating the test statistic for all of the plans that could have been generated by the randomization design. In frequentist inference, randomization allows inferences to be based on the randomization distribution rather than a subjective model, and this is important especially in survey sampling and design of experiments. Statistical inference from randomized studies is also more straightforward than many other situations. In Bayesian inference, randomization is also of importance: in survey sampling, use of sampling without replacement ensures the exchangeability of the sample with the population; in randomized experiments, randomization warrants a missing at random assumption for covariate information.

Objective randomization allows properly inductive procedures. Many statisticians prefer randomization-based analysis of data that was generated by well-defined randomization procedures. (However, it is true that in fields of science with developed theoretical knowledge and experimental control, randomized experiments may increase the costs of experimentation without improving the quality of inferences.) Similarly, results from randomized experiments are recommended by leading statistical authorities as allowing inferences with greater reliability than do observational studies of the same phenomena. However, a good observational study may be better than a bad randomized experiment.

The statistical analysis of a randomized experiment may be based on the randomization scheme stated in the experimental protocol and does not need a subjective model.

However, at any time, some hypotheses cannot be tested using objective statistical models, which accurately describe randomized experiments or random samples. In some cases, such randomized studies are uneconomical or unethical.

Model-based Analysis of Randomized Experiments

It is standard practice to refer to a statistical model, often a linear model, when analyzing data from randomized experiments. However, the randomization scheme guides the choice of a statistical model. It is not possible to choose an appropriate model without knowing the randomization scheme. Seriously misleading results can be obtained analyzing data from randomized experiments while ignoring the experimental protocol; common mistakes include forgetting the blocking used in an experiment and confusing repeated measurements on the same experimental unit with independent replicates of the treatment applied to different experimental units.

Paradigms for Inference

Different schools of statistical inference have become established. These schools—or "paradigms"—are not mutually exclusive, and methods that work well under one paradigm often have attractive interpretations under other paradigms.

Bandyopadhyay & Forster describe four paradigms: "(i) classical statistics or error statistics, (ii) Bayesian statistics, (iii) likelihood-based statistics, and (iv) the Akaikean-Information Criterion-based statistics". The classical (or frequentist) paradigm, the Bayesian paradigm, and the AIC-based paradigm are summarized below. The likelihood-based paradigm is essentially a sub-paradigm of the AIC-based paradigm.

Frequentist Inference

This paradigm calibrates the plausibility of propositions by considering (notional) repeated sampling of a population distribution to produce datasets similar to the one at hand. By considering the dataset's characteristics under repeated sampling, the frequentist properties of a statistical proposition can be quantified—although in practice this quantification may be challenging.

Examples of Frequentist Inference

- p-value

- Confidence interval

Frequentist Inference, Objectivity, and Decision Theory

One interpretation of frequentist inference (or classical inference) is that it is applicable only in terms of frequency probability; that is, in terms of repeated sampling from a population. However, the approach of Neyman develops these procedures in terms of pre-experiment probabilities. That is, before undertaking an experiment, one decides on a rule for coming to a conclusion such that the probability of being correct is controlled in a suitable way: such a probability need not have a frequentist or repeated sampling interpretation. In contrast, Bayesian inference works in terms of conditional probabilities (i.e. probabilities conditional on the observed data), compared to the marginal (but conditioned on unknown parameters) probabilities used in the frequentist approach.

The frequentist procedures of significance testing and confidence intervals can be constructed without regard to utility functions. However, some elements of frequentist statistics, such as statistical decision theory, do incorporate utility functions.In particular, frequentist developments of optimal inference (such as minimum-variance unbiased estimators, or uniformly most powerful testing) make use of loss functions, which play the role of (negative) utility functions. Loss functions need not be explicitly stated for statistical theorists to prove that a statistical procedure has an optimality property. However, loss-functions are often useful for stating optimality properties: for example, median-unbiased estimators are optimal under absolute value loss functions, in that they minimize expected loss, and least squares estimators are optimal under squared error loss functions, in that they minimize expected loss.

While statisticians using frequentist inference must choose for themselves the parameters of interest, and the estimators/test statistic to be used, the absence of obviously explicit utilities and prior distributions has helped frequentist procedures to become widely viewed as 'objective'.

Bayesian Inference

The Bayesian calculus describes degrees of belief using the 'language' of probability; beliefs are positive, integrate to one, and obey probability axioms. Bayesian inference uses the available posterior beliefs as the basis for making statistical propositions. There are several different justifications for using the Bayesian approach.

Examples of Bayesian Inference

- Credible interval for interval estimation

- Bayes factors for model comparison

Bayesian Inference, Subjectivity and Decision Theory

Many informal Bayesian inferences are based on "intuitively reasonable" summaries of the posterior. For example, the posterior mean, median and mode, highest posterior density intervals, and Bayes Factors can all be motivated in this way. While a user's utility function need not be stated for this sort of inference, these summaries do all depend (to some extent) on stated prior beliefs, and are generally viewed as subjective conclusions. (Methods of prior construction which do not require external input have been proposed but not yet fully developed.)

Formally, Bayesian inference is calibrated with reference to an explicitly stated utility, or loss function; the 'Bayes rule' is the one which maximizes expected utility, averaged over the posterior uncertainty. Formal Bayesian inference therefore automatically provides optimal decisions in a decision theoretic sense. Given assumptions, data and utility, Bayesian inference can be made for essentially any problem, although not every statistical inference need have a Bayesian interpretation. Analyses which are not formally Bayesian can be (logically) incoherent; a feature of Bayesian procedures which use proper priors (i.e. those integrable to one) is that they are guaranteed to be coherent. Some advocates of Bayesian inference assert that inference *must* take place in this decision-theoretic framework, and that Bayesian inference should not conclude with the evaluation and summarization of posterior beliefs.

Other Paradigms for Inference

Minimum Description Length

The minimum description length (MDL) principle has been developed from ideas in information theory and the theory of Kolmogorov complexity. The (MDL) principle selects statistical models that maximally compress the data; inference proceeds without assuming counterfactual or non-falsifiable "data-generating mechanisms" or probability models for the data, as might be done in frequentist or Bayesian approaches.

However, if a "data generating mechanism" does exist in reality, then according to Shannon's source coding theorem it provides the MDL description of the data, on average and asymptotically. In minimizing description length (or descriptive complexity), MDL estimation is similar to maximum likelihood estimation and maximum a posteriori estimation (using maximum-entropy Bayesian priors). However, MDL avoids assuming that the underlying probability model is known; the MDL principle can also be applied without assumptions that e.g. the data arose from independent sampling.

The MDL principle has been applied in communication-coding theory in information theory, in linear regression, and in data mining.

The evaluation of MDL-based inferential procedures often uses techniques or criteria from computational complexity theory.

Fiducial Inference

Fiducial inference was an approach to statistical inference based on fiducial probability, also known as a "fiducial distribution". In subsequent work, this approach has been called ill-defined, extremely limited in applicability, and even fallacious. However this argument is the same as that which shows that a so-called confidence distribution is not a valid probability distribution and, since this has not invalidated the application of confidence intervals, it does not necessarily invalidate conclusions drawn from fiducial arguments. An attempt was made to reinterpret the early work of Fisher's fiducial argument as a special case of an inference theory using Upper and lower probabilities.

Structural Inference

Developing ideas of Fisher and of Pitman from 1938 to 1939, George A. Barnard developed "structural inference" or "pivotal inference", an approach using invariant probabilities on group families. Barnard reformulated the arguments behind fiducial inference on a restricted class of models on which "fiducial" procedures would be well-defined and useful.

Inference Topics

The topics below are usually included in the area of statistical inference.

1. Statistical assumptions

2. Statistical decision theory

3. Estimation theory

4. Statistical hypothesis testing

5. Revising opinions in statistics

6. Design of experiments, the analysis of variance, and regression

7. Survey sampling

8. Summarizing statistical data

Bayesian Inference

Bayesian inference is a method of statistical inference in which Bayes' theorem is used to update the probability for a hypothesis as more evidence or information becomes available. Bayesian inference is an important technique in statistics, and especially in mathematical statistics. Bayesian updating is particularly important in the dynamic analysis of a sequence of data. Bayesian inference has found application in a wide range of activities, including science, engineering, philosophy, medicine, sport, and law. In the philosophy of decision theory, Bayesian inference is closely related to subjective probability, often called "Bayesian probability".

Introduction to Bayes' Rule

Formal

Bayesian inference derives the posterior probability as a consequence of two antecedents, a prior probability and a "likelihood function" derived from a statistical model for the observed data. Bayesian inference computes the posterior probability according to Bayes' theorem:

$$P(H \mid E) = \frac{P(E \mid H) \cdot P(H)}{P(E)}$$

where

- | means "event conditional on" (so that $(A \mid B)$ means *A given B*).

- *H* stands for any *hypothesis* whose probability may be affected by data (called *evidence* below). Often there are competing hypotheses, and the task is to determine which is the most probable.

- the *evidence E* corresponds to new data that were not used in computing the prior probability.

- $P(H)$, the *prior probability*, is the estimate of the probability of the hypothesis *H before* the data *E*, the current evidence, is observed.

- $P(H \mid E)$, the *posterior probability*, is the probability of *H given* , *E*, .e., *after E* is observed. This is what we want to know: the probability of a hypothesis *given* the observed evidence.

- $P(E \mid H)$ is the probability of observing *E given H*. As a function of *E* with *H* fixed, this is the *likelihood* – it indicates the compatibility of the evidence with the given hypothesis. The likelihood function is a function of the evidence, *E*, while the posterior probability is a function of the hypothesis, *H*.

- $P(E)$ is sometimes termed the marginal likelihood or "model evidence". This factor is the same for all possible hypotheses being considered (as is evident from the fact that the hypothesis *H* does not appear anywhere in the symbol, unlike for all the other factors), so this factor does not enter into determining the relative probabilities of different hypotheses.

For different values of *H*, only the factors $P(H)$ and $P(E \mid H)$, both in the numerator, affect the value of $P(H \mid E)$ – the posterior probability of a hypothesis is proportional to its prior probability (its inherent likeliness) and the newly acquired likelihood (its compatibility with the new observed evidence).

Bayes' rule can also be written as follows:

$$P(H \mid E) = \frac{P(E \mid H)}{P(E)} \cdot P(H)$$

where the factor $\frac{P(E \mid H)}{P(E)}$ can be interpreted as the impact of *E* on the probability of *H*.

Informal

If the evidence does not match up with a hypothesis, one should reject the hypothesis. But if a hypothesis is extremely unlikely *a priori*, one should also reject it, even if the evidence does appear to match up. For example, if one does not know whether the newborn baby next door is a boy or a girl, the color of decorations on the crib in front of the door may support the hypothesis of one gender or the other; but if in front of that door, instead of the crib, a dog kennel is found, the posterior probability that the family next door gave birth to a dog remains small in spite of the "evidence", since one's prior belief in such a hypothesis was already extremely small.

The critical point about Bayesian inference, then, is that it provides a principled way of combining new evidence with prior beliefs, through the application of Bayes' rule. (Contrast this with frequentist inference, which relies only on the evidence as a whole, with no reference to prior beliefs.)

Furthermore, Bayes' rule can be applied iteratively: after observing some evidence, the resulting posterior probability can then be treated as a prior probability, and a new posterior probability computed from new evidence. This allows for Bayesian principles to be applied to various kinds of evidence, whether viewed all at once or over time. This procedure is termed "Bayesian updating".

Alternatives to Bayesian Updating

Bayesian updating is widely used and computationally convenient. However, it is not the only updating rule that might be considered rational.

Ian Hacking noted that traditional "Dutch book" arguments did not specify Bayesian updating: they left open the possibility that non-Bayesian updating rules could avoid Dutch books. Hacking wrote "And neither the Dutch book argument, nor any other in the personalist arsenal of proofs of the probability axioms, entails the dynamic assumption. Not one entails Bayesianism. So the personalist requires the dynamic assumption to be Bayesian. It is true that in consistency a personalist could abandon the Bayesian model of learning from experience. Salt could lose its savour."

Indeed, there are non-Bayesian updating rules that also avoid Dutch books (as discussed in the literature on "probability kinematics" following the publication of Richard C. Jeffrey's rule, which applies Bayes' rule to the case where the evidence itself is assigned a probability. The additional hypotheses needed to uniquely require Bayesian updating have been deemed to be substantial, complicated, and unsatisfactory.

Formal Description of Bayesian Inference

Definitions

- x, a data point in general. This may in fact be a vector of values.

- θ, the parameter of the data point's distribution, i.e., $x \sim p(x \mid \theta)$. This may in fact be a vector of parameters.

- α, the hyperparameter of the parameter, i.e., $\theta \sim p(\theta \mid \alpha)$.. This may in fact be a vector of hyperparameters.

- **X**, a set of n observed data points, i.e., $x_1, \ldots, x_n$.

- $\tilde{x}$, a new data point whose distribution is to be predicted.

Bayesian Inference

- The prior distribution is the distribution of the parameter(s) before any data is observed, i.e. $p(\theta \mid \alpha)$.

- The prior distribution might not be easily determined. In this case, we can use the Jeffreys prior to obtain the posterior distribution before updating them with newer observations.

- The sampling distribution is the distribution of the observed data conditional on its parameters, i.e. $p(\mathbf{X} \mid \theta)$. This is also termed the likelihood, especially when viewed as a function of the parameter(s), sometimes written $L(\theta \mid \mathbf{X}) = p(\mathbf{X} \mid \theta)$..

- The marginal likelihood (sometimes also termed the *evidence*) is the distribution of the observed data marginalized over the parameter(s), i.e. $p(\mathbf{X} \mid \alpha) = \int_\theta p(\mathbf{X} \mid \theta) p(\theta \mid \alpha) \, d\theta$..

- The posterior distribution is the distribution of the parameter(s) after taking into account the observed data. This is determined by Bayes' rule, which forms the heart of Bayesian inference:

$$p(\theta \mid \mathbf{X}, \alpha) = \frac{p(\mathbf{X} \mid \theta) p(\theta \mid \alpha)}{p(\mathbf{X} \mid \alpha)} \propto p(\mathbf{X} \mid \theta) p(\theta \mid \alpha)$$

Note that this is expressed in words as "posterior is proportional to likelihood times prior", or sometimes as "posterior = likelihood times prior, over evidence".

Bayesian Prediction

- The posterior predictive distribution is the distribution of a new data point, marginalized over the posterior:

$$p(\tilde{x} \mid \mathbf{X}, \alpha) = \int_\theta p(\tilde{x} \mid \theta) p(\theta \mid \mathbf{X}, \alpha) \, d\theta$$

- The prior predictive distribution is the distribution of a new data point, marginalized over the prior:

$$p(\tilde{x} \mid \alpha) = \int_\theta p(\tilde{x} \mid \theta) p(\theta \mid \alpha) \, d\theta$$

Bayesian theory calls for the use of the posterior predictive distribution to do predictive inference, i.e., to predict the distribution of a new, unobserved data point. That is, instead of a fixed point as a prediction, a distribution over possible points is returned. Only this way is the entire posterior distribution of the parameter(s) used. By comparison, prediction in frequentist statistics often involves finding an optimum point estimate of the parameter(s)—e.g., by maximum likelihood or maximum a posteriori estimation (MAP)—and then plugging this estimate into the formula for the distribution of a data point. This has the disadvantage that it does not account for any uncertainty in the value of the parameter, and hence will underestimate the variance of the predictive distribution.

(In some instances, frequentist statistics can work around this problem. For example, confidence intervals and prediction intervals in frequentist statistics when constructed from a normal distribution with unknown mean and variance are constructed using a Student's t-distribution. This correctly estimates the variance, due to the fact that (1) the average of normally distributed random variables is also normally distributed; (2) the predictive distribution of a normally distributed data point with unknown mean and variance, using conjugate or uninformative priors, has a student's t-distribution. In Bayesian statistics, however, the posterior predictive distribution can always be determined exactly—or at least, to an arbitrary level of precision, when numerical methods are used.)

Note that both types of predictive distributions have the form of a compound probability distribution (as does the marginal likelihood). In fact, if the prior distribution is a conjugate prior, and hence the prior and posterior distributions come from the same family, it can easily be seen that both prior and posterior predictive distributions also come from the same family of compound distributions. The only difference is that the posterior predictive distribution uses the updated values of the hyperparameters (applying the Bayesian update rules given in the conjugate prior article), while the prior predictive distribution uses the values of the hyperparameters that appear in the prior distribution.

Inference Over Exclusive and Exhaustive Possibilities

If evidence is simultaneously used to update belief over a set of exclusive and exhaustive propositions, Bayesian inference may be thought of as acting on this belief distribution as a whole.

General Formulation

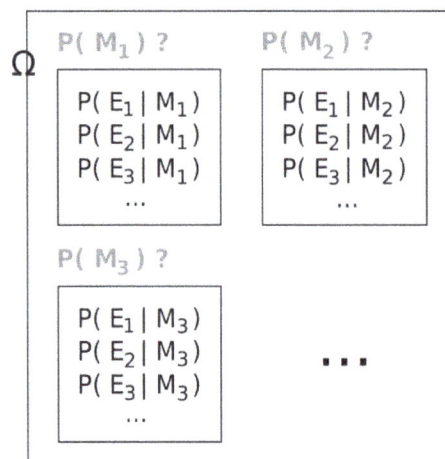

Diagram illustrating event space Ω in general formulation of Bayesian inference. Although this diagram shows discrete models and events, the continuous case may be visualized similarly using probability densities.

Suppose a process is generating independent and identically distributed events E_n, but the probability distribution is unknown. Let the event space Ω represent the current state of belief for this process. Each model is represented by event M_m. The conditional probabilities $P(E_n | M_m)$ are specified to define the models. $P(M_m)$ is the degree of belief in M_m. Before the first inference step, $\{P(M_m)\}$ is a set of *initial prior probabilities*. These must sum to 1, but are otherwise arbitrary.

Suppose that the process is observed to generate $E \in \{E_n\}$. For each $M \in \{M_m\}$, the prior $P(M)$ is updated to the posterior $P(M | E)$. From Bayes' theorem:

$$P(M | E) = \frac{P(E | M)}{\sum_m P(E | M_m) P(M_m)} \cdot P(M)$$

Upon observation of further evidence, this procedure may be repeated.

Multiple Observations

For a sequence of independent and identically distributed observations $\mathbf{E} = (e_1, \ldots, e_n)$, it can be shown by induction that repeated application of the above is equivalent to

$$P(M | \mathbf{E}) = \frac{P(\mathbf{E} | M)}{\sum_m P(\mathbf{E} | M_m) P(M_m)} \cdot P(M)$$

Where

$$P(\mathbf{E} | M) = \prod_k P(e_k | M).$$

Parametric formulation

By parameterizing the space of models, the belief in all models may be updated in a single step. The distribution of belief over the model space may then be thought of as a distribution of belief over the parameter space. The distributions in this section are expressed as continuous, represented by probability densities, as this is the usual situation. The technique is however equally applicable to discrete distributions.

Let the vector θ span the parameter space. Let the initial prior distribution over θ be $p(\theta | \alpha)$, where α is a set of parameters to the prior itself, or *hyperparameters*. Let $\mathbf{E} = (e_1, \ldots, e_n)$ be a sequence of independent and identically distributed event observations, where all e_i are distributed as $p(e | \theta)$ for some θ Bayes' theorem is applied to find the posterior distribution over θ:

$$p(\theta | E, \alpha) = \frac{p(E | \theta, \alpha)}{p(E | \alpha)} \cdot p(\theta | \alpha)$$

$$= \frac{p(E | \theta, \alpha)}{\int_\theta p(E | \theta, \alpha) p(\theta | \alpha) d\theta} \cdot p(\theta | \alpha)$$

Where

$$p(E | \theta, \alpha) = \prod_k p(e_k | \theta)$$

Mathematical Properties

Interpretation of Factor

$\frac{P(E \mid M)}{P(E)} > 1 \Rightarrow P(E \mid M) > P(E)$. That is, if the model were true, the evidence would be more likely than is predicted by the current state of belief. The reverse applies for a decrease in belief. If the belief does not change, $\frac{P(E \mid M)}{P(E)} = 1 \Rightarrow P(E \mid M) = P(E)$ That is, the evidence is independent of the model. If the model were true, the evidence would be exactly as likely as predicted by the current state of belief.

Cromwell's Rule

If $P(M) = 0$ then $P(M \mid E) = 0$. If $P(M) = 1$, then $P(M \mid E) = 1$. This can be interpreted to mean that hard convictions are insensitive to counter-evidence.

The former follows directly from Bayes' theorem. The latter can be derived by applying the first rule to the event "not M" in place of "M", yielding "if $1 - P(M) = 0$, then $1 - P(M \mid E) = 0$", from which the result immediately follows.

Asymptotic Behaviour of Posterior

Consider the behaviour of a belief distribution as it is updated a large number of times with independent and identically distributed trials. For sufficiently nice prior probabilities, the Bernstein-von Mises theorem gives that in the limit of infinite trials, the posterior converges to a Gaussian distribution independent of the initial prior under some conditions firstly outlined and rigorously proven by Joseph L. Doob in 1948, namely if the random variable in consideration has a finite probability space. The more general results were obtained later by the statistician David A. Freedman who published in two seminal research papers in 1963 and 1965 when and under what circumstances the asymptotic behaviour of posterior is guaranteed. His 1963 paper treats, like Doob (1949), the finite case and comes to a satisfactory conclusion. However, if the random variable has an infinite but countable probability space (i.e., corresponding to a die with infinite many faces) the 1965 paper demonstrates that for a dense subset of priors the Bernstein-von Mises theorem is not applicable. In this case there is almost surely no asymptotic convergence. Later in the 1980s and 1990s Freedman and Persi Diaconis continued to work on the case of infinite countable probability spaces. To summarise, there may be insufficient trials to suppress the effects of the initial choice, and especially for large (but finite) systems the convergence might be very slow.

Conjugate Priors

In parameterized form, the prior distribution is often assumed to come from a family of distributions called conjugate priors. The usefulness of a conjugate prior is that the corresponding posterior distribution will be in the same family, and the calculation may be expressed in closed form.

Estimates of Parameters and Predictions

It is often desired to use a posterior distribution to estimate a parameter or variable. Several methods of Bayesian estimation select measurements of central tendency from the posterior distribution.

For one-dimensional problems, a unique median exists for practical continuous problems. The posterior median is attractive as a robust estimator.

If there exists a finite mean for the posterior distribution, then the posterior mean is a method of estimation.

$$\tilde{\theta} = \mathrm{E}[\theta] = \int_\theta \theta \mathrm{P}(\theta \mid \mathrm{X}, \alpha) \mathrm{d}\theta$$

Taking a value with the greatest probability defines maximum *a posteriori* (MAP) estimates:

$$\{\theta_{MAP}\} \subset \arg\max_\theta \mathrm{p}(\theta \mid \mathrm{X}, \alpha).$$

There are examples where no maximum is attained, in which case the set of MAP estimates is empty.

There are other methods of estimation that minimize the posterior *risk* (expected-posterior loss) with respect to a loss function, and these are of interest to statistical decision theory using the sampling distribution ("frequentist statistics").

The posterior predictive distribution of a new observation $\tilde{x}$ (that is independent of previous observations) is determined by

$$\mathrm{p}(\tilde{x} \mid \mathrm{X}, \alpha) = \int_\theta \mathrm{p}(\tilde{x}, \theta \mid \mathrm{X}, \alpha) \mathrm{d}\theta = \int_\theta \mathrm{p}(\tilde{x} \mid \theta) \mathrm{p}(\theta \mid \mathrm{X}, \alpha) \mathrm{d}\theta.$$

Examples

Probability of a Hypothesis

Suppose there are two full bowls of cookies. Bowl #1 has 10 chocolate chip and 30 plain cookies, while bowl #2 has 20 of each. Our friend Fred picks a bowl at random, and then picks a cookie at random. We may assume there is no reason to believe Fred treats one bowl differently from another, likewise for the cookies. The cookie turns out to be a plain one. How probable is it that Fred picked it out of bowl #1?

Intuitively, it seems clear that the answer should be more than a half, since there are more plain cookies in bowl #1. The precise answer is given by Bayes' theorem. Let H_1 correspond to bowl #1, and H_2 to bowl #2. It is given that the bowls are identical from Fred's point of view, thus $P(H_1) = P(H_2)$, and the two must add up to 1, so both are equal to 0.5. The event E is the observation of a plain cookie. From the contents of the bowls, we know that $P(E \mid H_1) = 30/40 = 0.75$ and $P(E \mid H_2) = 20/40 = 0.5$. Bayes' formula then yields

$$\begin{aligned}
P(H_1 \mid E) &= \frac{P(E \mid H_1)P(H_1)}{P(E \mid H_1)P(H_1) + P(E \mid H_2)P(H_2)} \\
&= \frac{0.75 \times 0.5}{0.75 \times 0.5 + 0.5 \times 0.5} \\
&= 0.6
\end{aligned}$$

Before we observed the cookie, the probability we assigned for Fred having chosen bowl #1 was the prior probability, $P(H_1)$, which was 0.5. After observing the cookie, we must revise the probability to $P(H_1 | E)$, which is 0.6.

Making a Prediction

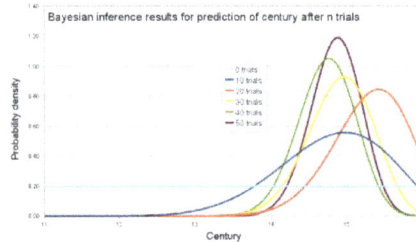

Example results for archaeology example. This simulation was generated using c=15.2.

An archaeologist is working at a site thought to be from the medieval period, between the 11th century to the 16th century. However, it is uncertain exactly when in this period the site was inhabited. Fragments of pottery are found, some of which are glazed and some of which are decorated. It is expected that if the site were inhabited during the early medieval period, then 1% of the pottery would be glazed and 50% of its area decorated, whereas if it had been inhabited in the late medieval period then 81% would be glazed and 5% of its area decorated. How confident can the archaeologist be in the date of inhabitation as fragments are unearthed?

The degree of belief in the continuous variable C (century) is to be calculated, with the discrete set of events $\{GD, G\bar{D}, \bar{G}D, \bar{G}\bar{D}\}$ as evidence. Assuming linear variation of glaze and decoration with time, and that these variables are independent,

$$P(E = GD \,|\, C = c) = (0.01 + \frac{0.81 - 0.01}{16 - 11}(c - 11))(0.5 - \frac{0.5 - 0.05}{16 - 11}(c - 11))$$

$$P(E = G\bar{D} \,|\, C = c) = (0.01 + \frac{0.81 - 0.01}{16 - 11}(c - 11))(0.5 + \frac{0.5 - 0.05}{16 - 11}(c - 11))$$

$$P(E = \bar{G}D \,|\, C = c) = ((1 - 0.01) - \frac{0.81 - 0.01}{16 - 11}(c - 11))(0.5 - \frac{0.5 - 0.05}{16 - 11}(c - 11))$$

$$P(E = \bar{G}\bar{D} \,|\, C = c) = ((1 - 0.01) - \frac{0.81 - 0.01}{16 - 11}(c - 11))(0.5 + \frac{0.5 - 0.05}{16 - 11}(c - 11))$$

Assume a uniform prior of $f_C(c) = 0.2$, and that trials are independent and identically distributed. When a new fragment of type e is discovered, Bayes' theorem is applied to update the degree of belief for each $b\, c$:

$$f_C(c \,|\, E = e) = \frac{P(E = e \,|\, C = c)}{P(E = e)} f_C(c) = \frac{P(E = e \,|\, C = c)}{\int_{11}^{16} P(E = e \,|\, C = c) f_C(c) \, dc} f_C(c)$$

A computer simulation of the changing belief as 50 fragments are unearthed is shown on the graph. In the simulation, the site was inhabited around 1420, or $c = 15.2$. By calculating the area under the relevant portion of the graph for 50 trials, the archaeologist can say that there is practically no chance the site was inhabited in the 11th and 12th centuries, about 1% chance that it was inhabited during the 13th century, 63% chance during the 14th century and 36% during the 15th century. Note that the Bernstein-von Mises theorem asserts here the asymptotic convergence to the "true" distribution because the probability space corresponding to the discrete set of events $\{GD, G\bar{D}, \bar{G}D, \bar{G}\bar{D}\}$ is finite.

In Frequentist Statistics and Decision Theory

A decision-theoretic justification of the use of Bayesian inference was given by Abraham Wald, who proved that every unique Bayesian procedure is admissible. Conversely, every admissible statistical procedure is either a Bayesian procedure or a limit of Bayesian procedures.

Wald characterized admissible procedures as Bayesian procedures (and limits of Bayesian procedures), making the Bayesian formalism a central technique in such areas of frequentist inference as parameter estimation, hypothesis testing, and computing confidence intervals. For example:

- "Under some conditions, all admissible procedures are either Bayes procedures or limits of Bayes procedures (in various senses). These remarkable results, at least in their original form, are due essentially to Wald. They are useful because the property of being Bayes is easier to analyze than admissibility."

- "In decision theory, a quite general method for proving admissibility consists in exhibiting a procedure as a unique Bayes solution."

- "In the first chapters of this work, prior distributions with finite support and the corresponding Bayes procedures were used to establish some of the main theorems relating to the comparison of experiments. Bayes procedures with respect to more general prior distributions have played a very important role in the development of statistics, including its asymptotic theory." "There are many problems where a glance at posterior distributions, for suitable priors, yields immediately interesting information. Also, this technique can hardly be avoided in sequential analysis."

- "A useful fact is that any Bayes decision rule obtained by taking a proper prior over the whole parameter space must be admissible"

- "An important area of investigation in the development of admissibility ideas has been that of conventional sampling-theory procedures, and many interesting results have been obtained."

Applications

Computer Applications

Bayesian inference has applications in artificial intelligence and expert systems. Bayesian inference techniques have been a fundamental part of computerized pattern recognition techniques since the late 1950s. There is also an ever growing connection between Bayesian methods and simulation-based Monte Carlo techniques since complex models cannot be processed in closed form by a Bayesian analysis, while a graphical model structure *may* allow for efficient simulation algorithms like the Gibbs sampling and other Metropolis–Hastings algorithm schemes. Recently Bayesian inference has gained popularity amongst the phylogenetics community for these reasons; a number of applications allow many demographic and evolutionary parameters to be estimated simultaneously.

As applied to statistical classification, Bayesian inference has been used in recent years to develop algorithms for identifying e-mail spam. Applications which make use of Bayesian inference for spam filtering include CRM114, DSPAM, Bogofilter, SpamAssassin, SpamBayes, Mozilla, XEAMS, and others. Spam classification is treated in more detail in the article on the naive Bayes classifier.

Solomonoff's Inductive inference is the theory of prediction based on observations; for example, predicting the next symbol based upon a given series of symbols. The only assumption is that the environment follows some unknown but computable probability distribution. It is a formal inductive framework that combines two well-studied principles of inductive inference: Bayesian statistics and Occam's Razor. Solomonoff's universal prior probability of any prefix p of a computable sequence x is the sum of the probabilities of all programs (for a universal computer) that compute something starting with p. Given some p and any computable but unknown probability distribution from which x is sampled, the universal prior and Bayes' theorem can be used to predict the yet unseen parts of x in optimal fashion.

In the Courtroom

Bayesian inference can be used by jurors to coherently accumulate the evidence for and against a defendant, and to see whether, in totality, it meets their personal threshold for 'beyond a reasonable doubt'. Bayes' theorem is applied successively to all evidence presented, with the posterior from one stage becoming the prior for the next. The benefit of a Bayesian approach is that it gives the juror an unbiased, rational mechanism for combining evidence. It may be appropriate to explain Bayes' theorem to jurors in odds form, as betting odds are more widely understood than probabilities. Alternatively, a logarithmic approach, replacing multiplication with addition, might be easier for a jury to handle.

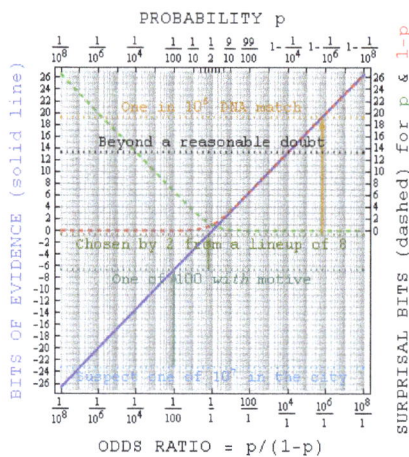

Adding up evidence.

If the existence of the crime is not in doubt, only the identity of the culprit, it has been suggested that the prior should be uniform over the qualifying population. For example, if 1,000 people could have committed the crime, the prior probability of guilt would be 1/1000.

The use of Bayes' theorem by jurors is controversial. In the United Kingdom, a defence expert witness explained Bayes' theorem to the jury in *R v Adams*. The jury convicted, but the case went to appeal on the basis that no means of accumulating evidence had been provided for jurors who did not wish to use Bayes' theorem. The Court of Appeal upheld the conviction, but it also gave the opinion that "To introduce Bayes' Theorem, or any similar method, into a criminal trial plunges the jury into inappropriate and unnecessary realms of theory and complexity, deflecting them from their proper task."

Gardner-Medwin argues that the criterion on which a verdict in a criminal trial should be based is *not* the probability of guilt, but rather the *probability of the evidence, given that the defendant is innocent* (akin to a frequentist p-value). He argues that if the posterior probability of guilt is to be computed by Bayes' theorem, the prior probability of guilt must be known. This will depend on the incidence of the crime, which is an unusual piece of evidence to consider in a criminal trial. Consider the following three propositions:

> A The known facts and testimony could have arisen if the defendant is guilty

> B The known facts and testimony could have arisen if the defendant is innocent

> C The defendant is guilty.

Gardner-Medwin argues that the jury should believe both A and not-B in order to convict. A and not-B implies the truth of C, but the reverse is not true. It is possible that B and C are both true, but in this case he argues that a jury should acquit, even though they know that they will be letting some guilty people go free.

Bayesian Epistemology

Bayesian epistemology is a movement that advocates for Bayesian inference as a means of justifying the rules of inductive logic.

Karl Popper and David Miller have rejected the alleged rationality of Bayesianism, i.e. using Bayes rule to make epistemological inferences: It is prone to the same vicious circle as any other justificationist epistemology, because it presupposes what it attempts to justify. According to this view, a rational interpretation of Bayesian inference would see it merely as a probabilistic version of falsification, rejecting the belief, commonly held by Bayesians, that high likelihood achieved by a series of Bayesian updates would prove the hypothesis beyond any reasonable doubt, or even with likelihood greater than 0.

Other

- The scientific method is sometimes interpreted as an application of Bayesian inference. In this view, Bayes' rule guides (or should guide) the updating of probabilities about hypotheses conditional on new observations or experiments.

- Bayesian search theory is used to search for lost objects.

- Bayesian inference in phylogeny

- Bayesian tool for methylation analysis

- Bayesian brain hypothesis says that the brain is a Bayesian mechanism.

- Bayesian inference in ecological studies

Bayes and Bayesian Inference

The problem considered by Bayes in Proposition 9 of his essay, "An Essay towards solving a Problem in the Doctrine of Chances", is the posterior distribution for the parameter a (the success rate) of the binomial distribution.

History

The term *Bayesian* refers to Thomas Bayes (1702–1761), who proved a special case of what is now called Bayes' theorem. However, it was Pierre-Simon Laplace (1749–1827) who introduced a general version of the theorem and used it to approach problems in celestial mechanics, medical statistics, reliability, and jurisprudence. Early Bayesian inference, which used uniform priors following Laplace's principle of insufficient reason, was called "inverse probability" (because it infers backwards from observations to parameters, or from effects to causes). After the 1920s, "inverse probability" was largely supplanted by a collection of methods that came to be called frequentist statistics.

In the 20th century, the ideas of Laplace were further developed in two different directions, giving rise to *objective* and *subjective* currents in Bayesian practice. In the objective or "non-informative" current, the statistical analysis depends on only the model assumed, the data analyzed, and the method assigning the prior, which differs from one objective Bayesian to another objective Bayesian. In the subjective or "informative" current, the specification of the prior depends on the belief (that is, propositions on which the analysis is prepared to act), which can summarize information from experts, previous studies, etc.

In the 1980s, there was a dramatic growth in research and applications of Bayesian methods, mostly attributed to the discovery of Markov chain Monte Carlo methods, which removed many of the computational problems, and an increasing interest in nonstandard, complex applications. Despite growth of Bayesian research, most undergraduate teaching is still based on frequentist statistics. Nonetheless, Bayesian methods are widely accepted and used, such as for example in the field of machine learning.

Asymptotic Theory

Asymptotic theory or large-sample theory is the branch of mathematics which studies asymptotic expansions.

An example of an asymptotic result is the prime number theorem: Let $\pi(x)$ be the number of prime numbers that are smaller than or equal to x. Then the limit

$$\lim_{x \to \infty} \frac{\pi(x) \ln(x)}{x}$$

exists, and it is equal to 1.

Asymptotic theory ("asymptotics") is used in several mathematical sciences. In statistics, asymptotic theory provides limiting approximations of the probability distribution of sample statistics, such as the likelihood ratio statistic and the expected value of the deviance. Asymptotic theory does not provide a method of evaluating the finite-sample distributions of sample statistics, however. Non-asymptotic bounds are provided by methods of approximation theory.

Asymptotic Distribution

In mathematics and statistics, an asymptotic distribution is a hypothetical distribution that is in a sense the "limiting" distribution of a sequence of distributions. A distribution is an ordered set of random variables

$$Z_i$$

for $i = 1$ to n for some positive integer n. An asymptotic distribution allows i to range without bound, that is, n is infinite.

A special case of an asymptotic distribution is when the late entries go to zero—that is, the Z_i go to 0 as i goes to infinity. Some instances of "asymptotic distribution" refer only to this special case.

This is based on the notion of an asymptotic function which cleanly approaches a constant value (the *asymptote*) as the independent variable goes to infinity; "clean" in this sense meaning that for any desired closeness epsilon there is some value of the independent variable after which the function never differs from the constant by more than epsilon.

An asymptote is a straight line that a curve approaches but never meets or crosses. Informally, one may speak of the curve meeting the asymptote "at infinity" although this is not a precise definition. In the equation

$$y = \frac{1}{x},$$

y becomes arbitrarily small in magnitude as x increases.

It is often used in time series analysis.

In mathematics an asymptotic expansion, asymptotic series or Poincaré expansion (after Henri Poincaré) is a formal series of functions which has the property that truncating the series after a finite number of terms provides an approximation to a given function as the argument of the function tends towards a particular, often infinite, point.

If φ_n is a sequence of continuous functions on some domain, and if L is a (possibly infinite) limit point of the domain, then the sequence constitutes an asymptotic scale if for every n,

$\varphi_{n+1}(x) = o(\varphi_n(x)) \, (x \to L)$. If f is a continuous function on the domain of the asymptotic scale, then an asymptotic expansion of f with respect to the scale is a formal series $\sum_{n=0}^{\infty} a_n \varphi_n(x)$ such that, for any fixed N,

$$f(x) = \sum_{n=0}^{N} a_n \varphi_n(x) + O(\varphi_{N+1}(x)) \, (x \to L).$$

In this case, we write

$$f(x) \sim \sum_{n=0}^{\infty} a_n \varphi_n(x) \, (x \to L).$$

The most common type of asymptotic expansion is a power series in either positive or negative terms. While a convergent Taylor series fits the definition as given, a non-convergent series is what is usually intended by the phrase. Methods of generating such expansions include the Euler–Maclaurin formula and integral transforms such as the Laplace and Mellin transforms. Repeated integration by parts will often lead to an asymptotic expansion.

Examples of Asymptotic Expansions

- Gamma function

$$\frac{e^x}{x^x \sqrt{2\pi x}} \Gamma(x+1) \sim 1 + \frac{1}{12x} + \frac{1}{288x^2} - \frac{139}{51840x^3} - \cdots (x \to \infty)$$

- Exponential integral

$$x e^x E_1(x) \sim \sum_{n=0}^{\infty} \frac{(-1)^n n!}{x^n} \, (x \to \infty)$$

- Riemann zeta function

$$\zeta(s) \sim \sum_{n=1}^{N-1} n^{-s} + \frac{N^{1-s}}{s-1} + N^{-s} \sum_{m=1}^{\infty} \frac{B_{2m} s^{\overline{2m-1}}}{(2m)! N^{2m-1}}$$

where B_{2m} are Bernoulli numbers and $s^{\overline{2m-1}}$ is a rising factorial. This expansion is valid for all complex s and is often used to compute the zeta function by using a large enough value of N, for instance $N > |s|$.

- Error function

$$\sqrt{\pi} x e^{x^2} \operatorname{erfc}(x) = 1 + \sum_{n=1}^{\infty} (-1)^n \frac{(2n)!}{n!(2x)^{2n}}.$$

Detailed Example

Asymptotic expansions often occur when an ordinary series is used in a formal expression that forces the taking of values outside of its domain of convergence. Thus, for example, one may start with the ordinary series

$$\frac{1}{1-w} = \sum_{n=0}^{\infty} w^n$$

The expression on the left is valid on the entire complex plane $w \neq 1$, while the right hand side converges only for $|w| < 1$. Multiplying by $e^{-w/t}$ and integrating both sides yields

$$\int_0^\infty \frac{e^{-w/t}}{1-w} dw = \sum_{n=0}^{\infty} t^{n+1} \int_0^\infty e^{-u} u^n du$$

The integral on the left hand side can be expressed in terms of the exponential integral. The integral on the right hand side, after the substitution $u = w/t$, , may be recognized as the gamma function. Evaluating both, one obtains the asymptotic expansion

$$e^{-1/t} \operatorname{Ei}\left(\frac{1}{t}\right) = \sum_{n=0}^{\infty} n! \, t^{n+1}$$

Here, the right hand side is clearly not convergent for any non-zero value of t. However, by keeping t small, and truncating the series on the right to a finite number of terms, one may obtain a fairly good approximation to the value of $\operatorname{Ei}(1/t)$. Substituting $x = -1/t$ and noting that $\operatorname{Ei}(x) = -E_1(-x)$ results in the asymptotic expansion given earlier in this article.

Estimation Theory

Estimation theory is a branch of statistics that deals with estimating the values of parameters based on measured empirical data that has a random component. The parameters describe an underlying physical setting in such a way that their value affects the distribution of the measured data. An estimator attempts to approximate the unknown parameters using the measurements.

For example, it is desired to estimate the proportion of a population of voters who will vote for a particular candidate. That proportion is the parameter sought; the estimate is based on a small random sample of voters.

Or, for example, in radar the goal is to estimate the range of objects (airplanes, boats, etc.) by analyzing the two-way transit timing of received echoes of transmitted pulses. Since the reflected pulses are unavoidably embedded in electrical noise, their measured values are randomly distributed, so that the transit time must be estimated.

In estimation theory, two approaches are generally considered.

- The probabilistic approach (described in this article) assumes that the measured data is random with probability distribution dependent on the parameters of interest

- The set-membership approach assumes that the measured data vector belongs to a set which depends on the parameter vector.

For example, in electrical communication theory, the measurements which contain information regarding the parameters of interest are often associated with a noisy signal. Without randomness, or noise, the problem would be deterministic and estimation would not be needed.

Basics

To build a model, several statistical "ingredients" need to be known. These are needed to ensure the estimator has some mathematical tractability.

The first is a set of statistical samples taken from a random vector (RV) of size N. Put into a vector,

$$\mathbf{x} = \begin{bmatrix} x[0] \\ x[1] \\ \vdots \\ x[N-1] \end{bmatrix}.$$

Secondly, there are the corresponding M parameters

$$\theta = \begin{bmatrix} \theta_1 \\ \theta_2 \\ \vdots \\ \theta_M \end{bmatrix},$$

which need to be established with their continuous probability density function (pdf) or its discrete counterpart, the probability mass function (pmf)

$$p(\mathbf{x} \mid \theta).$$

It is also possible for the parameters themselves to have a probability distribution (e.g., Bayesian statistics). It is then necessary to define the Bayesian probability

$$\pi(\theta).$$

After the model is formed, the goal is to estimate the parameters, commonly denoted $\tilde{\theta}$, where the "hat" indicates the estimate.

One common estimator is the minimum mean squared error estimator, which utilizes the error between the estimated parameters and the actual value of the parameters

$$e = \tilde{\theta} - \theta$$

as the basis for optimality. This error term is then squared and minimized for the MMSE estimator.

Estimators

Commonly used estimators and estimation methods, and topics related to them:

- Maximum likelihood estimators
- Bayes estimators
- Method of moments estimators
- Cramér–Rao bound
- Minimum mean squared error (MMSE), also known as Bayes least squared error (BLSE)
- Maximum a posteriori (MAP)
- Minimum variance unbiased estimator (MVUE)
- nonlinear system identification
- Best linear unbiased estimator (BLUE)
- Unbiased estimators
- Particle filter
- Markov chain Monte Carlo (MCMC)
- Kalman filter, and its various derivatives
- Wiener filter

Examples

Unknown Constant in Additive white Gaussian Noise

Consider a received discrete signal, $x[n]$, of N independent samples that consists of an unknown constant A with additive white Gaussian noise (AWGN) $w[n]$ with known variance σ^2 (*i.e.*, $\mathcal{N}(0,\sigma^2)$). Since the variance is known then the only unknown parameter is A.

The model for the signal is then

$$x[n] = A + w[n] \quad n = 0,1,\ldots,N-1$$

Two possible (of many) estimators for the parameter A are:

- $\hat{A}_1 = x[0]$
- $\hat{A}_2 = \dfrac{1}{N}\sum_{n=0}^{N-1} x[n]$ which is the sample mean

Both of these estimators have a mean of A, which can be shown through taking the expected value of each estimator

$$\mathrm{E}\left[\hat{A}_1\right] = \mathrm{E}[x[0]] = A$$

and

$$E\left[\hat{A}_2\right] = E\left[\frac{1}{N}\sum_{n=0}^{N-1}x[n]\right] = \frac{1}{N}\left[\sum_{n=0}^{N-1}E\left[x[n]\right]\right] = \frac{1}{N}\left[NA\right] = A$$

At this point, these two estimators would appear to perform the same. However, the difference between them becomes apparent when comparing the variances.

$$\mathrm{var}\left(\hat{A}_1\right) = \mathrm{var}\left(x[0]\right) = \sigma^2$$

and

$$\mathrm{var}\left(\hat{A}_2\right) = \mathrm{var}\left(\frac{1}{N}\sum_{n=0}^{N-1}x[n]\right) \overset{\text{independence}}{=} \frac{1}{N^2}\left[\sum_{n=0}^{N-1}\mathrm{var}(x[n])\right] = \frac{1}{N^2}\left[N\sigma^2\right] = \frac{\sigma^2}{N}$$

It would m that the sample mean is a better estimator since its variance is lower for every $N > 1$.

Maximum Likelihood

Continuing the example using the maximum likelihood estimator, the probability density function (pdf) of the noise for one sample $w[n]$ is

$$p(w[n]) = \frac{1}{\sigma\sqrt{2\pi}}\exp\left(-\frac{1}{2\sigma^2}w[n]^2\right)$$

and the probability of $x[n]$ becomes ($x[n]$ can be thought of a $\mathcal{N}(A,\sigma^2)$)

$$p(x[n]; A) = \frac{1}{\sigma\sqrt{2\pi}}\exp\left(-\frac{1}{2\sigma^2}(x[n]-A)^2\right)$$

By independence, the probability of $\mathbf{x}$ becomes

$$p(\mathbf{x}; A) = \prod_{n=0}^{N-1}p(x[n]; A) = \frac{1}{\left(\sigma\sqrt{2\pi}\right)^N}\exp\left(-\frac{1}{2\sigma^2}\sum_{n=0}^{N-1}(x[n]-A)^2\right)$$

Taking the natural logarithm of the pdf

$$\ln p(\mathbf{x}; A) = -N\ln\left(\sigma\sqrt{2\pi}\right) - \frac{1}{2\sigma^2}\sum_{n=0}^{N-1}(x[n]-A)^2$$

and the maximum likelihood estimator is

$$\hat{A} = \arg\max \ln p(\mathbf{x}; A)$$

Taking the first derivative of the log-likelihood function

$$\frac{\partial}{\partial A}\ln p(\mathbf{x};A) = \frac{1}{\sigma^2}\left[\sum_{n=0}^{N-1}(x[n]-A)\right] = \frac{1}{\sigma^2}\left[\sum_{n=0}^{N-1}x[n]-NA\right]$$

and setting it to zero

$$0 = \frac{1}{\sigma^2}\left[\sum_{n=0}^{N-1}x[n]-NA\right] = \sum_{n=0}^{N-1}x[n]-NA$$

This results in the maximum likelihood estimator

$$\hat{A} = \frac{1}{N}\sum_{n=0}^{N-1}x[n]$$

which is simply the sample mean. From this example, it was found that the sample mean is the maximum likelihood estimator for N samples of a fixed, unknown parameter corrupted by AWGN.

Cramér–Rao Lower Bound

To find the Cramér–Rao lower bound (CRLB) of the sample mean estimator, it is first necessary to find the Fisher information number

$$\mathcal{I}(A) = \mathrm{E}\left(\left[\frac{\partial}{\partial A}\ln p(\mathbf{x};A)\right]^2\right) = -\mathrm{E}\left[\frac{\partial^2}{\partial A^2}\ln p(\mathbf{x};A)\right]$$

and copying from above

$$\frac{\partial}{\partial A}\ln p(\mathbf{x};A) = \frac{1}{\sigma^2}\left[\sum_{n=0}^{N-1}x[n]-NA\right]$$

Taking the second derivative

$$\frac{\partial^2}{\partial A^2}\ln p(\mathbf{x};A) = \frac{1}{\sigma^2}(-N) = \frac{-N}{\sigma^2}$$

and finding the negative expected value is trivial since it is now a deterministic constant

$$-\mathrm{E}\left[\frac{\partial^2}{\partial A^2}\ln p(\mathbf{x};A)\right] = \frac{N}{\sigma^2}$$

Finally, putting the Fisher information into

$$\mathrm{var}\left(\hat{A}\right) \geq \frac{1}{\mathcal{I}}$$

results in

$$\mathrm{var}\left(\hat{A}\right) \geq \frac{\sigma^2}{N}$$

Comparing this to the variance of the sample mean (determined previously) shows that the sample mean is *equal to* the Cramér–Rao lower bound for all values of N and A. In other words, the sample mean is the (necessarily unique) efficient estimator, and thus also the minimum variance unbiased estimator (MVUE), in addition to being the maximum likelihood estimator.

Maximum of a Uniform Distribution

One of the simplest non-trivial examples of estimation is the estimation of the maximum of a uniform distribution. It is used as a hands-on classroom exercise and to illustrate basic principles of estimation theory. Further, in the case of estimation based on a single sample, it demonstrates philosophical issues and possible misunderstandings in the use of maximum likelihood estimators and likelihood functions.

Given a discrete uniform distribution $1, 2, \ldots, N$ with unknown maximum, the UMVU estimator for the maximum is given by

$$\frac{k+1}{k} m - 1 = m + \frac{m}{k} - 1$$

where m is the sample maximum and k is the sample size, sampling without replacement. This problem is commonly known as the German tank problem, due to application of maximum estimation to estimates of German tank production during World War II.

The formula may be understood intuitively as:

"The sample maximum plus the average gap between observations in the sample",

the gap being added to compensate for the negative bias of the sample maximum as an estimator for the population maximum.

This has a variance of

$$\frac{1}{k} \frac{(N-k)(N+1)}{(k+2)} \approx \frac{N^2}{k^2} \text{ for small samples } k \ll N$$

so a standard deviation of approximately $N >$, the (population) average size of a gap between samples; compare $\frac{m}{k}$ above. This can be seen as a very simple case of maximum spacing estimation.

The sample maximum is the maximum likelihood estimator for the population maximum, but, as discussed above, it is biased.

Applications

Numerous fields require the use of estimation theory. Some of these fields include (but are by no means limited to):

- Interpretation of scientific experiments
- Signal processing
- Clinical trials

- Opinion polls

- Quality control

- Telecommunications

- Project management

- Software engineering

- Control theory (in particular Adaptive control)

- Network intrusion detection system

- Orbit determination

Measured data are likely to be subject to noise or uncertainty and it is through statistical probability that optimal solutions are sought to extract as much information from the data as possible.

Statistical Hypothesis Testing

A statistical hypothesis is a hypothesis that is testable on the basis of observing a process that is modeled via a set of random variables. A statistical hypothesis test is a method of statistical inference. Commonly, two statistical data sets are compared, or a data set obtained by sampling is compared against a synthetic data set from an idealized model. A hypothesis is proposed for the statistical relationship between the two data sets, and this is compared as an alternative to an idealized null hypothesis that proposes no relationship between two data sets. The comparison is deemed *statistically significant* if the relationship between the data sets would be an unlikely realization of the null hypothesis according to a threshold probability—the significance level. Hypothesis tests are used in determining what outcomes of a study would lead to a rejection of the null hypothesis for a pre-specified level of significance. The process of distinguishing between the null hypothesis and the alternative hypothesis is aided by identifying two conceptual types of errors (type 1 & type 2), and by specifying parametric limits on e.g. how much type 1 error will be permitted.

An alternative framework for statistical hypothesis testing is to specify a set of statistical models, one for each candidate hypothesis, and then use model selection techniques to choose the most appropriate model. The most common selection techniques are based on either Akaike information criterion or Bayes factor.

Statistical hypothesis testing is sometimes called confirmatory data analysis. It can be contrasted with exploratory data analysis, which may not have pre-specified hypotheses.

Variations and Sub-classes

Statistical hypothesis testing is a key technique of both frequentist inference and Bayesian inference, although the two types of inference have notable differences. Statistical hypothesis tests define a procedure that controls (fixes) the probability of incorrectly *deciding* that a default position (null hypothesis) is incorrect. The procedure is based on how likely it would be for a set of observations to occur if the null hypothesis were true. Note that this probability of making an incorrect decision is *not* the probability that the null hypothesis is true, nor whether any specific alternative hypothesis is true. This contrasts with other possible techniques of decision theory in which the null and alternative hypothesis are treated on a more equal basis.

One naïve Bayesian approach to hypothesis testing is to base decisions on the posterior probability, but this fails when comparing point and continuous hypotheses. Other approaches to decision making, such as Bayesian decision theory, attempt to balance the consequences of incorrect decisions across all possibilities, rather than concentrating on a single null hypothesis. A number of other approaches to reaching a decision based on data are available via decision theory and optimal decisions, some of which have desirable properties. Hypothesis testing, though, is a dominant approach to data analysis in many fields of science. Extensions to the theory of hypothesis testing include the study of the power of tests, i.e. the probability of correctly rejecting the null hypothesis given that it is false. Such considerations can be used for the purpose of sample size determination prior to the collection of data.

The Testing Process

In the statistics literature, statistical hypothesis testing plays a fundamental role. The usual line of reasoning is as follows:

1. There is an initial research hypothesis of which the truth is unknown.

2. The first step is to state the relevant null and alternative hypotheses. This is important, as mis-stating the hypotheses will muddy the rest of the process.

3. The second step is to consider the statistical assumptions being made about the sample in doing the test; for example, assumptions about the statistical independence or about the form of the distributions of the observations. This is equally important as invalid assumptions will mean that the results of the test are invalid.

4. Decide which test is appropriate, and state the relevant test statistic T.

5. Derive the distribution of the test statistic under the null hypothesis from the assumptions. In standard cases this will be a well-known result. For example, the test statistic might follow a Student's t distribution or a normal distribution.

6. Select a significance level (a), a probability threshold below which the null hypothesis will be rejected. Common values are 5% and 1%.

7. The distribution of the test statistic under the null hypothesis partitions the possible values of T into those for which the null hypothesis is rejected—the so-called *critical region*—and those for which it is not. The probability of the critical region is a.

8. Compute from the observations the observed value t_{obs} of the test statistic T.

9. Decide to either reject the null hypothesis in favor of the alternative or not reject it. The decision rule is to reject the null hypothesis H_0 if the observed value t_{obs} is in the critical region, and to accept or "fail to reject" the hypothesis otherwise.

An alternative process is commonly used:

1. Compute from the observations the observed value t_{obs} of the test statistic T.

2. Calculate the p-value. This is the probability, under the null hypothesis, of sampling a test statistic at least as extreme as that which was observed.

3. Reject the null hypothesis, in favor of the alternative hypothesis, if and only if the p-value is less than the significance level (the selected probability) threshold.

The two processes are equivalent. The former process was advantageous in the past when only tables of test statistics at common probability thresholds were available. It allowed a decision to be made without the calculation of a probability. It was adequate for classwork and for operational use, but it was deficient for reporting results.

The latter process relied on extensive tables or on computational support not always available. The explicit calculation of a probability is useful for reporting. The calculations are now trivially performed with appropriate software.

The difference in the two processes applied to the Radioactive suitcase example (below):

- "The Geiger-counter reading is 10. The limit is 9. Check the suitcase."

- "The Geiger-counter reading is high; 97% of safe suitcases have lower readings. The limit is 95%. Check the suitcase."

The former report is adequate, the latter gives a more detailed explanation of the data and the reason why the suitcase is being checked.

It is important to note the difference between accepting the null hypothesis and simply failing to reject it. The "fail to reject" terminology highlights the fact that the null hypothesis is assumed to be true from the start of the test; if there is a lack of evidence against it, it simply continues to be assumed true. The phrase "accept the null hypothesis" may suggest it has been proved simply because it has not been disproved, a logical fallacy known as the argument from ignorance. Unless a test with particularly high power is used, the idea of "accepting" the null hypothesis may be dangerous. Nonetheless the terminology is prevalent throughout statistics, where the meaning actually intended is well understood.

The processes described here are perfectly adequate for computation. They seriously neglect the design of experiments considerations.

It is particularly critical that appropriate sample sizes be estimated before conducting the experiment.

The phrase "test of significance" was coined by statistician Ronald Fisher.

Interpretation

If the p-value is less than the required significance level (equivalently, if the observed test statistic is in the critical region), then we say the null hypothesis is rejected at the given level of significance.

Rejection of the null hypothesis is a conclusion. This is like a "guilty" verdict in a criminal trial: the evidence is sufficient to reject innocence, thus proving guilt. We might accept the alternative hypothesis (and the research hypothesis).

If the *p*-value is *not* less than the required significance level (equivalently, if the observed test statistic is outside the critical region), then the test has no result. The evidence is insufficient to support a conclusion. (This is like a jury that fails to reach a verdict.) The researcher typically gives extra consideration to those cases where the *p*-value is close to the significance level.

In the Lady tasting tea example (below), Fisher required the Lady to properly categorize all of the cups of tea to justify the conclusion that the result was unlikely to result from chance. He defined the critical region as that case alone. The region was defined by a probability (that the null hypothesis was correct) of less than 5%.

Whether rejection of the null hypothesis truly justifies acceptance of the research hypothesis depends on the structure of the hypotheses. Rejecting the hypothesis that a large paw print originated from a bear does not immediately prove the existence of Bigfoot. Hypothesis testing emphasizes the rejection, which is based on a probability, rather than the acceptance, which requires extra steps of logic.

"The probability of rejecting the null hypothesis is a function of five factors: whether the test is one- or two tailed, the level of significance, the standard deviation, the amount of deviation from the null hypothesis, and the number of observations." These factors are a source of criticism; factors under the control of the experimenter/analyst give the results an appearance of subjectivity.

Use and Importance

Statistics are helpful in analyzing most collections of data. This is equally true of hypothesis testing which can justify conclusions even when no scientific theory exists. In the Lady tasting tea example, it was "obvious" that no difference existed between (milk poured into tea) and (tea poured into milk). The data contradicted the "obvious".

Real world applications of hypothesis testing include:

- Testing whether more men than women suffer from nightmares
- Establishing authorship of documents
- Evaluating the effect of the full moon on behavior
- Determining the range at which a bat can detect an insect by echo
- Deciding whether hospital carpeting results in more infections
- Selecting the best means to stop smoking
- Checking whether bumper stickers reflect car owner behavior
- Testing the claims of handwriting analysts

Statistical hypothesis testing plays an important role in the whole of statistics and in statistical inference. For example, Lehmann (1992) in a review of the fundamental paper by Neyman and Pearson (1933) says: "Nevertheless, despite their shortcomings, the new paradigm formulated in the 1933 paper, and the many developments carried out within its framework continue to play a central role in both the theory and practice of statistics and can be expected to do so in the foreseeable future".

Significance testing has been the favored statistical tool in some experimental social sciences (over 90% of articles in the *Journal of Applied Psychology* during the early 1990s). Other fields have favored the estimation of parameters (e.g., effect size). Significance testing is used as a substitute for the traditional comparison of predicted value and experimental result at the core of the scientific method. When theory is only capable of predicting the sign of a relationship, a directional (one-sided) hypothesis test can be configured so that only a statistically significant result supports theory. This form of theory appraisal is the most heavily criticized application of hypothesis testing.

Cautions

"If the government required statistical procedures to carry warning labels like those on drugs, most inference methods would have long labels indeed." This caution applies to hypothesis tests and alternatives to them.

The successful hypothesis test is associated with a probability and a type-I error rate. The conclusion *might* be wrong.

The conclusion of the test is only as solid as the sample upon which it is based. The design of the experiment is critical. A number of unexpected effects have been observed including:

- The clever Hans effect. A horse appeared to be capable of doing simple arithmetic.

- The Hawthorne effect. Industrial workers were more productive in better illumination, and most productive in worse.

- The placebo effect. Pills with no medically active ingredients were remarkably effective.

A statistical analysis of misleading data produces misleading conclusions. The issue of data quality can be more subtle. In forecasting for example, there is no agreement on a measure of forecast accuracy. In the absence of a consensus measurement, no decision based on measurements will be without controversy.

The book *How to Lie with Statistics* is the most popular book on statistics ever published. It does not much consider hypothesis testing, but its cautions are applicable, including: Many claims are made on the basis of samples too small to convince. If a report does not mention sample size, be doubtful.

Hypothesis testing acts as a filter of statistical conclusions; only those results meeting a probability threshold are publishable. Economics also acts as a publication filter; only those results favorable to the author and funding source may be submitted for publication. The impact of filtering on publication is termed publication bias. A related problem is that of multiple testing (sometimes linked to data mining), in which a variety of tests for a variety of possible effects are applied to a single data set and only those yielding a significant result are reported. These are often dealt with by using multiplicity correction procedures that control the family wise error rate (FWER) or the false discovery rate (FDR).

Those making critical decisions based on the results of a hypothesis test are prudent to look at the details rather than the conclusion alone. In the physical sciences most results are fully accepted only when independently confirmed. The general advice concerning statistics is, "Figures never lie, but liars figure" (anonymous).

Examples

Lady Tasting Tea

In a famous example of hypothesis testing, known as the *Lady tasting tea*, a female colleague of Fisher claimed to be able to tell whether the tea or the milk was added first to a cup. Fisher proposed to give her eight cups, four of each variety, in random order. One could then ask what the probability was for her getting the number she got correct, but just by chance. The null hypothesis was that the Lady had no such ability. The test statistic was a simple count of the number of successes in selecting the 4 cups. The critical region was the single case of 4 successes of 4 possible based on a conventional probability criterion ($< 5\%$; 1 of 70 $\approx$ 1.4%). Fisher asserted that no alternative hypothesis was (ever) required. The lady correctly identified every cup, which would be considered a statistically significant result.

Courtroom Trial

A statistical test procedure is comparable to a criminal trial; a defendant is considered not guilty as long as his or her guilt is not proven. The prosecutor tries to prove the guilt of the defendant. Only when there is enough charging evidence the defendant is convicted.

In the start of the procedure, there are two hypotheses H_0 : "the defendant is not guilty", and H_1 : "the defendant is guilty". The first one is called *null hypothesis*, and is for the time being accepted. The second one is called *alternative (hypothesis)*. It is the hypothesis one hopes to support.

The hypothesis of innocence is only rejected when an error is very unlikely, because one doesn't want to convict an innocent defendant. Such an error is called *error of the first kind* (i.e., the conviction of an innocent person), and the occurrence of this error is controlled to be rare. As a consequence of this asymmetric behaviour, the *error of the second kind* (acquitting a person who committed the crime), is often rather large.

	H_0 is true Truly not guilty	H_1 is true Truly guilty
Accept null hypothesis Acquittal	Right decision	Wrong decision Type II Error
Reject null hypothesis Conviction	Wrong decision Type I Error	Right decision

A criminal trial can be regarded as either or both of two decision processes: guilty vs not guilty or evidence vs a threshold ("beyond a reasonable doubt"). In one view, the defendant is judged; in the other view the performance of the prosecution (which bears the burden of proof) is judged. A hypothesis test can be regarded as either a judgment of a hypothesis or as a judgment of evidence.

Philosopher's Beans

The following example was produced by a philosopher describing scientific methods generations before hypothesis testing was formalized and popularized.

Few beans of this handful are white. Most beans in this bag are white. Therefore: Probably, these beans were taken from another bag. This is an hypothetical inference.

The beans in the bag are the population. The handful are the sample. The null hypothesis is that the sample originated from the population. The criterion for rejecting the null-hypothesis is the "obvious" difference in appearance (an informal difference in the mean). The interesting result is that consideration of a real population and a real sample produced an imaginary bag. The philosopher was considering logic rather than probability. To be a real statistical hypothesis test, this example requires the formalities of a probability calculation and a comparison of that probability to a standard.

A simple generalization of the example considers a mixed bag of beans and a handful that contain either very few or very many white beans. The generalization considers both extremes. It requires more calculations and more comparisons to arrive at a formal answer, but the core philosophy is unchanged; If the composition of the handful is greatly different from that of the bag, then the sample probably originated from another bag. The original example is termed a one-sided or a one-tailed test while the generalization is termed a two-sided or two-tailed test.

The statement also relies on the inference that the sampling was random. If someone had been picking through the bag to find white beans, then it would explain why the handful had so many white beans, and also explain why the number of white beans in the bag was depleted (although the bag is probably intended to be assumed much larger than one's hand).

Clairvoyant Card Game

A person (the subject) is tested for clairvoyance. He is shown the reverse of a randomly chosen playing card 25 times and asked which of the four suits it belongs to. The number of hits, or correct answers, is called X.

As we try to find evidence of his clairvoyance, for the time being the null hypothesis is that the person is not clairvoyant. The alternative is, of course: the person is (more or less) clairvoyant.

If the null hypothesis is valid, the only thing the test person can do is guess. For every card, the probability (relative frequency) of any single suit appearing is 1/4. If the alternative is valid, the test subject will predict the suit correctly with probability greater than 1/4. We will call the probability of guessing correctly p. The hypotheses, then, are:

- null hypothesis : $H_0 : p = \frac{1}{4}$ (just guessing)

and

- alternative hypothesis : $H_1 : p > \frac{1}{4}$ (true clairvoyant).

When the test subject correctly predicts all 25 cards, we will consider him clairvoyant, and reject the null hypothesis. Thus also with 24 or 23 hits. With only 5 or 6 hits, on the other hand, there is no cause to consider him so. But what about 12 hits, or 17 hits? What is the critical number, c, of hits, at which point we consider the subject to be clairvoyant? How do we determine the critical value c? It is obvious that with the choice $c=25$ (i.e. we only accept clairvoyance when all cards are predicted correctly) we're more critical than with $c=10$. In the first case almost no test subjects will be recognized to be clairvoyant, in the second case, a certain number will pass the test. In practice, one decides how critical one will be. That is, one decides how often one accepts an error of the first kind – a false positive, or Type I error. With $c = 25$ the probability of such an error is:

$$P(reject\ \mathrm{H}_0 \mid \mathrm{H}_0\ is\ valid) = P(X = 25 \mid \mathrm{p} = \tfrac{1}{4}) = \left(\tfrac{1}{4}\right)^{25} \approx 10^{-15},$$

and hence, very small. The probability of a false positive is the probability of randomly guessing correctly all 25 times.

Being less critical, with $c=10$, gives:

$$P(reject\ \mathrm{H}_0 \mid \mathrm{H}_0\ is\ valid) = P(X \geq 10 \mid \mathrm{p} = \tfrac{1}{4}) = \sum_{k=10}^{25} P(X = k \mid \mathrm{p} = \tfrac{1}{4}) \approx 0.07.$$

Thus, $c = 10$ yields a much greater probability of false positive.

Before the test is actually performed, the maximum acceptable probability of a Type I error (α) is determined. Typically, values in the range of 1% to 5% are selected. (If the maximum acceptable error rate is zero, an infinite number of correct guesses is required.) Depending on this Type 1 error rate, the critical value c is calculated. For example, if we select an error rate of 1%, c is calculated thus:

$$P(reject\ \mathrm{H}_0 \mid \mathrm{H}_0\ is\ valid) = P(X \geq c \mid \mathrm{p} = \tfrac{1}{4}) \leq 0.01.$$

From all the numbers c, with this property, we choose the smallest, in order to minimize the probability of a Type II error, a false negative. For the above example, we select: $c = 13$.

Radioactive Suitcase

As an example, consider determining whether a suitcase contains some radioactive material. Placed under a Geiger counter, it produces 10 counts per minute. The null hypothesis is that no radioactive material is in the suitcase and that all measured counts are due to ambient radioactivity typical of the surrounding air and harmless objects. We can then calculate how likely it is that we would observe 10 counts per minute if the null hypothesis were true. If the null hypothesis predicts (say) on average 9 counts per minute, then according to the Poisson distribution typical for radioactive decay there is about 41% chance of recording 10 or more counts. Thus we can say that the suitcase is compatible with the null hypothesis (this does not guarantee that there is no radioactive material, just that we don't have enough evidence to suggest there is). On the other hand, if the null hypothesis predicts 3 counts per minute (for which the Poisson distribution predicts only 0.1% chance of recording 10 or more counts) then the suitcase is not compatible with the null hypothesis, and there are likely other factors responsible to produce the measurements.

The test does not directly assert the presence of radioactive material. A *successful* test asserts that the claim of no radioactive material present is unlikely given the reading (and therefore ...). The double negative (disproving the null hypothesis) of the method is confusing, but using a counter-example to disprove is standard mathematical practice. The attraction of the method is its practicality. We know (from experience) the expected range of counts with only ambient radioactivity present, so we can say that a measurement is *unusually* large. Statistics just formalizes the intuitive by using numbers instead of adjectives. We probably do not know the characteristics of the radioactive suitcases; We just assume that they produce larger readings.

To slightly formalize intuition: Radioactivity is suspected if the Geiger-count with the suitcase is among or exceeds the greatest (5% or 1%) of the Geiger-counts made with ambient radiation alone. This makes no assumptions about the distribution of counts. Many ambient radiation observations are required to obtain good probability estimates for rare events.

The test described here is more fully the null-hypothesis statistical significance test. The null hypothesis represents what we would believe by default, before seeing any evidence. Statistical significance is a possible finding of the test, declared when the observed sample is unlikely to have occurred by chance if the null hypothesis were true. The name of the test describes its formulation and its possible outcome. One characteristic of the test is its crisp decision: to reject or not reject the null hypothesis. A calculated value is compared to a threshold, which is determined from the tolerable risk of error.

Definition of Terms

The following definitions are mainly based on the exposition in the book by Lehmann and Romano:

Statistical hypothesis

> A statement about the parameters describing a population (not a sample).

Statistic

> A value calculated from a sample, often to summarize the sample for comparison purposes.

Simple hypothesis

> Any hypothesis which specifies the population distribution completely.

Composite hypothesis

> Any hypothesis which does *not* specify the population distribution completely.

Null hypothesis (H_0)

> A simple hypothesis associated with a contradiction to a theory one would like to prove.

Alternative hypothesis (H_1)

A hypothesis (often composite) associated with a theory one would like to prove.

Statistical test

A procedure whose inputs are samples and whose result is a hypothesis.

Region of acceptance

The set of values of the test statistic for which we fail to reject the null hypothesis.

Region of rejection / Critical region

The set of values of the test statistic for which the null hypothesis is rejected.

Critical value

The threshold value delimiting the regions of acceptance and rejection for the test statistic.

Power of a test $(1 - \beta)$

The test's probability of correctly rejecting the null hypothesis. The complement of the false negative rate, β. Power is termed sensitivity in biostatistics. ("This is a sensitive test. Because the result is negative, we can confidently say that the patient does not have the condition.")

Size

For simple hypotheses, this is the test's probability of *incorrectly* rejecting the null hypothesis. The false positive rate. For composite hypotheses this is the supremum of the probability of rejecting the null hypothesis over all cases covered by the null hypothesis. The complement of the false positive rate is termed specificity in biostatistics. ("This is a specific test. Because the result is positive, we can confidently say that the patient has the condition.")

Significance level of a test (α)

It is the upper bound imposed on the size of a test. Its value is chosen by the statistician prior to looking at the data or choosing any particular test to be used. It is the maximum exposure to erroneously rejecting H_0 he/she is ready to accept. Testing H_0 at significance level α means testing H_0 with a test whose size does not exceed α. In most cases, one uses tests whose size is equal to the significance level.

p-value

The probability, assuming the null hypothesis is true, of observing a result at least as extreme as the test statistic.

Statistical significance test

A predecessor to the statistical hypothesis test. An experimental result was said to be statistically significant if a sample was sufficiently inconsistent with the (null) hypothesis. This was variously considered common sense, a pragmatic heuristic for identifying meaningful experimental results, a convention establishing a threshold of statistical evidence or a method for drawing conclusions from data. The statistical hypothesis test added mathematical rigor and philosophical consistency to the concept by making the alternative hypothesis explicit. The term is loosely used to describe the modern version which is now part of statistical hypothesis testing.

Conservative test

> A test is conservative if, when constructed for a given nominal significance level, the true probability of *incorrectly* rejecting the null hypothesis is never greater than the nominal level.

Exact test

> A test in which the significance level or critical value can be computed exactly, i.e., without any approximation. In some contexts this term is restricted to tests applied to categorical data and to permutation tests, in which computations are carried out by complete enumeration of all possible outcomes and their probabilities.

A statistical hypothesis test compares a test statistic (*z* or *t* for examples) to a threshold. The test statistic (the formula found in the table below) is based on optimality. For a fixed level of Type I error rate, use of these statistics minimizes Type II error rates (equivalent to maximizing power). The following terms describe tests in terms of such optimality:

Most powerful test

> For a given *size* or *significance level*, the test with the greatest power (probability of rejection) for a given value of the parameter(s) being tested, contained in the alternative hypothesis.

Uniformly most powerful test (UMP)

> A test with the greatest *power* for all values of the parameter(s) being tested, contained in the alternative hypothesis.

Common Test Statistics

One-sample tests are appropriate when a sample is being compared to the population from a hypothesis. The population characteristics are known from theory or are calculated from the population.

Two-sample tests are appropriate for comparing two samples, typically experimental and control samples from a scientifically controlled experiment.

Paired tests are appropriate for comparing two samples where it is impossible to control important variables. Rather than comparing two sets, members are paired between samples so the difference between the members becomes the sample. Typically the mean of the differences is then compared to zero. The common example scenario for when a paired difference test is appropriate is when a single set of test subjects has something applied to them and the test is intended to check for an effect.

Z-tests are appropriate for comparing means under stringent conditions regarding normality and a known standard deviation.

A *t*-test is appropriate for comparing means under relaxed conditions (less is assumed).

Tests of proportions are analogous to tests of means (the 50% proportion).

Chi-squared tests use the same calculations and the same probability distribution for different applications:

- Chi-squared tests for variance are used to determine whether a normal population has a specified variance. The null hypothesis is that it does.

- Chi-squared tests of independence are used for deciding whether two variables are associated or are independent. The variables are categorical rather than numeric. It can be used to decide whether left-handedness is correlated with libertarian politics (or not). The null hypothesis is that the variables are independent. The numbers used in the calculation are the observed and expected frequencies of occurrence (from contingency tables).

- Chi-squared goodness of fit tests are used to determine the adequacy of curves fit to data. The null hypothesis is that the curve fit is adequate. It is common to determine curve shapes to minimize the mean square error, so it is appropriate that the goodness-of-fit calculation sums the squared errors.

F-tests (analysis of variance, ANOVA) are commonly used when deciding whether groupings of data by category are meaningful. If the variance of test scores of the left-handed in a class is much smaller than the variance of the whole class, then it may be useful to study lefties as a group. The null hypothesis is that two variances are the same – so the proposed grouping is not meaningful.

In the table below, the symbols used are defined at the bottom of the table. Many other tests can be found in other articles. Proofs exist that the test statistics are appropriate.

Name	Formula	Assumptions or notes
One-sample z-test	$z = \dfrac{\bar{x} - \mu_0}{(\sigma / \sqrt{n})}$	(Normal population **or** $n > 30$ **and** σ known. (z is the distance from the mean in relation to the standard deviation of the mean). For non-normal distributions it is possible to calculate a minimum proportion of a population that falls within k standard deviations for any k.
Two-sample z-test	$z = \dfrac{(\bar{x}_1 - \bar{x}_2) - d_0}{\sqrt{\dfrac{\sigma_1^2}{n_1} + \dfrac{\sigma_2^2}{n_2}}}$	Normal population **and** independent observations **and** σ_1 and σ_2 are known
One-sample t-test	$t = \dfrac{\bar{x} - \mu_0}{(s / \sqrt{n})},$ $df = n-1$	(Normal population **or** $n > 30$) **and** unknown

Paired t-test	$$t = \frac{\bar{d} - d_0}{(s_d / \sqrt{n})},$$ $$df = n - 1$$	(Normal population of differences **or** $n > 30$) **and** unknown or small sample size $n < 30$
Two-sample pooled t-test, equal variances	$$t = \frac{(\bar{x}_1 - \bar{x}_2) - d_0}{s_p\sqrt{\frac{1}{n_1} + \frac{1}{n_2}}},$$ $$df = n_1 + n_2 - 2$$ $$s_p^2 = \frac{(n_1 - 1)s_1^2 + (n_2 - 1)s_2^2}{n_1 + n_2 - 2},$$	(Normal populations **or** $n_1 + n_2 > 40$) **and** independent observations **and** $\sigma_1 = \sigma_2$ unknown
Two-sample unpooled t-test, unequal variances (Welch's t-test)	$$df = \frac{\left(\frac{s_1^2}{n_1} + \frac{s_2^2}{n_2}\right)^2}{\frac{\left(\frac{s_1^2}{n_1}\right)^2}{n_1 - 1} + \frac{\left(\frac{s_2^2}{n_2}\right)^2}{n_2 - 1}}$$ $$t = \frac{(\bar{x}_1 - \bar{x}_2) - d_0}{\sqrt{\frac{s_1^2}{n_1} + \frac{s_2^2}{n_2}}},$$	(Normal populations **or** $n_1 + n_2 > 40$) **and** independent observations **and** $\sigma_1 \neq \sigma_2$ both unknown
One-proportion z-test	$$z = \frac{\hat{p} - p_0}{\sqrt{p_0(1 - p_0)}}\sqrt{n}$$	$n\,p_o > 10$ **and** $n\,(1 - p_o) > 10$ **and** it is a SRS (Simple Random Sample), see notes.
Two-proportion z-test, pooled for $H_0 : p_1 = p_2$	$$z = \frac{(\hat{p}_1 - \hat{p}_2)}{\sqrt{\hat{p}(1 - \hat{p})(\frac{1}{n_1} + \frac{1}{n_2})}}$$ $$\hat{p} = \frac{x_1 + x_2}{n_1 + n_2}$$	$n_1 p_1 > 5$ **and** $n_1(1 - p_1) > 5$ **and** $n_2 p_2 > 5$ **and** $n_2(1 - p_2) > 5$ **and** independent observations.
Two-proportion z-test, unpooled for $\lvert d_0 \rvert > 0$	$$z = \frac{(\hat{p}_1 - \hat{p}_2) - d_0}{\sqrt{\frac{\hat{p}_1(1 - \hat{p}_1)}{n_1} + \frac{\hat{p}_2(1 - \hat{p}_2)}{n_2}}}$$	$n_1 p_1 > 5$ **and** $n_1(1 - p_1) > 5$ **and** $n_2 p_2 > 5$ **and** $n_2(1 - p_2) > 5$ **and** independent observations.

Chi-squared test for variance	$\chi^2 = (n-1)\dfrac{s^2}{\sigma_0^2}$	Normal population
Chi-squared test for goodness of fit	$\chi^2 = \sum^k \dfrac{(\text{observed}-\text{expected})^2}{\text{expected}}$	$df = k - 1 - \#\ parameters\ estimated$, and one of these must hold. • All expected counts are at least 5. • All expected counts are > 1 and no more than 20% of expected counts are less than 5
Two-sample F test for equality of variances	$F = \dfrac{s_1^2}{s_2^2}$	Normal populations Arrange so $s_1^2 \geq s_2^2$ and reject H_0 for $F > F(\alpha/2, n_1-1, n_2-1)^[$
Regression t-test of $H_0: R^2 = 0$.	$t = \sqrt{\dfrac{R^2(n-k-1^*)}{1-R^2}}$	Reject H_0 for $t > t(\alpha/2, n-k-1^*)^[$ *Subtract 1 for intercept; k terms contain independent variables.

Origins and Early Controversy

Significance testing is largely the product of Karl Pearson (p-value, Pearson's chi-squared test), William Sealy Gosset (Student's t-distribution), and Ronald Fisher ("null hypothesis", analysis of variance, "significance test"), while hypothesis testing was developed by Jerzy Neyman and Egon Pearson (son of Karl). Ronald Fisher began his life in statistics as a Bayesian (Zabell 1992), but Fisher soon grew disenchanted with the subjectivity involved (namely use of the principle of indifference when determining prior probabilities), and sought to provide a more "objective" approach to inductive inference.

Fisher was an agricultural statistician who emphasized rigorous experimental design and methods to extract a result from few samples assuming Gaussian distributions. Neyman (who teamed with the younger Pearson) emphasized mathematical rigor and methods to obtain more results from many samples and a wider range of distributions. Modern hypothesis testing is an inconsistent hybrid of the Fisher vs Neyman/Pearson formulation, methods and terminology developed in the early 20th century. While hypothesis testing was popularized early in the 20th century, evidence of its use can be found much earlier. In the 1770s Laplace considered the statistics of almost half a million births. The statistics showed an excess of boys compared to girls. He concluded by calculation of a p-value that the excess was a real, but unexplained, effect.

Fisher popularized the "significance test". He required a null-hypothesis (corresponding to a population frequency distribution) and a sample. His (now familiar) calculations determined whether to reject the null-hypothesis or not. Significance testing did not utilize an alternative hypothesis so there was no concept of a Type II error.

The p-value was devised as an informal, but objective, index meant to help a researcher determine (based on other knowledge) whether to modify future experiments or strengthen one's faith in the null hypothesis. Hypothesis testing (and Type I/II errors) was devised by Neyman and Pearson as a more objective alternative to Fisher's p-value, also meant to determine researcher behaviour, but without requiring any inductive inference by the researcher.

Neyman & Pearson considered a different problem (which they called "hypothesis testing"). They initially considered two simple hypotheses (both with frequency distributions). They calculated two probabilities and typically selected the hypothesis associated with the higher probability (the hypothesis more likely to have generated the sample). Their method always selected a hypothesis. It also allowed the calculation of both types of error probabilities.

Fisher and Neyman/Pearson clashed bitterly. Neyman/Pearson considered their formulation to be an improved generalization of significance testing.(The defining paper was abstract. Mathematicians have generalized and refined the theory for decades.) Fisher thought that it was not applicable to scientific research because often, during the course of the experiment, it is discovered that the initial assumptions about the null hypothesis are questionable due to unexpected sources of error. He believed that the use of rigid reject/accept decisions based on models formulated before data is collected was incompatible with this common scenario faced by scientists and attempts to apply this method to scientific research would lead to mass confusion.

The dispute between Fisher and Neyman–Pearson was waged on philosophical grounds, characterized by a philosopher as a dispute over the proper role of models in statistical inference.

Events intervened: Neyman accepted a position in the western hemisphere, breaking his partnership with Pearson and separating disputants (who had occupied the same building) by much of the planetary diameter. World War II provided an intermission in the debate. The dispute between Fisher and Neyman terminated (unresolved after 27 years) with Fisher's death in 1962. Neyman wrote a well-regarded eulogy. Some of Neyman's later publications reported p-values and significance levels.

The modern version of hypothesis testing is a hybrid of the two approaches that resulted from confusion by writers of statistical textbooks (as predicted by Fisher) beginning in the 1940s. (But signal detection, for example, still uses the Neyman/Pearson formulation.) Great conceptual differences and many caveats in addition to those mentioned above were ignored. Neyman and Pearson provided the stronger terminology, the more rigorous mathematics and the more consistent philosophy, but the subject taught today in introductory statistics has more similarities with Fisher's method than theirs. This history explains the inconsistent terminology (example: the null hypothesis is never accepted, but there is a region of acceptance).

Sometime around 1940, in an apparent effort to provide researchers with a "non-controversial" way to have their cake and eat it too, the authors of statistical text books began anonymously combining these two strategies by using the p-value in place of the test statistic (or data) to test against the Neyman–Pearson "significance level". Thus, researchers were encouraged to infer the strength of their data against some null hypothesis using p-values, while also thinking they are retaining the post-data collection objectivity provided by hypothesis testing. It then became customary for the null hypothesis, which was originally some realistic research hypothesis, to be used almost solely as a strawman "nil" hypothesis (one where a treatment has no effect, regardless of the context).

A Comparison Between Fisherian, Frequentist (Neyman–Pearson)

Fisher's null hypothesis testing	Neyman–Pearson decision theory
1. Set up a statistical null hypothesis. The null need not be a nil hypothesis (i.e., zero difference).	1. Set up two statistical hypotheses, H1 and H2, and decide about α, β, and sample size before the experiment, based on subjective cost-benefit considerations. These define a rejection region for each hypothesis.
2. Report the exact level of significance (e.g., p = 0.051 or p = 0.049). Do not use a conventional 5% level, and do not talk about accepting or rejecting hypotheses. If the result is "not significant", draw no conclusions and make no decisions, but suspend judgement until further data is available.	2. If the data falls into the rejection region of H1, accept H2; otherwise accept H1. Note that accepting a hypothesis does not mean that you believe in it, but only that you act as if it were true.
3. Use this procedure only if little is known about the problem at hand, and only to draw provisional conclusions in the context of an attempt to understand the experimental situation.	3. The usefulness of the procedure is limited among others to situations where you have a disjunction of hypotheses (e.g., either μ1 = 8 or μ2 = 10 is true) and where you can make meaningful cost-benefit trade-offs for choosing alpha and beta.

Early Choices of Null Hypothesis

Paul Meehl has argued that the epistemological importance of the choice of null hypothesis has gone largely unacknowledged. When the null hypothesis is predicted by theory, a more precise experiment will be a more severe test of the underlying theory. When the null hypothesis defaults to "no difference" or "no effect", a more precise experiment is a less severe test of the theory that motivated performing the experiment. An examination of the origins of the latter practice may therefore be useful:

1778: Pierre Laplace compares the birthrates of boys and girls in multiple European cities. He states: "it is natural to conclude that these possibilities are very nearly in the same ratio". Thus Laplace's null hypothesis that the birthrates of boys and girls should be equal given "conventional wisdom".

1900: Karl Pearson develops the chi squared test to determine "whether a given form of frequency curve will effectively describe the samples drawn from a given population." Thus the null hypothesis is that a population is described by some distribution predicted by theory. He uses as an example the numbers of five and sixes in the Weldon dice throw data.

1904: Karl Pearson develops the concept of "contingency" in order to determine whether outcomes are independent of a given categorical factor. Here the null hypothesis is by default that two things are unrelated (e.g. scar formation and death rates from smallpox). The null hypothesis in this case is no longer predicted by theory or conventional wisdom, but is instead the principle of indifference that lead Fisher and others to dismiss the use of "inverse probabilities".

Versus Null Hypothesis Statistical Significance Testing

An example of Neyman–Pearson hypothesis testing can be made by a change to the radioactive suitcase example. If the "suitcase" is actually a shielded container for the transportation of radioactive material, then a test might be used to select among three hypotheses: no radioactive source present, one present, two (all) present. The test could be required for safety, with actions required in each case. The Neyman–Pearson lemma of hypothesis testing says that a good criterion for the selection of hypotheses is the ratio of their probabilities (a likelihood ratio). A simple method of solution is to select the hypothesis with the highest probability for the Geiger counts observed. The typical result matches intuition: few counts imply no source, many counts imply two sources and intermediate counts imply one source.

Neyman–Pearson theory can accommodate both prior probabilities and the costs of actions resulting from decisions. The former allows each test to consider the results of earlier tests (unlike Fisher's significance tests). The latter allows the consideration of economic issues (for example) as well as probabilities. A likelihood ratio remains a good criterion for selecting among hypotheses.

The two forms of hypothesis testing are based on different problem formulations. The original test is analogous to a true/false question; the Neyman–Pearson test is more like multiple choice. In the view of Tukey the former produces a conclusion on the basis of only strong evidence while the latter produces a decision on the basis of available evidence. While the two tests seem quite different both mathematically and philosophically, later developments lead to the opposite claim. Consider many tiny radioactive sources. The hypotheses become 0,1,2,3... grains of radioactive sand. There is little distinction between none or some radiation (Fisher) and 0 grains of radioactive sand versus all of the alternatives (Neyman–Pearson). The major Neyman–Pearson paper of 1933 also considered composite hypotheses (ones whose distribution includes an unknown parameter). An example proved the optimality of the (Student's) t-test, "there can be no better test for the hypothesis under consideration" (p 321). Neyman–Pearson theory was proving the optimality of Fisherian methods from its inception.

Fisher's significance testing has proven a popular flexible statistical tool in application with little mathematical growth potential. Neyman–Pearson hypothesis testing is claimed as a pillar of mathematical statistics, creating a new paradigm for the field. It also stimulated new applications in Statistical process control, detection theory, decision theory and game theory. Both formulations have been successful, but the successes have been of a different character.

The dispute over formulations is unresolved. Science primarily uses Fisher's (slightly modified) formulation as taught in introductory statistics. Statisticians study Neyman–Pearson theory in graduate school. Mathematicians are proud of uniting the formulations. Philosophers consider them separately. Learned opinions deem the formulations variously competitive (Fisher vs Neyman), incompatible or complementary. The dispute has become more complex since Bayesian inference has achieved respectability.

The terminology is inconsistent. Hypothesis testing can mean any mixture of two formulations that both changed with time. Any discussion of significance testing vs hypothesis testing is doubly vulnerable to confusion.

Fisher thought that hypothesis testing was a useful strategy for performing industrial quality control, however, he strongly disagreed that hypothesis testing could be useful for scientists. Hypothesis testing provides a means of finding test statistics used in significance testing. The concept of power is useful in explaining the consequences of adjusting the significance level and is heavily used in sample size determination. The two methods remain philosophically distinct. They usually (but *not always*) produce the same mathematical answer. The preferred answer is context dependent. While the existing merger of Fisher and Neyman–Pearson theories has been heavily criticized, modifying the merger to achieve Bayesian goals has been considered.

Criticism

Criticism of statistical hypothesis testing fills volumes citing 300–400 primary references. Much of the criticism can be summarized by the following issues:

- The interpretation of a *p*-value is dependent upon stopping rule and definition of multiple comparison. The former often changes during the course of a study and the latter is unavoidably ambiguous. (i.e. "p values depend on both the (data) observed and on the other possible (data) that might have been observed but weren't").

- Confusion resulting (in part) from combining the methods of Fisher and Neyman–Pearson which are conceptually distinct.

- Emphasis on statistical significance to the exclusion of estimation and confirmation by repeated experiments.

- Rigidly requiring statistical significance as a criterion for publication, resulting in publication bias. Most of the criticism is indirect. Rather than being wrong, statistical hypothesis testing is misunderstood, overused and misused.

- When used to detect whether a difference exists between groups, a paradox arises. As improvements are made to experimental design (e.g., increased precision of measurement and sample size), the test becomes more lenient. Unless one accepts the absurd assumption that all sources of noise in the data cancel out completely, the chance of finding statistical significance in either direction approaches 100%.

- Layers of philosophical concerns. The probability of statistical significance is a function of decisions made by experimenters/analysts. If the decisions are based on convention they are termed arbitrary or mindless while those not so based may be termed subjective. To minimize type II errors, large samples are recommended. In psychology practically all null hypotheses are claimed to be false for sufficiently large samples so "...it is usually nonsensical to perform an experiment with the *sole* aim of rejecting the null hypothesis.". "Statistically significant findings are often misleading" in psychology. Statistical significance does not imply practical significance and correlation does not imply causation. Casting doubt on the null hypothesis is thus far from directly supporting the research hypothesis.

- "[I]t does not tell us what we want to know". Lists of dozens of complaints are available.

Critics and supporters are largely in factual agreement regarding the characteristics of null hypothesis significance testing (NHST): While it can provide critical information, it is *inadequate as the sole tool for statistical analysis. Successfully rejecting the null hypothesis may offer no support for the research hypothesis*. The continuing controversy concerns the selection of the best statistical practices for the near-term future given the (often poor) existing practices. Critics would prefer to ban NHST completely, forcing a complete departure from those practices, while supporters suggest a less absolute change.

Controversy over significance testing, and its effects on publication bias in particular, has produced several results. The American Psychological Association has strengthened its statistical reporting requirements after review, medical journal publishers have recognized the obligation to publish some results that are not statistically significant to combat publication bias and a journal (*Journal of Articles in Support of the Null Hypothesis*) has been created to publish such results exclusively. Textbooks have added some cautions and increased coverage of the tools necessary to estimate the size of the sample required to produce significant results. Major organizations have not abandoned use of significance tests although some have discussed doing so.

Alternatives

The numerous criticisms of significance testing do not lead to a single alternative. A unifying position of critics is that statistics should not lead to a conclusion or a decision but to a probability or to an estimated value with a confidence interval rather than to an accept-reject decision regarding a particular hypothesis. It is unlikely that the controversy surrounding significance testing will be resolved in the near future. Its supposed flaws and unpopularity do not eliminate the need for an objective and transparent means of reaching conclusions regarding studies that produce statistical results. Critics have not unified around an alternative. Other forms of reporting confidence or uncertainty could probably grow in popularity. One strong critic of significance testing suggested a list of reporting alternatives: effect sizes for importance, prediction intervals for confidence, replications and extensions for replicability, meta-analyses for generality. None of these suggested alternatives produces a conclusion/decision. Lehmann said that hypothesis testing theory can be presented in terms of conclusions/decisions, probabilities, or confidence intervals. "The distinction between the ... approaches is largely one of reporting and interpretation."

On one "alternative" there is no disagreement: Fisher himself said, "In relation to the test of significance, we may say that a phenomenon is experimentally demonstrable when we know how to conduct an experiment which will rarely fail to give us a statistically significant result." Cohen, an influential critic of significance testing, concurred, "... don't look for a magic alternative to NHST *[null hypothesis significance testing]* ... It doesn't exist." "... given the problems of statistical induction, we must finally rely, as have the older sciences, on replication." The "alternative" to significance testing is repeated testing. The easiest way to decrease statistical uncertainty is by obtaining more data, whether by increased sample size or by repeated tests. Nickerson claimed to have never seen the publication of a literally replicated experiment in psychology. An indirect approach to replication is meta-analysis.

Bayesian inference is one proposed alternative to significance testing. (Nickerson cited 10 sources suggesting it, including Rozeboom (1960)). For example, Bayesian parameter estimation can provide rich information about the data from which researchers can draw inferences, while using uncertain priors that exert only minimal influence on the results when enough data is available. Psychologist John K. Kruschke has suggested Bayesian estimation as an alternative for the *t*-test. Alternatively two competing models/hypothesis can be compared using Bayes factors. Bayesian methods could be criticized for requiring information that is seldom available in the cases where significance testing is most heavily used. Neither the prior probabilities nor the probability distribution of the test statistic under the alternative hypothesis are often available in the social sciences.

Advocates of a Bayesian approach sometimes claim that the goal of a researcher is most often to objectively assess the probability that a hypothesis is true based on the data they have collected. Neither Fisher's significance testing, nor Neyman–Pearson hypothesis testing can provide this information, and do not claim to. The probability a hypothesis is true can only be derived from use of Bayes' Theorem, which was unsatisfactory to both the Fisher and Neyman–Pearson camps due to the explicit use of subjectivity in the form of the prior probability. Fisher's strategy is to sidestep this with the p-value (an objective *index* based on the data alone) followed by *inductive inference*, while Neyman–Pearson devised their approach of *inductive behaviour*.

Philosophy

Hypothesis testing and philosophy intersect. Inferential statistics, which includes hypothesis testing, is applied probability. Both probability and its application are intertwined with philosophy. Philosopher David Hume wrote, "All knowledge degenerates into probability." Competing practical definitions of probability reflect philosophical differences. The most common application of hypothesis testing is in the scientific interpretation of experimental data, which is naturally studied by the philosophy of science.

Fisher and Neyman opposed the subjectivity of probability. Their views contributed to the objective definitions. The core of their historical disagreement was philosophical.

Many of the philosophical criticisms of hypothesis testing are discussed by statisticians in other contexts, particularly correlation does not imply causation and the design of experiments. Hypothesis testing is of continuing interest to philosophers.

Education

Statistics is increasingly being taught in schools with hypothesis testing being one of the elements taught. Many conclusions reported in the popular press (political opinion polls to medical studies) are based on statistics. An informed public should understand the limitations of statistical conclusions and many college fields of study require a course in statistics for the same reason. An introductory college statistics class places much emphasis on hypothesis testing – perhaps half of the course. Such fields as literature and divinity now include findings based on statistical analysis. An introductory statistics class teaches hypothesis testing as a cookbook process. Hypothesis testing is also taught at the postgraduate level. Statisticians learn how to create good statistical test procedures (like z, Student's t, F and chi-squared). Statistical hypothesis testing is considered a mature area within statistics, but a limited amount of development continues.

The cookbook method of teaching introductory statistics leaves no time for history, philosophy or controversy. Hypothesis testing has been taught as received unified method. Surveys showed that graduates of the class were filled with philosophical misconceptions (on all aspects of statistical inference) that persisted among instructors. While the problem was addressed more than a decade ago, and calls for educational reform continue, students still graduate from statistics classes holding fundamental misconceptions about hypothesis testing. Ideas for improving the teaching of hypothesis testing include encouraging students to search for statistical errors in published papers, teaching the history of statistics and emphasizing the controversy in a generally dry subject.

References

- Bickel, Peter J.; Doksum, Kjell A. (2001). Mathematical statistics: Basic and selected topics. 1 (Second (updated printing 2007) ed.). Prentice Hall. ISBN 0-13-850363-X. MR 443141.

- Freedman, D. A. (2009). Statistical models: Theory and practice (revised ed.). Cambridge University Press. pp. xiv+442 pp. ISBN 978-0-521-74385-3. MR 2489600.

- Hinkelmann, Klaus; Kempthorne, Oscar (2008). Introduction to Experimental Design (Second ed.). Wiley. ISBN 978-0-471-72756-9.

- Peirce, C. S. (1883), "A Theory of Probable Inference", Studies in Logic, pp. 126-181, Little, Brown, and Company. (Reprinted 1983, John Benjamins Publishing Company, ISBN 90-272-3271-7)

- Pfanzagl, Johann; with the assistance of R. Hamböker (1994). Parametric Statistical Theory. Berlin: Walter de Gruyter. ISBN 3-11-013863-8. MR 1291393.

- Rissanen, Jorma (1989). Stochastic Complexity in Statistical Inquiry. Series in computer science. 15. Singapore: World Scientific. ISBN 9971-5-0859-1. MR 1082556.

- Traub, Joseph F.; Wasilkowski, G. W.; Wozniakowski, H. (1988). Information-Based Complexity. Academic Press. ISBN 0-12-697545-0.

- Aster, Richard; Borchers, Brian, and Thurber, Clifford (2012). Parameter Estimation and Inverse Problems, Second Edition, Elsevier. ISBN 0123850487, ISBN 978-0123850485

- Bickel, Peter J. & Doksum, Kjell A. (2001). Mathematical Statistics, Volume 1: Basic and Selected Topics (Second (updated printing 2007) ed.). Pearson Prentice–Hall. ISBN 0-13-850363-X.

- Jaynes E. T. (2003) Probability Theory: The Logic of Science, CUP. ISBN 978-0-521-59271-0 (Link to Fragmentary Edition of March 1996).

- Howson, C. & Urbach, P. (2005). Scientific Reasoning: the Bayesian Approach (3rd ed.). Open Court Publishing Company. ISBN 978-0-8126-9578-6.

- Lehmann, E. L.; Romano, Joseph P. (2005). Testing Statistical Hypotheses (3E ed.). New York: Springer. ISBN 0-387-98864-5.

- Hinkelmann, Klaus and Kempthorne, Oscar (2008). Design and Analysis of Experiments. I and II (Second ed.). Wiley. ISBN 978-0-470-38551-7.

Evolution of Probability and Statistics

The meaning of statistics has involved in the past few decades. Initially the meaning of the term was limited to information about states. The meaning has since then also included all the collections of information of all types and analysis and interpretation of such data. The evolution of probability and statistics helps the reader in understanding the growth of the subject.

History of Statistics

The history of statistics can be said to start around 1749 although, over time, there have been changes to the interpretation of the word *statistics*. In early times, the meaning was restricted to information about states. This was later extended to include all collections of information of all types, and later still it was extended to include the analysis and interpretation of such data. In modern terms, "statistics" means both sets of collected information, as in national accounts and temperature records, and analytical work which requires statistical inference.

Statistical activities are often associated with models expressed using probabilities, and require probability theory for them to be put on a firm theoretical basis.

A number of statistical concepts have an important impact on a wide range of sciences. These include the design of experiments and approaches to statistical inference such as Bayesian inference, each of which can be considered to have their own sequence in the development of the ideas underlying modern statistics.

Introduction

By the 18th century, the term "statistics" designated the systematic collection of demographic and economic data by states. For at least two millennia, these data were mainly tabulations of human and material resources that might be taxed or put to military use. In the early 19th century, collection intensified, and the meaning of "statistics" broadened to include the discipline concerned with the collection, summary, and analysis of data. Today, data are collected and statistics are computed and widely distributed in government, business, most of the sciences and sports, and even for many pastimes. Electronic computers have expedited more elaborate statistical computation even as they have facilitated the collection and aggregation of data. A single data analyst may have available a set of data-files with millions of records, each with dozens or hundreds of separate measurements. These were collected over time from computer activity (for example, a stock exchange) or from computerized sensors, point-of-sale registers, and so on. Computers then produce simple, accurate summaries, and allow more tedious analyses, such as those that require inverting a large matrix or perform hundreds of steps of iteration, that would never be attempted by hand. Faster computing has allowed statisticians to develop "computer-intensive" methods which may look at all permutations, or use randomization to look at 10,000 permutations of a problem, to estimate answers that are not easy to quantify by theory alone.

The term "mathematical statistics" designates the mathematical theories of probability and statistical inference, which are used in statistical practice. The relation between statistics and probability theory developed rather late, however. In the 19th century, statistics increasingly used probability theory, whose initial results were found in the 17th and 18th centuries, particularly in the analysis of games of chance (gambling). By 1800, astronomy used probability models and statistical theories, particularly the method of least squares. Early probability theory and statistics was systematized in the 19th century and statistical reasoning and probability models were used by social scientists to advance the new sciences of experimental psychology and sociology, and by physical scientists in thermodynamics and statistical mechanics. The development of statistical reasoning was closely associated with the development of inductive logic and the scientific method, which are concerns that move statisticians away from the narrower area of mathematical statistics. Much of the theoretical work was readily available by the time computers were available to exploit them. By the 1970s, Johnson and Kotz produced a four-volume Compendium on Statistical Distributions (First Edition 1969-1972), which is still an invaluable resource.

Applied statistics can be regarded as not a field of mathematics but an autonomous mathematical science, like computer science and operations research. Unlike mathematics, statistics had its origins in public administration. Applications arose early in demography and economics; large areas of micro- and macro-economics today are "statistics" with an emphasis on time-series analyses. With its emphasis on learning from data and making best predictions, statistics also has been shaped by areas of academic research including psychological testing, medicine and epidemiology. The ideas of statistical testing have considerable overlap with decision science. With its concerns with searching and effectively presenting data, statistics has overlap with information science and computer science.

Etymology

The term *statistics* is ultimately derived from the New Latin *statisticum collegium* ("council of state") and the Italian word *statista* ("statesman" or "politician"). The German *Statistik*, first introduced by Gottfried Achenwall (1749), originally designated the analysis of data about the state, signifying the "science of state" (then called *political arithmetic* in English). It acquired the meaning of the collection and classification of data generally in the early 19th century. It was introduced into English in 1791 by Sir John Sinclair when he published the first of 21 volumes titled *Statistical Account of Scotland*.

Thus, the original principal purpose of *Statistik* was data to be used by governmental and (often centralized) administrative bodies. The collection of data about states and localities continues, largely through national and international statistical services. In particular, censuses provide frequently updated information about the population.

The first book to have 'statistics' in its title was "Contributions to Vital Statistics" (1845) by Francis GP Neison, actuary to the Medical Invalid and General Life Office.

Origins in Probability Theory

Basic forms of statistics have been used since the beginning of civilization. Early empires often collated censuses of the population or recorded the trade in various commodities. The Roman Empire was one of the first states to extensively gather data on the size of the empire's population, geographical area and wealth.

The use of statistical methods dates back to least to the 5th century BCE. The historian Thucydides in his *History of the Peloponnesian War* describes how the Athenians calculated the height of the wall of Platea by counting the number of bricks in an unplastered section of the wall sufficiently near them to be able to count them. The count was repeated several times by a number of soldiers. The most frequent value (in modern terminology - the mode) so determined was taken to be the most likely value of the number of bricks. Multiplying this value by the height of the bricks used in the wall allowed the Athenians to determine the height of the ladders necessary to scale the walls.

In the Indian epic - the Mahabharata (Book 3: The Story of Nala) - King Rtuparna estimated the number of fruit and leaves (2095 fruit and 50,000,000 - five crores - leaves) on two great branches of a Vibhitaka tree by counting them on a single twig. This number was then multiplied by the number of twigs on the branches. This estimate was later checked and found to be very close to the actual number. With knowledge of this method Nala was subsequently able to regain his kingdom.

The earliest writing on statistics was found in a 9th-century book entitled: "Manuscript on Deciphering Cryptographic Messages", written by Al-Kindi (801–873 CE). In his book, Al-Kindi gave a detailed description of how to use statistics and frequency analysis to decipher encrypted messages. This text arguably gave rise to the birth of both statistics and cryptanalysis.

The Trial of the Pyx is a test of the purity of the coinage of the Royal Mint which has been held on a regular basis since the 12th century. The Trial itself is based on statistical sampling methods. After minting a series of coins - originally from ten pounds of silver - a single coin was placed in the Pyx - a box in Westminster Abbey. After a given period now once a year - the coins are removed and weighed. A sample of coins removed from the box are then tested for purity.

The *Nuova Cronica*, a 14th-century history of Florence by the Florentine banker and official Giovanni Villani, includes much statistical information on population, ordinances, commerce and trade, education, and religious facilities and has been described as the first introduction of statistics as a positive element in history, though neither the term nor the concept of statistics as a specific field yet existed. But this was proven to be incorrect after the rediscovery of Al-Kindi's book on frequency analysis.

The arithmetic mean, although a concept known to the Greeks, was not generalised to more than two values until the 16th century. The invention of the decimal system by Simon Stevin in 1585 seems likely to have facilitated these calculations. This method was first adopted in astronomy by Tycho Brahe who was attempting to reduce the errors in his estimates of the locations of various celestial bodies.

The idea of the median originated in Edward Wright's book on navigation (*Certaine Errors in Navigation*) in 1599 in a section concerning the determination of location with a compass. Wright felt that this value was the most likely to be the correct value in a series of observations.

Sir William Petty, a 17th-century economist who used early statistical methods to analyse demographic data.

The birth of statistics is often dated to 1662, when John Graunt, along with William Petty, developed early human statistical and census methods that provided a framework for modern demography. He produced the first life table, giving probabilities of survival to each age. His book *Natural and Political Observations Made upon the Bills of Mortality* used analysis of the mortality rolls to make the first statistically based estimation of the population of London. He knew that there were around 13,000 funerals per year in London and that three people died per eleven families per year. He estimated from the parish records that the average family size was 8 and calculated that the population of London was about 384,000. Laplace in 1802 estimated the population of France with a similar method.

Although the original scope of statistics was limited to data useful for governance, the approach was extended to many fields of a scientific or commercial nature during the 19th century. The mathematical foundations for the subject heavily drew on the new probability theory, pioneered in the 16th century in the correspondence amongst Gerolamo Cardano, Pierre de Fermat and Blaise Pascal. Christiaan Huygens (1657) gave the earliest known scientific treatment of the subject. Jakob Bernoulli's *Ars Conjectandi* (posthumous, 1713) and Abraham de Moivre's *The Doctrine of Chances* (1718) treated the subject as a branch of mathematics. In his book Bernoulli introduced the idea of representing complete certainty as one and probability as a number between zero and one.

The formal study of theory of errors may be traced back to Roger Cotes' *Opera Miscellanea* (posthumous, 1722), but a memoir prepared by Thomas Simpson in 1755 (printed 1756) first applied the theory to the discussion of errors of observation. The reprint (1757) of this memoir lays down the axioms that positive and negative errors are equally probable, and that there are certain assignable limits within which all errors may be supposed to fall; continuous errors are discussed and a probability curve is given. Simpson discussed several possible distributions of error. He first considered the uniform distribution and then the discrete symmetric triangular distribution followed by the continuous symmetric triangle distribution. Tobias Mayer, in his study of the libration of the moon (*Kosmographische Nachrichten*, Nuremberg, 1750), invented the first formal method for estimating the unknown quantities by generalized the averaging of observations under identical circumstances to the averaging of groups of similar equations.

Ruder Boškovic in 1755 based in his work on the shape of the earth proposed in his book *De Litteraria expeditione per pontificiam ditionem ad dimetiendos duos meridiani gradus a PP. Maire et Boscovicli* that the true value of a series of observations would be that which minimises the sum of absolute errors. In modern terminology this value is the median. The first example of what later became known as the normal curve was studied by Abraham de Moivre who plotted this curve on November 12, 1733. de Moivre was studying the number of heads that occurred when a 'fair' coin was tossed.

In 1761 Thomas Bayes proved Bayes' theorem and in 1765 Joseph Priestley invented the first timeline charts.

Johann Heinrich Lambert in his 1765 book *Anlage zur Architectonic* proposed the semicircle as a distribution of errors:

$$f(x) = \frac{1}{2}\sqrt{(1-x^2)}$$

with -1 < *x* < 1.

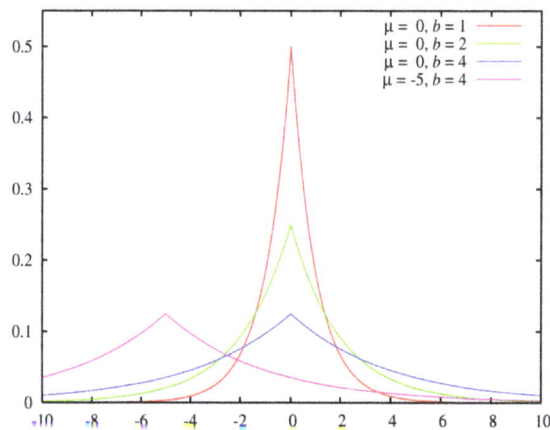

Probability density plots for the Laplace distribution.

Pierre-Simon Laplace (1774) made the first attempt to deduce a rule for the combination of observations from the principles of the theory of probabilities. He represented the law of probability of errors by a curve and deduced a formula for the mean of three observations.

Laplace in 1774 noted that the frequency of an error could be expressed as an exponential function of its magnitude once its sign was disregarded. This distribution is now known as the Laplace distribution. Lagrange proposed a parabolic distribution of errors in 1776.

Laplace in 1778 published his second law of errors wherein he noted that the frequency of an error was proportional to the exponential of the square of its magnitude. This was subsequently rediscovered by Gauss (possibly in 1795) and is now best known as the normal distribution which is of central importance in statistics. This distribution was first referred to as the *normal* distribution by Pierce in 1873 who was studying measurement errors when an object was dropped onto a wooden base. He chose the term *normal* because of its frequent occurrence in naturally occurring variables.

Lagrange also suggested in 1781 two other distributions for errors - a Raised cosine distribution and a logarithmic distribution.

Laplace gave (1781) a formula for the law of facility of error (a term due to Joseph Louis Lagrange, 1774), but one which led to unmanageable equations. Daniel Bernoulli (1778) introduced the principle of the maximum product of the probabilities of a system of concurrent errors.

In 1786 William Playfair (1759-1823) introduced the idea of graphical representation into statistics. He invented the line chart, bar chart and histogram and incorporated them into his works on economics, the *Commercial and Political Atlas*. This was followed in 1795 by his invention of the pie chart and circle chart which he used to display the evolution of England's imports and exports. These latter charts came to general attention when he published examples in his *Statistical Breviary* in 1801.

Laplace, in an investigation of the motions of Saturn and Jupiter in 1787, generalized Mayer's method by using different linear combinations of a single group of equations.

In 1802 Laplace estimated the population of France to be 28,328,612. He calculated this figure using the number of births in the previous year and census data for three communities. The census data of these communities showed that they had 2,037,615 persons and that the number of births were 71,866. Assuming that these samples were representative of France, Laplace produced his estimate for the entire population.

The method of least squares, which was used to minimize errors in data measurement, was published independently by Adrien-Marie Legendre (1805), Robert Adrain (1808), and Carl Friedrich Gauss (1809). Gauss had used the method in his famous 1801 prediction of the location of the dwarf planet Ceres. The observations that Gauss based his calculations on were made by the Italian monk Piazzi.

The term *probable error (der wahrscheinliche Fehler)* - the median deviation from the mean - was introduced in 1815 by the German astronomer Frederik Wilhelm Bessel. Antoine Augustin Cournot in 1843 was the first to use the term *median (valeur médiane)* for the value that divides a probability distribution into two equal halves.

Other contributors to the theory of errors were Ellis (1844), De Morgan (1864), Glaisher (1872), and Giovanni Schiaparelli (1875).Peters's (1856) formula for , the "probable error" of a single observation was widely used and inspired early robust statistics.

In the 19th century authors on statistical theory included Laplace, S. Lacroix (1816), Littrow (1833), Dedekind (1860), Helmert (1872), Laurent (1873), Liagre, Didion, De Morgan and Boole.

Gustav Theodor Fechner used the median (*Centralwerth*) in sociological and psychological phenomena. It had earlier been used only in astronomy and related fields. Francis Galton used the English term *median* for the first time in 1881 having earlier used the terms *middle-most value* in 1869 and the *medium* in 1880.

Adolphe Quetelet (1796–1874), another important founder of statistics, introduced the notion of the "average man" (*l'homme moyen*) as a means of understanding complex social phenomena such as crime rates, marriage rates, and suicide rates.

The first tests of the normal distribution were invented by the German statistician Wilhelm Lexis in the 1870s. The only data sets available to him that he was able to show were normally distributed were birth rates.

Development of Modern Statistics

Although the origins of statistical theory lie in the 18th century advances in probability, the modern field of statistics only emerged in the late 19th and early 20th century in three stages. The first wave, at the turn of the century, was led by the work of Francis Galton and Karl Pearson, who transformed statistics into a rigorous mathematical discipline used for analysis, not just in science, but in industry and politics as well. The second wave of the 1910s and 20s was initiated by William Gosset, and reached its culmination in the insights of Ronald Fisher. This involved the development of better design of experiments models, hypothesis testing and techniques for use with small data samples. The final wave, which mainly saw the refinement and expansion of earlier developments, emerged from the collaborative work between Egon Pearson and Jerzy Neyman in the 1930s. Today, statistical methods are applied in all fields that involve decision making, for making accurate inferences from a collated body of data and for making decisions in the face of uncertainty based on statistical methodology.

The original logo of the Royal Statistical Society, founded in 1834.

The first statistical bodies were established in the early 19th century. The Royal Statistical Society was founded in 1834 and Florence Nightingale, its first female member, pioneered the application of statistical analysis to health problems for the furtherance of epidemiological understanding and public health practice. However, the methods then used would not be considered as modern statistics today.

The Oxford scholar Francis Ysidro Edgeworth's book, *Metretike: or The Method of Measuring Probability and Utility* (1887) dealt with probability as the basis of inductive reasoning, and his later works focused on the 'philosophy of chance'. His first paper on statistics (1883) explored the law of error (normal distribution), and his *Methods of Statistics* (1885) introduced an early version of the t distribution, the Edgeworth expansion, the Edgeworth series, the method of variate transformation and the asymptotic theory of maximum likelihood estimates.

The Norwegian Anders Nicolai Kiær introduced the concept of stratified sampling in 1895. Arthur Lyon Bowley introduced new methods of data sampling in 1906 when working on social statistics. Although statistical surveys of social conditions had started with Charles Booth's "Life and Labour of the People in London" (1889-1903) and Seebohm Rowntree's "Poverty, A Study of Town Life" (1901), Bowley's, key innovation consisted of the use of random sampling techniques. His efforts culminated in his *New Survey of London Life and Labour*.

Francis Galton is credited as one of the principal founders of statistical theory. His contributions to the field included introducing the concepts of standard deviation, correlation, regression and the application of these methods to the study of the variety of human characteristics - height, weight, eyelash length among others. He found that many of these could be fitted to a normal curve distribution.

Galton submitted a paper to *Nature* in 1907 on the usefulness of the median. He examined the accuracy of 787 guesses of the weight of an ox at a country fair. The actual weight was 1208 pounds: the median guess was 1198. The guesses were markedly non-normally distributed.

Galton's publication of *Natural Inheritance* in 1889 sparked the interest of a brilliant mathematician, Karl Pearson, then working at University College London, and he went on to found the discipline of mathematical statistics. He emphasised the statistical foundation of scientific laws and promoted its study and his laboratory attracted students from around the world attracted by his new methods of analysis, including Udny Yule. His work grew to encompass the fields of biology, epidemiology, anthropometry, medicine and social history. In 1901, with Walter Weldon, founder of biometry, and Galton, he founded the journal *Biometrika* as the first journal of mathematical statistics and biometry.

His work, and that of Galton's, underpins many of the 'classical' statistical methods which are in common use today, including the Correlation coefficient, defined as a product-moment; the method of moments for the fitting of distributions to samples; Pearson's system of continuous curves that forms the basis of the now conventional continuous probability distributions; Chi distance a precursor and special case of the Mahalanobis distance and P-value, defined as the probability measure of the complement of the ball with the hypothesized value as center point and chi distance as radius. He also introduced the term 'standard deviation'.

He also founded the statistical hypothesis testing theory, Pearson's chi-squared test and principal component analysis. In 1911 he founded the world's first university statistics department at University College London.

The second wave of mathematical statistics was pioneered by Ronald Fisher who wrote two textbooks, *Statistical Methods for Research Workers*, published in 1925 and *The Design of Experiments* in 1935, that were to define the academic discipline in universities around the world. He also systematized previous results, putting them on a firm mathematical footing. In his 1918 seminal paper *The Correlation between Relatives on the Supposition of Mendelian Inheritance*, the first use to use the statistical term, variance. In 1919, at Rothamsted Experimental Station he started a major study of the extensive collections of data recorded over many years. This resulted in a series of reports under the general title *Studies in Crop Variation*. In 1930 he published *The Genetical Theory of Natural Selection* where he applied statistics to evolution.

Over the next seven years, he pioneered the principles of the design of experiments and elaborated his studies of analysis of variance. He furthered his studies of the statistics of small samples. Perhaps even more important, he began his systematic approach of the analysis of real data as the springboard for the development of new statistical methods. He developed computational algorithms for analyzing data from his balanced experimental designs. In 1925, this work resulted in the publication of his first book, *Statistical Methods for Research Workers*. This book went through many editions and translations in later years, and it became the standard reference work for scientists in many disciplines. In 1935, this book was followed by *The Design of Experiments*, which was also widely used.

Ronald Fisher, "A genius who almost single-handedly created the foundations for modern statistical science",

In addition to analysis of variance, Fisher named and promoted the method of maximum likelihood estimation. Fisher also originated the concepts of sufficiency, ancillary statistics, Fisher's linear discriminator and Fisher information. His article *On a distribution yielding the error functions of several well known statistics* (1924) presented Pearson's chi-squared test and William Gosset's t in the same framework as the Gaussian distribution, and his own parameter in the analysis of variance Fisher's z-distribution (more commonly used decades later in the form of the F distribution). The 5% level of significance appears to have been introduced by Fisher in 1925. Fisher stated that deviations exceeding twice the standard deviation are regarded as significant. Before this deviations exceeding three times the probable error were considered significant. For a symmetrical distribution the probable error is half the interquartile range. For a normal distribution the probable error is approximately 2/3 the standard deviation. It appears that Fisher's 5% criterion was rooted in previous practice.

Other important contributions at this time included Charles Spearman's rank correlation coefficient that was a useful extension of the Pearson correlation coefficient. William Sealy Gosset, the English statistician better known under his pseudonym of *Student*, introduced Student's t-distribution, a continuous probability distribution useful in situations where the sample size is small and population standard deviation is unknown.

Egon Pearson (Karl's son) and Jerzy Neyman introduced the concepts of "Type II" error, power of a test and confidence intervals. Jerzy Neyman in 1934 showed that stratified random sampling was in general a better method of estimation than purposive (quota) sampling.

Design of Experiments

In 1747, while serving as surgeon on HM Bark *Salisbury*, James Lind carried out a controlled experiment to develop a cure for scurvy. In this study his subjects' cases "were as similar as I could have them", that is he provided strict entry requirements to reduce extraneous variation. The men were paired, which provided blocking. From a modern perspective, the main thing that is missing is randomized allocation of subjects to treatments.

James Lind carried out the first ever clinical trial in 1747, in an effort to find a treatment for scurvy.

James Lind is today often described as a one-factor-at-a-time experimenter. Similar one-factor-at-a-time (OFAT) experimentation was performed at the Rothamsted Research Station in the 1840s by Sir John Lawes to determine the optimal inorganic fertilizer for use on wheat.

A theory of statistical inference was developed by Charles S. Peirce in "Illustrations of the Logic of Science" (1877–1878) and "A Theory of Probable Inference" (1883), two publications that emphasized the importance of randomization-based inference in statistics. In another study, Peirce randomly assigned volunteers to a blinded, repeated-measures design to evaluate their ability to discriminate weights.

Peirce's experiment inspired other researchers in psychology and education, which developed a research tradition of randomized experiments in laboratories and specialized textbooks in the 1800s. Peirce also contributed the first English-language publication on an optimal design for regression-models in 1876. A pioneering optimal design for polynomial regression was suggested by Gergonne in 1815. In 1918 Kirstine Smith published optimal designs for polynomials of degree six (and less).

The use of a sequence of experiments, where the design of each may depend on the results of previous experiments, including the possible decision to stop experimenting, was pioneered by Abraham Wald in the context of sequential tests of statistical hypotheses. Surveys are available of optimal sequential designs, and of adaptive designs. One specific type of sequential design is the "two-armed bandit", generalized to the multi-armed bandit, on which early work was done by Herbert Robbins in 1952.

The term "design of experiments" (DOE) derives from early statistical work performed by Sir Ronald Fisher. He was described by Anders Hald as "a genius who almost single-handedly created the foundations for modern statistical science." Fisher initiated the principles of design of experiments and elaborated on his studies of "analysis of variance". Perhaps even more important, Fisher began his systematic approach to the analysis of real data as the springboard for the development of new statistical methods. He began to pay particular attention to the labour involved in the necessary computations performed by hand, and developed methods that were as practical as they were founded in rigour. In 1925, this work culminated in the publication of his first book, *Statistical*

Methods for Research Workers. This went into many editions and translations in later years, and became a standard reference work for scientists in many disciplines.

A methodology for designing experiments was proposed by Ronald A. Fisher, in his innovative book *The Design of Experiments* (1935) which also became a standard. As an example, he described how to test the hypothesis that a certain lady could distinguish by flavour alone whether the milk or the tea was first placed in the cup. While this sounds like a frivolous application, it allowed him to illustrate the most important ideas of experimental design.

Agricultural science advances served to meet the combination of larger city populations and fewer farms. But for crop scientists to take due account of widely differing geographical growing climates and needs, it was important to differentiate local growing conditions. To extrapolate experiments on local crops to a national scale, they had to extend crop sample testing economically to overall populations. As statistical methods advanced (primarily the efficacy of designed experiments instead of one-factor-at-a-time experimentation), representative factorial design of experiments began to enable the meaningful extension, by inference, of experimental sampling results to the population as a whole. But it was hard to decide how representative was the crop sample chosen. Factorial design methodology showed how to estimate and correct for any random variation within in the sample and also in the data collection procedures.

Bayesian Statistics

Pierre-Simon, marquis de Laplace, one of the main early developers of Bayesian statistics.

The term *Bayesian* refers to Thomas Bayes (1702–1761), who proved a special case of what is now called Bayes' theorem. However it was Pierre-Simon Laplace (1749–1827) who introduced a general version of the theorem and applied it to celestial mechanics, medical statistics, reliability, and jurisprudence. When insufficient knowledge was available to specify an informed prior, Laplace used uniform priors, according to his "principle of insufficient reason". Laplace assumed uniform priors for mathematical simplicity rather than for philosophical reasons. Laplace also introduced-primitive versions of conjugate priors and the theorem of von Mises and Bernstein, according to which the posteriors corresponding to initially differing priors ultimately agree, as the number of observations increases. This early Bayesian inference, which used uniform priors following Laplace's principle of insufficient reason, was called "inverse probability" (because it infers backwards from observations to parameters, or from effects to causes).

After the 1920s, inverse probability was largely supplanted by a collection of methods that were developed by Ronald A. Fisher, Jerzy Neyman and Egon Pearson. Their methods came to be called frequentist statistics. Fisher rejected the Bayesian view, writing that "the theory of inverse probability is founded upon an error, and must be wholly rejected". At the end of his life, however, Fisher expressed greater respect for the essay of Bayes, which Fisher believed to have anticipated his own, fiducial approach to probability; Fisher still maintained that Laplace's views on probability were "fallacious rubbish". Neyman started out as a "quasi-Bayesian", but subsequently developed confidence intervals (a key method in frequentist statistics) because "the whole theory would look nicer if it were built from the start without reference to Bayesianism and priors". The word *Bayesian* appeared around 1950, and by the 1960s it became the term preferred by those dissatisfied with the limitations of frequentist statistics.

In the 20th century, the ideas of Laplace were further developed in two different directions, giving rise to *objective* and *subjective* currents in Bayesian practice. In the objectivist stream, the statistical analysis depends on only the model assumed and the data analysed. No subjective decisions need to be involved. In contrast, "subjectivist" statisticians deny the possibility of fully objective analysis for the general case.

In the further development of Laplace's ideas, subjective ideas predate objectivist positions. The idea that 'probability' should be interpreted as 'subjective degree of belief in a proposition' was proposed, for example, by John Maynard Keynes in the early 1920s. This idea was taken further by Bruno de Finetti in Italy (*Fondamenti Logici del Ragionamento Probabilistico*, 1930) and Frank Ramsey in Cambridge (*The Foundations of Mathematics*, 1931). The approach was devised to solve problems with the frequentist definition of probability but also with the earlier, objectivist approach of Laplace. The subjective Bayesian methods were further developed and popularized in the 1950s by L.J. Savage.

Objective Bayesian inference was further developed by Harold Jeffreys at the University of Cambridge. His seminal book "Theory of probability" first appeared in 1939 and played an important role in the revival of the Bayesian view of probability. In 1957, Edwin Jaynes promoted the concept of maximum entropy for constructing priors, which is an important principle in the formulation of objective methods, mainly for discrete problems. In 1965, Dennis Lindley's 2-volume work "Introduction to Probability and Statistics from a Bayesian Viewpoint" brought Bayesian methods to a wide audience. In 1979, José-Miguel Bernardo introduced reference analysis, which offers a general applicable framework for objective analysis. Other well-known proponents of Bayesian probability theory include I.J. Good, B.O. Koopman, Howard Raiffa, Robert Schlaifer and Alan Turing.

In the 1980s, there was a dramatic growth in research and applications of Bayesian methods, mostly attributed to the discovery of Markov chain Monte Carlo methods, which removed many of the computational problems, and an increasing interest in nonstandard, complex applications. Despite growth of Bayesian research, most undergraduate teaching is still based on frequentist statistics. Nonetheless, Bayesian methods are widely accepted and used, such as for example in the field of machine learning.

History of Probability

Probability has a dual aspect: on the one hand the likelihood of hypotheses given the evidence for them, and on the other hand the behavior of stochastic processes such as the throwing of dice or coins. The study of the former is historically older in, for example, the law of evidence, while the mathematical treatment of dice began with the work of Cardano, Pascal and Fermat between the 16th and 17th century.

Probability is distinguished from statistics. While statistics deals with data and inferences from it, (stochastic) probability deals with the stochastic (random) processes which lie behind data or outcomes.

Origins

Ancient and medieval law of evidence developed a grading of degrees of proof, probabilities, presumptions and half-proof to deal with the uncertainties of evidence in court. In Renaissance times, betting was discussed in terms of odds such as "ten to one" and maritime insurance premiums were estimated based on intuitive risks, but there was no theory on how to calculate such odds or premiums.

The mathematical methods of probability arose in the correspondence of Gerolamo Cardano, Pierre de Fermat and Blaise Pascal (1654) on such questions as the fair division of the stake in an interrupted game of chance. Christiaan Huygens (1657) gave a comprehensive treatment of the subject.

From *Games, Gods and Gambling* ISBN 978-0-85264-171-2 by F. N. David:

> In ancient times there were games played using astragali, or Talus bone. The Pottery of ancient Greece was evidence to show that there was a circle drawn on the floor and the astragali were tossed into this circle, much like playing marbles. In Egypt, excavators of tombs found a game they called "Hounds and Jackals", which closely resembles the modern game "Snakes and Ladders". It seems that this is the early stages of the creation of dice.

> First dice game mentioned in literature of the Christian era was called Hazard. Played with 2 or 3 dice. Thought to have been brought to Europe by the knights returning from the Crusades.

> Dante Alighieri (1265-1321) mentions this game. A commentor of Dante puts further thought into this game: the thought was that with 3 dice, the lowest number you can get is 3, an ace for every die. Achieving a 4 can be done with 3 die by having a two on one die and aces on the other two dice.

> Cardano also thought about the throwing of three die. 3 dice are thrown: there are the same number of ways to throw a 9 as there are a 10. For a 9:(621) (531) (522) (441) (432) (333) and for 10: (631) (622) (541) (532) (442) (433). From this, Cardano found that the probability of throwing a 9 is less than that of throwing a 10. He also demonstrated the efficacy of defining odds as the ratio of favourable to unfavourable outcomes (which implies that the probability of an event is given by the ratio of favourable outcomes to the total number of possible outcomes).

In addition, the famous Galileo wrote about die-throwing sometime between 1613 and 1623. Essentially thought about Cardano's problem, about the probability of throwing a 9 is less than throwing a 10. Galileo had the following to say: Certain numbers have the ability to be thrown because there are more ways to create that number. Although 9 and 10 have the same number of ways to be created, 10 is considered by dice players to be more common than 9.

Eighteenth Century

Jacob Bernoulli's *Ars Conjectandi* (posthumous, 1713) and Abraham de Moivre's *The Doctrine of Chances* (1718) put probability on a sound mathematical footing, showing how to calculate a wide range of complex probabilities. Bernoulli proved a version of the fundamental law of large numbers, which states that in a large number of trials, the average of the outcomes is likely to be very close to the expected value - for example, in 1000 throws of a fair coin, it is likely that there are close to 500 heads (and the larger the number of throws, the closer to half-and-half the proportion is likely to be).

Nineteenth Century

The power of probabilistic methods in dealing with uncertainty was shown by Gauss's determination of the orbit of Ceres from a few observations. The theory of errors used the method of least squares to correct error-prone observations, especially in astronomy, based on the assumption of a normal distribution of errors to determine the most likely true value. In 1812, Laplace issued his *Théorie analytique des probabilités* in which he consolidated and laid down many fundamental results in probability and statistics such as the moment generating function, method of least squares, inductive probability, and hypothesis testing.

Towards the end of the nineteenth century, a major success of explanation in terms of probabilities was the Statistical mechanics of Ludwig Boltzmann and J. Willard Gibbs which explained properties of gases such as temperature in terms of the random motions of large numbers of particles.

The field of the history of probability itself was established by Isaac Todhunter's monumental *A History of the Mathematical Theory of Probability from the Time of Pascal to that of Laplace* (1865).

Twentieth Century

Probability and statistics became closely connected through the work on hypothesis testing of R. A. Fisher and Jerzy Neyman, which is now widely applied in biological and psychological experiments and in clinical trials of drugs, as well as in economics and elsewhere. A hypothesis, for example that a drug is usually effective, gives rise to a probability distribution that would be observed if the hypothesis is true. If observations approximately agree with the hypothesis, it is confirmed, if not, the hypothesis is rejected.

The theory of stochastic processes broadened into such areas as Markov processes and Brownian motion, the random movement of tiny particles suspended in a fluid. That provided a model for the study of random fluctuations in stock markets, leading to the use of sophisticated probability models in mathematical finance, including such successes as the widely used Black–Scholes formula for the valuation of options.

The twentieth century also saw long-running disputes on the interpretations of probability. In the mid-century frequentism was dominant, holding that probability means long-run relative frequency in a large number of trials. At the end of the century there was some revival of the Bayesian view, according to which the fundamental notion of probability is how well a proposition is supported by the evidence for it.

The mathematical treatment of probabilities, especially when there are infinitely many possible outcomes, was facilitated by Kolmogorov's axioms (1933).

References

- Cochran W.G. (1978) "Laplace's ratio estimators". pp 3-10. In David H.A., (ed). Contributions to Survey Sampling and Applied Statistics: papers in honor of H. O. Hartley. Academic Press, New York ISBN 978-1483237930

- Singh, Simon (2000). The code book : the science of secrecy from ancient Egypt to quantum cryptography (1st Anchor Books ed.). New York: Anchor Books. ISBN 0-385-49532-3.

- Bernardo J (2005). "Reference analysis". Handbook of statistics. Handbook of Statistics. 25: 17–90. doi:10.1016/S0169-7161(05)25002-2. ISBN 9780444515391.

- Daston, Lorraine (1988). Classical Probability in the Enlightenment. Princeton: Princeton University Press. ISBN 0-691-08497-1.

- Franklin, James (2001). The Science of Conjecture: Evidence and Probability Before Pascal. Baltimore, MD: Johns Hopkins University Press. ISBN 0-8018-6569-7.

- Hacking, Ian (2006). The Emergence of Probability (2nd ed.). New York: Cambridge University Press. ISBN 978-0-521-86655-2.

- Hald, Anders (2003). A History of Probability and Statistics and Their Applications before 1750. Hoboken, NJ: Wiley. ISBN 0-471-47129-1.

- von Plato, Jan (1994). Creating Modern Probability: Its Mathematics, Physics and Philosophy in Historical Perspective. New York: Cambridge University Press. ISBN 978-0-521-59735-7.

- Salsburg, David (2001). The Lady Tasting Tea: How Statistics Revolutionized Science in the Twentieth Century. ISBN 0-7167-4106-7

- Stigler, Stephen M. (1990). The History of Statistics: The Measurement of Uncertainty before 1900. Belknap Press/Harvard University Press. ISBN 0-674-40341-X.

Permissions

Index

A

Actuarial Science, 16, 21

Anova, 10, 287, 346

Approximate Distributions, 310

B

Bayesian Inference, 10, 70, 148, 151, 153, 166, 170, 186-188, 260, 307-308, 310, 312-318, 323-326, 335, 351, 353, 356, 366-367

Bayesian Models, 15, 296

Bernoulli Trial, 35, 38, 69, 291, 306

Binomial Distribution, 62, 69-70, 78, 111, 157, 162, 224-225, 273, 276, 279, 285-286, 291-292, 299, 306-307, 326

Bivariate Analysis, 295

Borel Set, 41, 236, 239, 241, 243

C

Calculation, 85, 95, 117, 124, 158-159, 161, 169, 186, 213, 320, 337, 341, 346, 349

Census, 1, 3-4, 255, 359, 361

Conditionally Independent, 41, 141, 157, 159, 164-166

Continuous Distribution, 38, 67-68, 82, 88, 90, 106, 109-111, 119, 142, 291, 304-305

Continuous Random Variable, 23, 27, 30, 33, 63, 66, 78, 81, 88-90, 97, 100, 106, 109, 127, 135-136, 139, 142, 265, 301, 303

Corrected Sample Standard Deviation, 120

Cumulative Distribution Function, 27, 31-32, 41, 61-62, 64, 66-67, 73, 75, 77, 79-83, 88-90, 92-95, 101, 117, 278, 299-301, 303-304

D

Data Set, 2, 5, 12, 112, 179, 248, 255, 265-266, 290, 294, 308, 335, 340

Data Type, 6

Demography, 16, 357, 359

Descriptive Statistics, 1-3, 14, 108, 290, 292, 294-296, 307-309

Discrete Probability Distribution, 38, 61, 64, 72, 77, 89, 299, 301

Discrete Random Variable, 27-29, 63, 76-78, 81-82, 89, 97-98, 105, 109, 112, 116, 135-136, 301

Discrete Weighted Variance, 110

Dispersion, 2, 63, 91, 98, 113, 273-274, 279, 285, 294-295, 301

Distribution, 1-2, 6, 9, 13, 15, 18-19, 25-29, 31-35, 37-39, 41-42, 61-64, 66-83, 86-98, 100-101, 105-106, 108-111, 115-121, 123, 126, 133-134, 138, 141-144, 157-160, 162, 164-166, 176, 187, 189, 193, 195, 198-199, 204-205, 207-208, 212, 215-219, 222, 224-232, 234-235, 243-245, 248-252, 257, 259-263, 265-266, 269, 272-282, 285-288, 291-293, 295-301, 303-311, 314, 316-321, 323-324, 326-327, 329-330, 334, 336, 342-343, 346, 350-351, 353, 359-364, 370

E

Estimation, 3-4, 6, 8, 14, 104, 118-120, 169, 221, 251-252, 255, 263, 266-268, 272, 280-281, 288, 293, 297, 308, 312-314, 317, 320-321, 323, 329-331, 334, 339, 352-353, 355, 359, 364

Experimental Protocol, 5, 311

Exponential Distribution, 34, 69-70, 95, 111, 230, 245, 292, 306-307

F

Failure Probability, 21

Fiducial Inference, 314

Formula, 20, 34, 36, 45, 66, 83, 85, 88, 94, 98, 100, 104-106, 112-116, 121, 135, 138, 141, 151, 157, 187, 196, 198, 204, 214, 218, 267, 270-271, 283, 286, 303, 317, 321, 328, 334, 345-346, 360-361, 370

Foundations Of Statistics, 190

Frequentist Inference, 179, 186, 310-312, 316, 323, 335

G

Gaussian Distribution, 68, 207, 249-250, 291, 305, 320, 364

H

Harmonic Mean, 122

Hawthorne Plant, 5

Huygens, 18-19, 71, 107-108, 359, 369

Hypothesis Testing, 3, 10, 70, 90, 108, 169, 179-180, 257, 298, 307-309, 311, 313-315, 317, 319, 321, 323, 325, 327, 329, 331, 333, 335-341, 343, 345, 347, 349-355, 362-363, 370

I

Independent Identically Distributed, 6

Integration By Parts, 100, 106, 328

Interquartile, 364

Interquartile Range, 364

Interval Estimation, 8, 312

K

Kolmogorov, 20-21, 57-59, 67, 71, 93, 127-128, 131, 135-

136, 170, 181, 193-194, 225, 227, 298, 304, 310, 313, 368, 370

L

Least Absolute Deviations, 8, 259

Least Squares, 7-8, 13, 20, 57, 180, 251-256, 258-259, 264, 266, 268-269, 272, 276, 280-281, 293, 312, 357, 361, 370

Linear Least Squares, 8, 258

M

Mann-whitney U, 10, 298

Markov Process, 15, 207, 229

Mathematical Treatment, 21, 27, 186, 368, 370

Mean, 1-3, 6-8, 10, 18, 39, 47, 63-64, 70, 76, 79, 82, 91, 94, 97, 104, 108-110, 113-119, 122-125, 134, 141-142, 167-168, 170, 173, 197-199, 207-208, 213, 218-219, 227-229, 232-233, 235-237, 241, 243, 249-250, 252, 254, 256, 261-262, 265, 272-276, 278-281, 285-286, 292-296, 299-301, 307, 309-310, 313, 318, 320-321, 330-334, 336, 341, 344-346, 350-351, 358, 360-361, 368

Mean Square Weighted Deviation, 10

Mean Squared Error, 7-8, 118, 330-331

Measure Theoretic Formulation, 78

Median, 63, 79, 91, 94-95, 122, 287, 294-295, 298, 301, 309, 312-313, 321, 358, 360-361, 363

Median-unbiased Estimators, 312

Modeling, 3, 21, 28, 125, 218, 251, 254-255, 259, 261, 292, 296, 298, 308-309

Multivariate Statistics, 17

N

Newtonian Mechanics, 24

Non-linear Least Squares, 8, 258

Normal Distribution, 1, 19, 34, 37-38, 62, 68, 70, 73-74, 76, 79, 81-82, 88, 91, 93, 95, 97, 110, 118, 120-121, 198, 215, 217-219, 257, 261-262, 266, 269, 272, 277, 281, 288, 291, 295-296, 299, 305, 307, 310, 318, 336, 360-362, 364, 370

Null Hypothesis, 2, 7, 9-10, 13, 335-338, 340-346, 349-350, 352-353

O

Objective Analysis, 187, 367

Optimality Property, 312

P

P-value, 9-10, 90, 312, 325, 337-338, 344, 349, 352, 354, 363

Poisson Distribution, 69-70, 72, 111, 224-228, 243, 273, 278-279, 285, 292, 306-307, 342

Population Variance, 109, 117-120

Primary Analysis, 5

Probability Distribution, 6, 9, 18, 26-28, 31, 38-39, 61-64, 66-68, 71-74, 77, 79-82, 89, 93-95, 98, 100, 105, 109, 119-120, 123, 126, 133-134, 141-143, 158-159, 164-166, 176, 187, 189, 193, 195, 208, 212, 216, 218, 229, 248, 250-251, 260, 273, 279, 291, 293, 296-297, 299-301, 303-305, 309, 314, 318, 324, 327, 329-330, 346, 353, 361, 364, 370

Probability Distributions, 14, 31, 37, 61-63, 66-69, 71-75, 77, 79, 81-83, 85, 87, 89, 91, 93, 95, 97, 101, 103, 105, 107, 109, 111, 113, 115, 117, 119, 121, 123, 125, 144, 147, 168, 183, 189, 193, 248-249, 273, 291, 293, 295, 298-300, 303-305, 309-310, 363

Probability Mass Function, 26-30, 61-64, 76-79, 81, 94, 109, 111, 142, 273, 281, 291, 299-301, 330

Probability Spaces, 37, 44, 58, 320

R

Random Number Generation, 67, 92, 304

Random Variable, 6, 9, 23, 26-35, 37-38, 41-42, 47, 61-64, 66-67, 73-78, 80-85, 88-90, 92-93, 97-98, 100, 102-106, 108-109, 111-112, 116-117, 121, 124-125, 127-128, 134-145, 159, 165, 204, 224, 226-229, 232-233, 235-238, 243, 245-246, 254, 259, 261, 265-266, 272, 291, 299-301, 303-304, 320

Random Variable of Type, 27

Real-valued Random Variables, 28, 31-32, 195

Regressions, 252

Residual Sum Of Squares, 8

S

Sample Variance, 6, 70, 116-121, 292, 307

Sampling Distribution, 260, 298, 317, 321

Samuelson's Inequality, 121

Significance Testing, 180, 312, 339, 349-354

Standard Deviation, 2-3, 7, 13, 32, 63, 91, 108, 117-118, 120, 123, 219, 253, 294-295, 301, 338, 345-346, 362-364

Statistical Computing, 14, 272

Statistical Inference, 4, 6, 14, 18, 25, 108, 126, 130, 148, 166, 169-170, 181, 247-249, 251, 291-293, 307-311, 313-315, 317, 319, 321, 323, 325, 327, 329, 331, 333, 335, 337, 339, 341, 343, 345, 347, 349, 351, 353-357, 365

Statistical Mechanics, 16, 24, 71, 357, 370

Statistical Process Control, 17, 351

Structural Inference, 314

T

Topological Space, 31

Total Probability, 26, 45, 47, 77, 129, 132, 154-155, 159

U

Umvue, 7

Univariate Analysis, 295

Univariate Distributions, 68, 80, 305

Unknown Parameters, 6, 251-254, 273-274, 282-283, 293, 309, 312, 329

V

Variance, 6-8, 10, 13, 27, 32, 63, 70, 76, 82, 98, 104-105, 108-124, 141-142, 197-199, 203, 207-208, 217-218, 238,

249-250, 254, 256, 261-262, 264-267, 272-274, 277-279, 281, 285-286, 292-296, 298, 301, 307, 309, 312, 314, 318, 331-332, 334, 346, 363-365

W

Weighted Average, 63, 97-98, 301

Weighted Least Squares, 254, 281

www.ingramcontent.com/pod-product-compliance
Lightning Source LLC
Chambersburg PA
CBHW061320190326
41458CB00011B/3844